AF597681

Computer-Enhanced Analytical Spectroscopy

Volume 2

MODERN ANALYTICAL CHEMISTRY

Series Editor: David M. Hercules
University of Pittsburgh

ANALYTICAL ATOMIC SPECTROSCOPY
William G. Schrenk

APPLIED ATOMIC SPECTROSCOPY
Volumes 1 and 2
Edited by E. L. Grove

CHEMICAL DERIVATIZATION IN ANALYTICAL CHEMISTRY
Edited by R. W. Frei and J. F. Lawrence
Volume 1: Chromatography
Volume 2: Separation and Continuous Flow Techniques

COMPUTER-ENHANCED ANALYTICAL SPECTROSCOPY
Volume 1: Edited by Henk L. C. Meuzelaar and Thomas L. Isenhour
Volume 2: Edited by Henk L. C. Meuzelaar

ION CHROMATOGRAPHY
Hamish Small

ION-SELECTIVE ELECTRODES IN ANALYTICAL CHEMISTRY
Volumes 1 and 2
Edited by Henry Freiser

LIQUID CHROMATOGRAPHY/MASS SPECTROMETRY
Techniques and Applications
Alfred L. Yergey, Charles G. Edmonds, Ivor A. S. Lewis, and Marvin L. Vestal

MODERN FLUORESCENCE SPECTROSCOPY
Volumes 1–4
Edited by E. L. Wehry

PHOTOELECTRON AND AUGER SPECTROSCOPY
Thomas A. Carlson

PRINCIPLES OF CHEMICAL SENSORS
Jiří Janata

TRANSFORM TECHNIQUES IN CHEMISTRY
Edited by Peter R. Griffiths

A Continuation Order Plan is available for this series. A continuation order will bring delivery of each new volume immediately upon publication. Volumes are billed only upon actual shipment. For further information please contact the publisher.

Computer-Enhanced Analytical Spectroscopy

Volume 2

Edited by

Henk L. C. Meuzelaar

The University of Utah
Salt Lake City, Utah

Plenum Press • New York and London

Library of Congress Cataloging in Publication Data

(Revised for. vol. 2)

Computer-enhanced analytical spectroscopy.

(Modern analytical chemistry)
Vol. 2: edited by Henk L. C. Meuzelaar.
Vol. 2: Papers from the Second Hidden Peak Symposium on Computer-Enhanced Analytical Spectroscopy, held in June 1988 at Snowbird Resort, Utah.
Includes bibliographies and indexes.
1. Spectrum analysis—Data processing. I. Meuzelaar, Henk L. C. II. Isenhour, Thomas L. III. Hidden Peak Symposium (1st: 1986: Snowbird, Utah) IV. Hidden Peak Symposium on Computer-Enhanced Analytical Spectroscopy (2nd: 1988: Snowbird, Utah) V. Series.
QD95.C6323 1987 543′.0858 87-15883
ISBN 0-306-43276-5

A Division of Plenum Publishing Corporation
233 Spring Street, New York, N.Y. 10013

Printed in the United States of America

Contributors

Joseph E. Bonelli, Health Science Center, University of Colorado, Denver, Colorado

Scott E. Carpenter, Department of Chemistry, The University of Iowa, Iowa City, Iowa

Bo Curry, Hewlett-Packard Laboratories, Palo Alto, California

John T. Ditillo, U.S. Army Chemical Research, Development, and Engineering Center, Aberdeen Proving Ground, Maryland

Steven L. Durfee, Somatogen, Inc., Broomfield, Colorado

Cynthia S. Firnhaber, Somatogen, Inc., Broomfield, Colorado

Stephen D. Frans, Department of Chemistry, University of Utah, Salt Lake City, Utah

Paul J. Gemperline, Department of Chemistry, East Carolina University, Greenville, North Carolina

Alexander F. H. Goetz, Center for the Study of Earth from Space, Cooperative Institute for Research in Environmental Sciences, University of Colorado, Boulder, Colorado

J. Craig Hamilton, Department of Chemistry, East Carolina University, Greenville, North Carolina

Peter B. Harrington, Department of Chemistry and Geochemistry, Colorado School of Mines, Golden, Colorado

Joel M. Harris, Department of Chemistry, University of Utah, Salt Lake City, Utah

Stephen Hoffman, Somatogen, Inc., Broomfield, Colorado

Erkki J. Karjalainen, Department of Clinical Chemistry, University of Helsinki, Helsinki, Finland

Robert T. Kroutil, U.S. Army Chemical Research, Development, and Engineering Center, Aberdeen Proving Ground, Maryland

Craig A. Lehmann, Department of Medical Technology, School of Allied Health Professions, State University of New York at Stony Brook, Health Sciences Center, Stony Brook, New York

Steven P. Levine, School of Public Health, University of Michigan, Ann Arbor, Michigan

George C. Levy, N.I.H. Resource and CASE Center, Syracuse University, Syracuse, New York

Ying Li-shi, School of Public Health, University of Michigan, Ann Arbor, Michigan

Stanton Y. Loh, Department of Chemistry, Baker Laboratory, Cornell University, Ithaca, New York

Gregory L. McClure, Perkin-Elmer Corporation, Norwalk, Connecticut

Malcolm K. McIntyre, Department of Chemistry, The University of Iowa, Iowa City, Iowa

Fred W. McLafferty, Department of Chemistry, Baker Laboratory, Cornell University, Ithaca, New York

Pete E. Poston, Department of Chemistry, University of Utah, Salt Lake City, Utah

Gary W. Small, Department of Chemistry, University of Iowa, Iowa City, Iowa

Douglas B. Stauffer, Department of Chemistry, Baker Laboratory, Cornell University, Ithaca, New York

Thomas E. Street, Department of Chemistry and Geochemistry, Colorado School of Mines, Golden, Colorado

Sterling A. Tomellini, Department of Chemistry, University of New Hampshire, Durham, New Hampshire

Kent J. Voorhees, Department of Chemistry and Geochemistry, Colorado School of Mines, Golden, Colorado

Andrew L. Wong, Department of Chemistry, University of Utah, Salt Lake City, Utah

Preface

The Second Hidden Peak Symposium on Computer-Enhanced Analytical Spectroscopy, held in June, 1988, at the Snowbird Resort (Salt Lake City, Utah), centered around twelve keynote lectures delivered by some of the foremost experts and pioneers in this rapidly expanding field. The editor is highly indebted to each of these colleagues for contributing a chapter to the second volume of *Computer-Enhanced Analytical Spectroscopy*. The primary objective of this volume is to present a representative cross-section of current activities in the field while balancing out the lighter coverage of some topics and areas in Volume 1.

An exciting new topic, remote IR sensing, is covered in Chapters 4 and 5. Deconvolution and signal-processing methods have now been extended to UV/VIS (Chapter 1) and GC/MS (Chapter 3) applications. Furthermore, the development and testing of novel factor analysis techniques in the areas of UV/VIS and IR spectroscopy are discussed in Chapters 2 and 12, respectively. Fundamental aspects of library search techniques are presented in Chapters 7 (MS) and 9 (NMR). Chapters 6, 10, and 11 cover selected uses of expert systems in NMR, IR, and MS, respectively. Finally, an integrated expert system approach to the interpretation of GC/IR/MS data is outlined in Chapter 8.

In an attempt to facilitate access to the various topics for the newcomer to the field, the twelve chapters have been organized into two main parts: Unsupervised Methods: Spectral Enhancement, Deconvolution, and Data Reduction, and Supervised Methods: Expert Systems, Modeling, and Quantitation.

All in all, the second volume is intended to constitute a logical complement to Volume 1. Consequently, for a more comprehensive update of developments in the field over the past five years or so, the reader is encouraged to consult both Volume 1 and Volume 2.

Henk L. C. Meuzelaar

Salt Lake City

Contents

PART I: UNSUPERVISED METHODS: Spectral Enhancement, Deconvolution, and Data Reduction

Chapter 1

Advances in Regression: Use of Models in Spectroscopic Data Analysis

Joel M. Harris, Stephen D. Frans, Pete E. Poston, and Andrew L. Wong

Chapter 2

Factor Analysis of Spectro-Chromatographic Data

Paul J. Gemperline and J. Craig Hamilton

Chapter 3

Isolation of Pure Spectra in GC/MS by Mathematical Chromatography: Entropy Considerations

Erkki J. Karjalainen

Chapter 4

Signal Processing Techniques for Remote Infrared Chemical Sensing

Robert T. Kroutil, John T. Ditillo, and Gary W. Small

Chapter 5

Imaging Spectrometry of the Earth: A Breakthrough and a Nightmare

Alexander F. H. Goetz

Chapter 6

Computer-Enhanced Nuclear Magnetic Resonance Spectroscopy

George C. Levy

PART II: SUPERVISED METHODS: Expert Systems, Modeling, and Quantitation

Chapter 7

Computer Identification of Mass Spectra

Fred W. McLafferty, Stanton Y. Loh, and Douglas B. Stauffer

Chapter 8

A Distributed Expert System for Interpretation of GC/IR/MS Data

Bo Curry

Chapter 9

Computer-Aided Solutions to ^{13}C NMR Spectral Interpretation Problems

Gary W. Small, Scott E. Carpenter, and Malcolm K. McIntyre

Chapter 12

Theory and Application of the CIRCOM Software for Quantitative Spectroscopic Analysis

Gregory L. McClure and Craig A. Lehmann

Part I

Unsupervised Methods: Spectral Enhancement, Deconvolution, and Data Reduction

Chapter 1

Advances in Regression: Use of Models in Spectroscopic Data Analysis

Joel M. Harris, Stephen D. Frans, Pete E. Poston, and Andrew L. Wong

1.1. INTRODUCTION

Redundant or correlated behavior of spectroscopic data can be costly to the number of degrees of freedom of a measurement and thereby limit the information content of a spectroscopic method of analysis.[1,2] Correlations can take on many forms, including the band shapes of spectroscopic lines, reproducible patterns of spectral features and time-dependent signals, or the predictable variation of component concentrations in hyphenated spectroscopic methods.[3] While such correlated behavior limits a spectroscopic technique in its ability to provide information, our knowledge of this behavior represents valuable prior information that may be brought to bear on the analysis of the data. Expectations of correlations, in the form of a model, can be used to selectively filter out random fluctuations in data and extract meaningful information in the presence of noise.

Joel M. Harris, Stephen D. Frans, Pete E. Poston, and Andrew L. Wong • Department of Chemistry, University of Utah, Salt Lake City, Utah 84112.

Regression analysis,[(4)] or the method of least squares, is one of the oldest methods of statistical data analysis, having been first developed in the early 1800s by Gauss and independently by Legendre. The method provides a general approach to extracting underlying relationships from data, including the parameters that describe the relationship between points and the uncertainties in those parameters. Regression methods have their roots in the method of maximum likelihood,[(5)] which assures that the parameter estimates are unbiased and efficient. In this chapter, regression analysis of spectroscopic data will be presented, with the emphasis on using models to describe the correlations expected in the data, proper weighting of observations, and determining uncertainties in estimated parameters. While this approach is particularly powerful for multidimensional spectroscopic methods (e.g., time-resolved fluorescence, GC/MS, LC/UV), the theory of regression methods will first be developed with examples from measurements of lower dimensionality. The basis of regression methods for multidimensional data in the simple statisics of estimating a mean and standard deviation provides an intuitive basis for understanding more powerful analysis procedures while extending our background in simple statistics into methods for manipulating spectroscopic data.

1.2. ANALYSIS OF ZERO-DIMENSIONAL SPECTROSCOPIC DATA (NUMBERS)

1.2.1. Method of Maximum Likelihood

The simplest of spectroscopic measurements provide an outcome that is only a number, the variation of which—with an independent variable such as wavelength—is not considered. An example of such a measurement is "colorimetric" analysis, where a sample is reacted with a chromogenic reagent and the absorbance of the product is determined at a single wavelength. Let us assume that we have made a series of N such measurements, x_i, each drawn from a population described by a normal distribution having a mean μ and a standard deviation σ_i, which can vary with the measurement. Given these results, we wish to determine the "maximum likelihood" estimate of the mean $\hat{m}$, which is an estimate of the mean of the underlying distribution that maximizes the probability we will observe these results.[(5)] The probability of observing a series of events is the product of probabilities for observing the individual events; thus, the probability of having observed the N measurements, x_i, is

$$P_N = \prod_{i=1}^{N} P_i(x_i, m, \sigma_i) \tag{1.1}$$

where P_i is normally distributed:

$$P_i = \frac{1}{\sigma_i\sqrt{2\pi}} \exp\left[-\frac{1}{2}\left(\frac{x_i - m}{\sigma_i}\right)^2\right] \tag{1.2}$$

Taking the product of Gaussian probabilities as a summation within the exponential gives the following expression for the probability of having observed the N results:

$$P_N = \left(\prod_{i=1}^{N} \frac{1}{\sigma_i\sqrt{2\pi}}\right) \exp\left[-\frac{1}{2}\sum_{i=1}^{N}\left(\frac{x_i - m}{\sigma_i}\right)^2\right] \tag{1.3}$$

To maximize P_N, we minimize the argument of the exponent with respect to $\hat{m}$:

$$\frac{\partial}{\partial m}\left[-\frac{1}{2}\sum_{i=1}^{N}\left(\frac{x_i - m}{\sigma_i}\right)^2\right] = 0 = \sum_{i=1}^{N}\left(\frac{x_i - \hat{m}}{\sigma_i^2}\right) \tag{1.4}$$

Note that equation (1.4) is a least squares expression that minimizes the squared deviations between the mean estimate $\hat{m}$ and the observed data x_i, weighted by the inverse of the variance of the observations, σ_i^2. Solving equation (1.4) for $\hat{m}$ gives the estimate of the mean of the underlying distribution, which maximizes the probability that we observed the particular series of N measurements:

$$\hat{m} \equiv \hat{x} = \left[\sum_{i=1}^{N} (x_i/\sigma_i^2)\right] \Big/ \sum_{i=1}^{N} (1/\sigma_i^2) \tag{1.5}$$

If the uncertainty of each of the measurements is constant, $\sigma_i = \sigma$, then $1/\sigma^2$ can be factored out of the summation, and the maximum likelihood estimate of the mean given by equation (1.5) is simply the average of the N measurements.

The uncertainty of the mean estimate determined by equation (1.5) can be found from a propagation of errors[4,5] applied to this expression:

$$\sigma_m^2 = \sum_{i=1}^{N}\left(\frac{\partial m}{\partial x_i}\right)^2 \sigma_i^2 = 1 \Big/ \sum_{i=1}^{N} (1/\sigma_i)^2 \tag{1.6}$$

Again, if the uncertainty of each measurement is constant, that is, $\sigma_i = \sigma$, then equation (1.6) predicts that the variance of the average of N measurements is $1/N$ times smaller than the variance of the individual measurements.

1.2.2. Maximum Likelihood Quantitative Estimates for Peaks

A common goal in analytical spectroscopy is to estimate the concentration of a sample responsible for an observed peak that rises from the baseline as a function of wavelength, frequency, or time. For such data, a number of strategies may be implemented to estimate the sample concentration, including measurements of peak height or peak area. The maximum likelihood method, developed above, provides an optimum method of data analysis for such cases.[6] To apply this method, consider N measurements of a spectroscopic signal across a peak $z_i = cg_i + e_i$, where c is the true sample concentration, g_i is a model peak shape function, and e_i is the error in the measured signal. Under these conditions, each data point provides a measure of sample concentration,

$$c_i = z_i/g_i \tag{1.7}$$

the uncertainty of which depends on the nature of the errors e_i in the measured signal.

If the magnitude of the noise in the signal is constant, independent of signal amplitude $\sigma_{z_i} = \sigma_z$, then the standard deviation of the concentration estimate varies inversely with the peak shape function $\sigma_{c_i} = \sigma_z/g_i$. Substituting this uncertainty relation and equation (1.7) into equation (1.5) provides a maximum likelihood estimate of the sample concentration $\hat{c}$,

$$\hat{c} = \left(\sum_i z_i g_i\right) \Big/ \sum_i g_i^2 \tag{1.8}$$

This result corresponds to calculating the zero-displacement value of the cross-correlation between the signal and the shape function, and is identical to a "matched filter" estimate.[7]

When the uncertainty of the signal depends on the signal magnitude (as in the case of shot noise or proportional noise), then one need only substitute the uncertainty relationship into equation (1.5) to obtain the appropriate maximum likelihood expression. For example, when the predominant noise present in a signal arises from fluctuations in measurement sensitivity, such as excitation source flicker, the standard deviation of the signal increases in proportion to signal size. Since the signal and its uncertainty are proportional to g_i, the uncertainty in concentration is constant, independent of i. Substituting this relationship into equation (1.5) gives the following maximum likelihood expression for estimating the concentration:

$$\hat{c} = (1/N) \sum_i (z_i/g_i) \tag{1.9}$$

For the case of shot noise, where the standard deviation of the signal varies with the square root of the signal magnitude, the maximum likelihood estimate of the sample concentration is given by the peak area.

1.2.3. Application to Photoacoustic Spectroscopy

Absorption of radiation from a pulsed laser and nonradiative relaxation of the excited states produces a rapid temperature rise in the sample, which in turn generates a pressure wave that can be detected by a piezoelectric transducer.[8] Reflections of the acoustic wave within the sample and transducer result in a reproducible high-frequency signal that persists for over 50 μs. While the peak compression signal at the start of the wave can be used for quantifying the sample absorbance,[8] the entire acoustic wave carries amplitude information that can be used to provide a more precise determination. The model of the peak shape, g_i, can be obtained from well-averaged photoacoustic transients obtained from more concentrated samples.

To compute a maximum likelihood estimate of the sample absorbance from such data, the relationship between signal errors and signal size must also be determined from replicate measurements. A plot of signal variance versus the square of the signal amplitude from such measurements is generally linear,[6] indicating a strong proportional noise component arising primarily from pulse-to-pulse variation in laser energy. The intercept of this plot is not zero, showing a constant noise source (detector noise) at low signal amplitudes. Since the noise sources are uncorrelated, their variances add so that the overall signal variance is given by

$$\sigma_{z_i}^2 = k^2 z_i^2 + \sigma_z^2 \tag{1.10}$$

where k is the coefficient for proportional noise and σ_z^2 is the constant noise variance. Substituting this mixed proportional and constant noise model into equation (1.5) along with equation (1.7) gives a maximum likelihood estimate that weights large signals by $1/g_i$ and small signals by g_i, according to

$$\hat{c} = \frac{\sum_i [z_i g_i / (k^2 z_i^2 + \sigma_z^2)]}{\sum_i [g_i^2 / (k^2 z_i^2 + \sigma_z^2)]} \tag{1.11}$$

Application of this equation to determining the sample concentration or absorbance from photoacoustic transients is illustrated in Fig. 1.1. The capability of a maximum likelihood estimate to extract quantitative

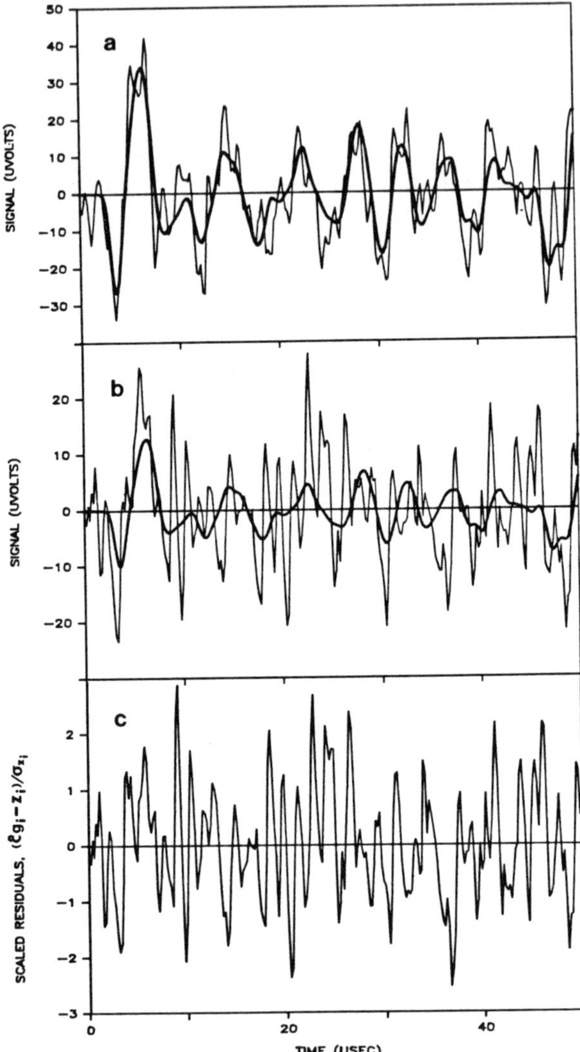

FIGURE 1.1. Maximum likelihood absorbance estimates from photoacoustic signals.[6] Photoacoustic transients from azulene in carbon tetrachloride in parts (a) and (b) have absorbances $A = 4.7 \times 10^{-5}\,\text{cm}^{-1}$ and $1.9 \times 10^{-5}\,\text{cm}^{-1}$, respectively. The best fits to the data, $\hat{c}g_i$, are shown as heavy lines. Part (c) shows the residual error for part (b), scaled by the expected error.

information from noisy data is illustrated by these results. For the $1.9 \times 10^{-5}\,\mathrm{cm}^{-1}$ absorbance sample in Figure 1.1(b), for example, the maximum likelihood estimate of absorbance is in error by only 20% of the true value, despite the largest peak in the transient being comparable to the noise. The scaled residuals, shown in Fig. 1.1(c), are random and of a magnitude expected for the experimental error, indicating that most of the quantitative information has been successfully extracted.

The limit of detection[(9)] of the maximum likelihood photoacoustic absorbance measurement, determined from replicates, was $A_{\min} = 7 \times 10^{-6}\,\mathrm{cm}^{-1}$, which represents a 5(±2)-fold improvement over single-point measurements at the peak of the waveform. This observed result is indistinguishable from the 5.6(±0.2)-fold improvement predicted from a propagation of errors through equation (1.11). One can conclude from these results that the method of maximum likelihood, an optimum technique for combining measurements of differing uncertainty, is appropriate for the quantitative interpretation of signal peaks where the peak shape is known in advance or is reproducible and can be measured precisely. Propagation of errors through the maximum likelihood estimate expressions allows one to predict both the uncertainty in the quantitative results and the improvement in precision relative to other methods.

1.3. ANALYSIS OF ONE-DIMENSIONAL SPECTROSCOPIC DATA (SPECTRA, WAVEFORMS)

1.3.1. Spectrophotometry of Mixtures

Acquisition of spectroscopic signals as a function of an independent variable (e.g., wavelength or time) increases the information content of the measurement[(1,2)] over zero-dimensional results and allows the identity and composition of complex mixtures to be determined. The linear relationship between absorbance and concentration given by the Beer–Lambert law makes the use of linear regression analysis appropriate for determining the concentration of individual components in the mixture. The measured absorptivity (in cm^{-1}), a_i, of an n-component sample at wavelength i, can be written as the sum of the absorptivity contributions of each component,

$$a_i = k_{i1}c_1 + \cdots + k_{ij}c_j + \cdots + k_{in}c_n + r_i \tag{1.12}$$

where k_{ij} is the molar absorptivity of component j at wavelength i, c_j is the molar concentration of component j, and r_i represents the residual error in the measurement. A series of absorbance measurements at m different wavelengths generates a system of m equations, each having the form of equation (1.12). It is convenient to express this system of equations in matrix form,

$$\mathbf{A} = \mathbf{KC} + \mathbf{R} \tag{1.13}$$

where $\mathbf{A}$ is a vector of absorbances of the mixture, measured at m different wavelengths (the mixture spectrum), $\mathbf{K}$ is an $m \times n$ matrix of standard spectra of the n components measured at each of the m wavelengths, $\mathbf{C}$ is a vector containing the n unknown concentrations, and $\mathbf{R}$ is an m-element vector containing the residual error in the measured absorption spectrum of the mixture. Note that $\mathbf{K}$ is a model of the wavelength variation we expect to observe in the mixture spectrum, where each of the columns of $\mathbf{K}$ is weighted by an element in $\mathbf{C}$, which is the concentration of the particular component in the mixture.

Given such a model for mixture spectrums where the concentrations of the components are unknown and an excess of degrees of freedom $(m > n)$, one obtains a maximum likelihood estimate of component concentrations by first calculating the sum of the squared residuals with respect to the n concentrations (divided by the constant measurement variance $\sigma_i^2 = \sigma^2$):

$$\chi^2 = \sum_{i=1}^{m} r_i^2/\sigma_i^2 \tag{1.14a}$$

$$= (1/\sigma^2)\mathbf{R}^T\mathbf{R} \tag{1.14b}$$

To obtain and optimize the estimate, this chi-squared statistic[5] is minimized with respect to each of the unknown concentrations by solving a series of n "normal" equations of the form

$$\partial\chi^2/\partial c_j = 0 \tag{1.15}$$

Note that equation (1.15) is simply a multivariate form of equation (1.4), which was derived for maximum likelihood estimation of a single variable.

This series of n simultaneous equations has a simple linear algebra solution[4,10] given by

$$\hat{\mathbf{C}} = (\mathbf{K}^T\mathbf{K})^{-1}\mathbf{K}^T\mathbf{A} \tag{1.16}$$

where the product of the first three terms on the right-hand side of equation (1.16) is often termed a pseudoinverse, or least squares inverse, of the matrix $\mathbf{K}$. A check on the validity of equation (1.16) can be made by allowing the residuals to vanish, $\mathbf{R} = 0$, where the measured absorbance is exactly predicted by the model $\mathbf{A} = \mathbf{KC}$. Substituting $\mathbf{KC}$ for $\mathbf{A}$ in equation (1.16) shows that, under these ideal conditions, the maximum likelihood estimate for the concentrations $\hat{\mathbf{C}}$ is identical to the true concentration vector $\mathbf{C}$.

1.3.2. Analysis of Errors in Linear Regression

The uncertainty or variance associated with the concentration estimates extracted from a mixture spectrum can be found by a propagation of errors analysis of equation (1.16).[4] The results of such analysis include both the variance of each of the extracted parameters as well as the covariance between parameters. These terms are collected into an $n \times n$ variance–covariance matrix $\mathbf{V}$, with the parameter variances appearing on the diagonal and the covariance terms as the off-diagonal elements. The linear algebra expression for the analysis has a remarkably simple form[4] given by

$$\mathbf{V} = (\mathbf{K}^T\mathbf{K})^{-1}\sigma^2 \tag{1.17}$$

where σ^2 is the variance associated with the measured absorbance of the mixture spectrum.

The concentration errors associated with such a spectrophotometric determination of a mixture arise from the product of two terms on the right-hand side of equation (1.17), which have entirely different origins.[11] The variance term σ^2 depends only on the precision with which the mixture spectrum is measured, and is thus related to the characteristics of both the instrument and experimental methods. The second term $(\mathbf{K}^T\mathbf{K})^{-1}$ is an $n \times n$ matrix that serves to amplify the measurement error in the estimation of concentrations. The magnitudes of the elements in this matrix, f_{jj}, depend on differences between the spectra of the components and the wavelengths chosen to measure absorbance, the latter being an exercise in experimental design.[11] The greater the similarity between two standard spectra, which appear in the columns of $\mathbf{K}$, the larger will be the inverse of $\mathbf{K}^T\mathbf{K}$, particularly the diagonal elements corresponding to the similar spectra and the off-diagonal elements between them.

1.3.3. Selecting "Analytical" Wavelengths

Choosing a set of wavelengths at which to gather absorbance data in order to estimate the concentration of components in a mixture is a historic problem in spectrophotometric analysis.[12–14] Since minimizing the parameter variance (least squares) returns a value of the parameter that maximizes the likelihood of having observed the data, one would optimally design quantitative, spectrophotometric experiments by selecting "analytical" wavelengths that minimize the elements of the variance–covariance matrix. Since the design only affects $(\mathbf{K}^T\mathbf{K})^{-1}$ in equation (1.16), one need not consider the measurement precision contribution to $\mathbf{V}$ when selecting wavelengths at which to gather data.

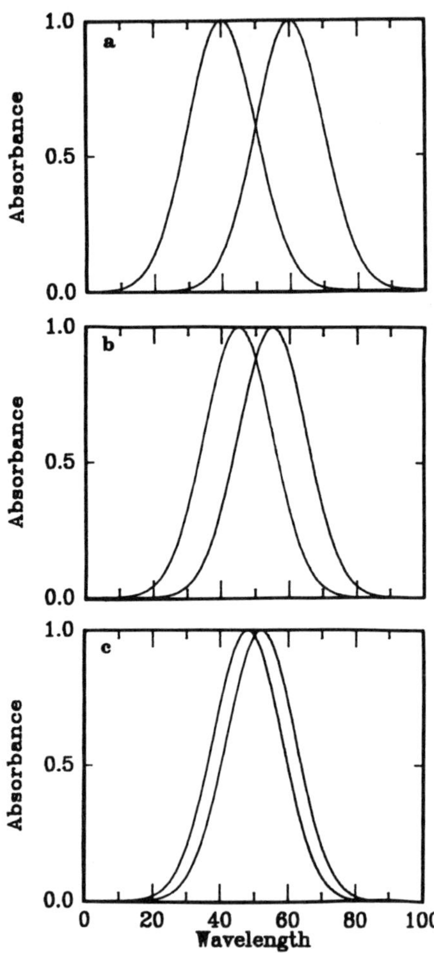

FIGURE 1.2. Model spectra for two-component mixtures: (a) $\lambda_{max} = (40, 60)$; resolution, $R_s = 0.5$. (b) $\lambda_{max} = (45, 55)$; resolution, $R_s = 0.25$. (c) $\lambda_{max} = (48, 52)$; resolution, $R_s = 0.1$. The width (standard deviation) of the Gaussian peaks is 10 wavelength units. R_s is the difference in the means of the two Gaussian peaks divided by four times the standard deviation.

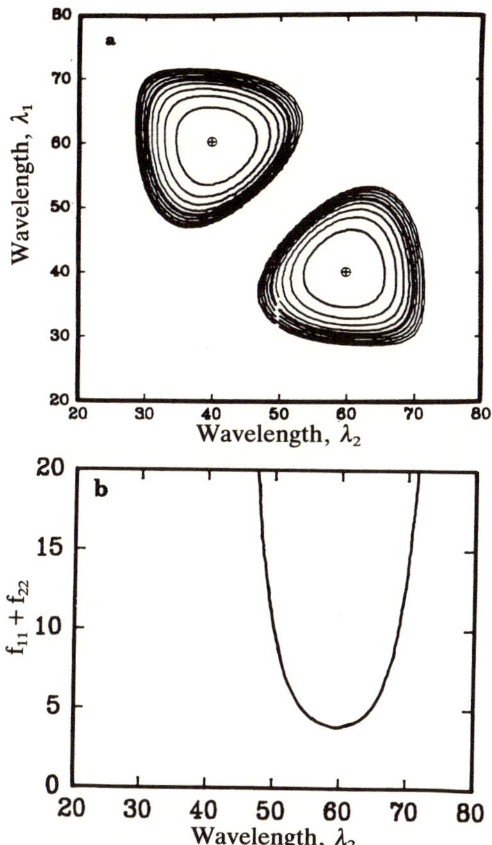

FIGURE 1.3. Variance in concentration estimates versus two analytical wavelengths for spectra from Fig. 1.2(b). (a) Contour plot of $f_{11} + f_{22}$; increment between contours is 1.73. (b) Horizontal slice through (a) at $\lambda_1 = 40$.

This regression-based concept for selecting "analytical" wavelengths has been evaluated for two-component mixtures.[15] Spectra were modeled as Gaussians as shown in Fig. 1.2, where the distance between the means was varied. To assess the predicted error in the estimated concentrations as a function of wavelengths in the design matrix **K**, the sum of the diagonal elements of $(\mathbf{K}^T\mathbf{K})^{-1}$, $f_{11} + f_{22}$, is plotted versus the first two wavelengths chosen. The results are shown as a contour plot in Fig. 1.3 for the spectra in Fig. 1.2(b). The error surface has equivalent minima at $(\lambda_1, \lambda_2) = (40, 60)$ and $(60, 40)$, where $f_{11} + f_{22} = 3.90$. Since the standard spectra are symmetric, $f_{11} = f_{22}$, the minimum value of $f_{11} + f_{22}$ indicates that replicate determinations of the concentrations of the components by a two-wavelength measurement at the best wavelengths would exhibit a variance that is 1.95 times larger than the variance of the absorbance measurements [see equation (1.17)]. The optimum set of wavelengths represents a distinct minimum, as shown in the slice through the error surface in Fig. 1.2(b).

Generating optimum data at more than two wavelengths presents a larger error minimization problem; an effective approach to dealing[(15)] with this problem is to fix the first two wavelengths at the above values and to vary the next pair. The concentration variance that results from this approach is plotted in Fig. 1.4. The results indicate that the same pair of wavelengths are optimal for the third and fourth measurements as for the first two. The concentration variance arising from two measurements,

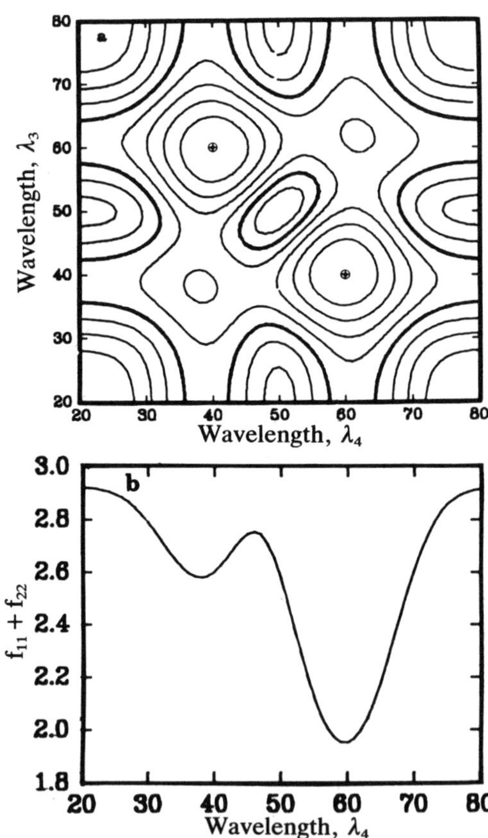

FIGURE 1.4. Variance in concentration estimates for a four-wavelength design. Wavelengths (40, 60) are preselected according to Fig. 1.3, and the next two wavelengths are varied. Minima at (40, 60) and (60, 40) are indicated, where $f_{11} + f_{22} = 1.95$. (a) Contour plot of $f_{11} + f_{22}$. (b) Horizontal slice at $\lambda_3 = 40$.

each at the same pair of wavelengths, is exactly one-half that observed for the two-wavelength design of Fig. 1.3. This is not surprising because the design has not changed except to double the number of measurements, which improves the measurement variance by a factor of two. While the optimal wavelengths in a design remain the same as the number of measurement is increased, the minima in the error surface become much less distinct; this trend is apparent in comparing Fig. 1.3 with Fig. 1.4, and thus flattening of the error surface continues as m increases. The penalty for measuring at less than the optimum wavelengths is smaller once measurements at or near the optimum region are included in the design.

To test this trend and its effects on concentration precision, a 50-fold replicated, 2-wavelength experiment design was compared to a measurement of a complete 100-wavelength spectrum with wavelengths spread uniformly over the range of Fig. 1.2. For the resolution of component spectra $R_s = 0.5$, 0.25, and 0.1, the concentration variance improved by a factor of 3.1, 3.7, and 4.1, respectively, when using the optimum 2-wavelength design instead of measuring a complete spectrum. The improvement in precision was greatest when the spectra were poorly resolved, and the magnitude of the improvement was rather modest when the number of measurements was large.

This modest gain in precision provided by a replicated n-wavelength design for an n-component determination is offset by a significant penalty—an insensitivity to model error. In the analysis of a mixture spectrum modeled by equation (1.13), we assume that all the components in the mixture are represented among the standard spectra in the matrix $\mathbf{K}$. If this is not the case, due to an unexpected contamination, for example, the vector of residual error, $\mathbf{R}$, which is the difference between the best fit $\mathbf{K}\tilde{\mathbf{C}}$ and the measured spectrum $\mathbf{A}$, will generally show structure due to the spectral variation not accommodated by the model; furthermore, the magnitude of the residuals will exceed the expected measurement error. In the case of a replicated n-wavelength design for an n-component determination, the residuals are not sensitive to model error and will never exceed the measurement precision. This situation is analogous to fitting calibration data to a straight line by acquiring replicate measurements of the dependent variable at only two points along the x-axis. If these points are at the origin and the extremum of the x-axis, then the precision of estimating the intercept and slope are optimized. On the other hand, this choice of measurements along the x-axis gives no hint as to whether the data should be fit to a straight line, that is, whether the assumed model is correct. For large numbers of measurements, acquiring data over the entire range of the x-axis returns slightly poorer precision in the estimated slope and intercept, but allows a nonlinear response to be detected in the residuals.

1.3.4. Weighting Observations in One-Dimensional Linear Regression

In deriving the concentration estimates that maximize the likelihood of observing a particular mixture spectrum, we have thus far assumed that the measurement variance is constant, as in equation (1.14a). The estimated concentrations $\hat{\mathbf{C}}$ given by equation (1.16) are the values that minimize the sum of the unweighted squared residuals $\mathbf{R}^T\mathbf{R}$. If error of measurement does not satisfy this assumption, then the residuals $\mathbf{R}$ are drawn from populations of differing variance, and the $1/\sigma_i^2$ factors in the summation that defines chi-square in equation (1.14) cannot be equated and brought outside the sum. The normal equations [equation (1.15)] under these conditions must therefore minimize the sum of the squared residuals with each residual weighted by the inverse of the expected variance; this is analogous to equation (1.4) for zero-dimensional data.

A convenient algebraic approach to achieving this goal is to multiply equation (1.13) by a weighting factor that makes the elements of the residual vector have the same variance. A factor that will accomplish this goal is an $m \times m$ diagonal matrix $\mathbf{W}$, where the elements $w_{ii} = 1/\sigma_i$.[4] Multiplying both sides of equation (1.13) by this matrix results in the following identity:

$$\mathbf{WA} = \mathbf{WKC} + \mathbf{WR} \tag{1.18}$$

The elements of the weighted vector of residuals, $\mathbf{WR}$, are drawn from a population having the same variance (i.e., equal to unity). The weighted definition of chi-square has a simple linear algebra form given by

$$\begin{aligned}\chi^2 &= (\mathbf{WR})^T(\mathbf{WR}) \\ &= \mathbf{R}^T\mathbf{W}^T\mathbf{WR} = \sum_{i=1}^{m} r_i^2/\sigma_i^2 \end{aligned} \tag{1.19}$$

where $\mathbf{W}^T\mathbf{W}$ is a diagonal matrix whose elements are $1/\sigma_i^2$. Since $\mathbf{WR}$ of equation (1.18) has uniform variance, one can obtain the maximum likelihood concentration estimates for this equation using the linear algebra solution to the normal equations for the uniform variance case, equation (1.16). Multiplying the weighted mixture spectrum $\mathbf{WA}$ on the left by the pseudoinverse or least squares inverse of $\mathbf{WK}$ gives the concentrations that minimize chi-square of equation (1.19):

$$\begin{aligned}\hat{\mathbf{C}} &= [(\mathbf{WK})^T(\mathbf{WK})]^{-1}(\mathbf{WK})^T(\mathbf{WA}) \\ &= (\mathbf{K}^T\mathbf{W}^T\mathbf{WK})^{-1}\mathbf{K}^T\mathbf{W}^T\mathbf{WA} \end{aligned} \tag{1.20}$$

The variance–covariance matrix for the estimated concentrations from this weighted least squares analysis arises from the inverse term of equation (1.20), which is analogous to equation (1.17) for the unweighted case:

$$\mathbf{V} = (\mathbf{K}^T\mathbf{W}^T\mathbf{W}\mathbf{K})^{-1} \tag{1.21}$$

1.3.5. Application to Time-Resolved Fluorescence Spectroscopy

Fluorescence measurements made with photon-counting detection and stable excitation sources are generally dominated by shot noise, which is characterized by a Poisson error distribution.[5] As a result, the residual errors arising from fitting spectra or time-decay curves have a variance equal to the mean number of counts detected, which for large number of counts (>100) may be approximated (with <20% error) by the number of observed counts, $\sigma_i^2 \sim a_i$. Substituting this approximation into $\mathbf{W}$ results in the diagonal elements of $\mathbf{W}^T\mathbf{W}$ having the value $1/a_i$, so that the product $\mathbf{W}^T\mathbf{W}\mathbf{A}$ in equation (1.21) is an m-element column vector where all elements are unity. Interestingly, all of the information about the measured fluorescence spectrum or time-decay curve used to estimate the concentrations in equation (1.21) resides in $\mathbf{W}^T\mathbf{W}$, which is within the inverse.

Recently, this method of data analysis has been applied to quantitative resolution fluorescence decay curves where the lifetimes of the two components are similar.[16] The form of the one-dimensional data is shown in Fig. 1.5 where naphthalene in cyclohexane is repetitively excited with a pulsed laser, and the decay of fluorescence intensity is collected as a time-histogram of single-photon arrivals following the excitation pulse.[17] The decay of the excited-state population is governed by first-order kinetics. This single component transient can therefore be fit to a single exponential decay of the form $a_i = (c_j/\tau_j)\exp(i\Delta t/\tau_j)$ by minimizing the chi-square of equation (1.19) with respect to c_j and τ_j. While the magnitude of the unweighted residuals shows a dependence on signal intensity, as shown in Fig. 1.5(a), the residuals weighted by the inverse of the expected shot noise, shown in Fig. (1.5b), are random and have the same variance.

Time-resolved fluorescence spectroscopy can be used for multicomponent determinations; the technique is especially useful, for example, as an in situ spectroscopic probe of molecular environments.[18] Since detection of fluorescence intensity is linear, the decay curve of intensity from a multicomponent sample can be modeled according to equations (1.12) and (1.13), where $\mathbf{K}$ contains normalized decay curves of the components $k_{ij} = (1/\tau_j)\exp(i\Delta t/\tau_j)$, and the vector $\mathbf{C}$ contains the total number of photon counts (the preexponential factors) that are propor-

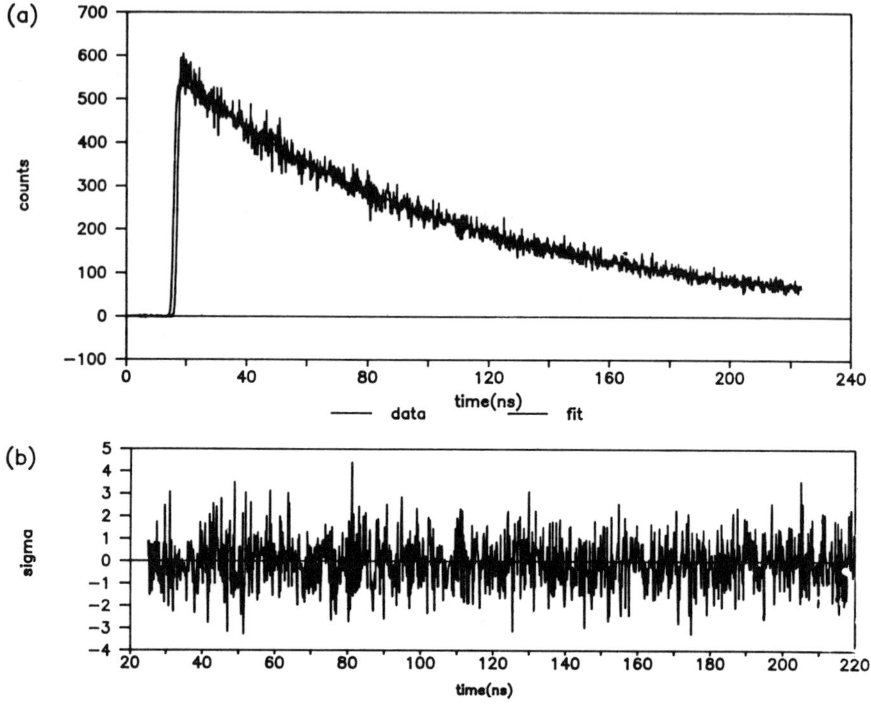

FIGURE 1.5. Fluorescence decay curve for naphthalene in cyclohexane. (a) The data are fit to a single exponential decay. (b) The residuals are weighted by the predicted error $r_i/(a_i)^{1/2}$.

tional to concentration. This approach to the quantitative analysis of mixed decay curves allows the use of equation (1.20) to efficiently determine the statistically optimal set of preexponential factors. If the fluorescence lifetimes of the components are known in advance, then the known vectors that comprise the matrix **K** can be used to extract concentrations in one step.

The advantage of having a physical model for a spectroscopic process being measured, and thereby a functional form for the data, is that **K** need not be known in advance. By varying the n fluorescence lifetimes, τ_j, which define the vectors in **K**, one can determine the particular lifetime values that minimize chi-square according to equation (1.19) and thus obtain the best estimate of the matrix **K** for a measured decay curve. While the preexponentials and fluorescence lifetimes can both be determined by a search of parameters, more precise parameter

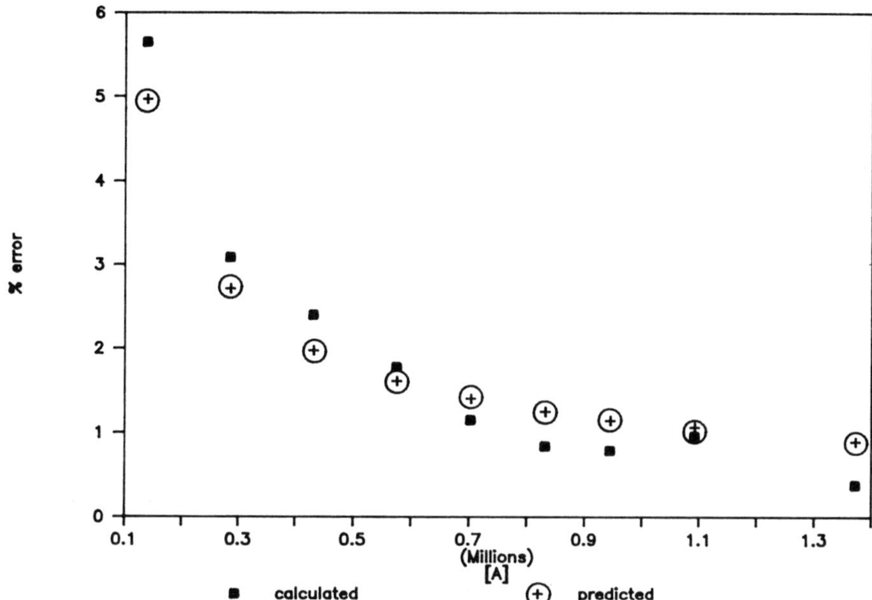

FIGURE 1.6. Observed and predicted errors in quantitative analysis of two-component fluorescence decay curves. The lifetimes are $\tau_1 = 107.7$ ns and $\tau_2 = 85.0$ ns. Solid squares are the relative standard deviation predicted for the shorter-lived component using the variance–covariance matrix; circled symbols are the observed relative standard deviation for this component. The x-axis indicates the total photons counted in the measurement, $a_1 + a_2$.

estimates are obtained faster by incorporating the linear least squares determination of $\hat{\mathbf{C}}$ using equation (1.20) into the search for the n nonlinear parameters τ_j.[18] A second advantage of using a weighted, linear regression step to estimate the component amplitudes is that the uncertainty of the estimates can be predicted from first principles. Using the relative standard deviations derived from the variance–covariance matrix for equation (1.20), $\mathbf{V} = (\mathbf{K}^T\mathbf{W}^T\mathbf{W}\mathbf{K})^{-1}$, the errors in determining the component amplitudes from a series of measured fluorescence decay curves were predicted and compared with the observed precision found in replicate measurements. As shown in Fig. 1.6, the error predictions of the variance–covariance matrix follow the observed results over a wide range of total photon counts in the data.

1.4. TWO-DIMENSIONAL SPECTROSCOPIC MEASUREMENTS

1.4.1. Combinations of Correlated and Uncorrelated Dimensions

The exponential decay of intensity in a time-resolved fluorescence experiment provides an excellent example of a correlated measurement dimension. While fluorescence intensity is measured at hundreds of points in time in such an experiment, the intensity channels are not independent but are related by the functional form of the decay of the components. It is prior knowledge of this relationship that allows the **K** matrix to be determined by fitting only one parameter per component in the sample. While such correlated behavior is valuable for resolving overlapped data from mixtures, the number of degrees of freedom in such a measurement is much smaller than in data that are less predictable and therefore more informative.[(2)]

Among the most powerful spectroscopic methods for resolving and identifying components in complex mixtures are combinations[(3)] of correlated and uncorrelated measurement dimensions. Examples include GC/MS, LC/UV, GC/IR, and time-resolved fluorescence spectroscopy. In these methods, the correlated measurement dimension (generally the time dependence, as in chromatography/spectroscopy combinations) can be used to resolve overlap between the components, either by using a physical model of the response[(19–21)] or by seeking correlations in the time dependence with factor analysis.[(22–24)] The spectra of components, thus resolved, generally show much richer variation and thus contain more information for identification. In the absence of any prior knowledge of what possible components are present, the unpredictable nature of the spectra makes mixture analysis with only this dimension impossible. Therefore, the combination of predictable and unpredictable measurement dimensions is ideally suited to determining the composition of a complex sample.

1.4.2. Modeling the Correlated Dimension: pH/UV Data Analysis

The use of physical models and regression methods for resolving component behavior in a correlated dimension is greatly assisted by measurements taken along a second spectroscopic dimension that is less predictable. Differences in the components along the information-rich spectroscopic variable aid in the convergence of the model. An example of this benefit has been demonstrated for analyzing spectrophotometric titrations by measuring a complete UV/VIS absorption spectrum as a function of pH. A synthetic example of such a data set is shown in Fig. 1.7, where the absorption of the mixture of two monoprotic acid–base

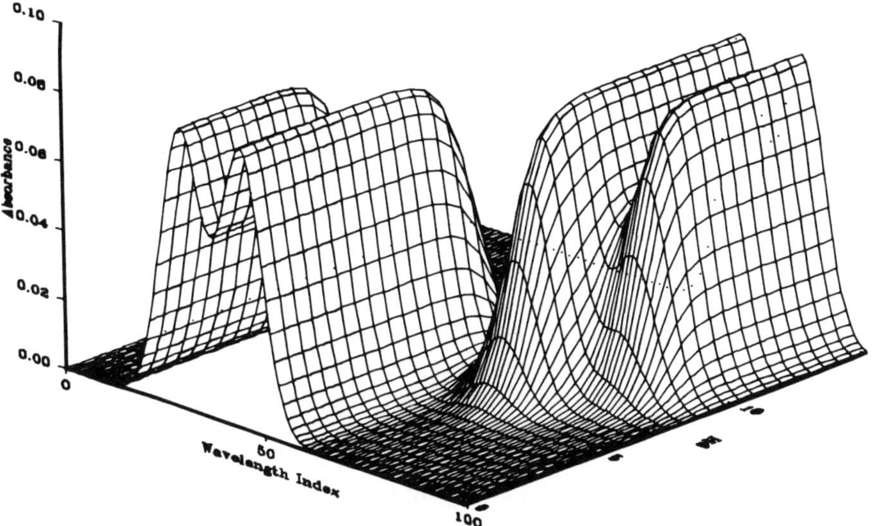

FIGURE 1.7. Synthetic data for a spectrophotometric titration of a binary mixture of monoprotic acids.

pairs shows that the dominance of the acid forms at low pH, and the transformation to base forms at higher pH. Since the absorbances of the components in the mixture are additive, the absorbance at any wavelength i and pH j is the sum of the contribution of each of the components in the mixture. The data for a given sample are no longer a single vector but rather a matrix of absorption spectra as a function of pH.

If the spectra of the components are independent of pH where the acid and base forms change only in relative proportion, the response can be modeled according to equations (1.12) and (1.13). The matrix $\mathbf{A}$ contains the absorption spectra versus pH for the mixture, where i is the index of wavelength down the rows and j is the index of pH across the columns. Like the one-dimensional case, $\mathbf{K}$ contains the spectra of the n components in its columns; however, $\mathbf{C}$ is now a matrix containing the pH dependent distribution curves in its rows. For a given measurement of $\mathbf{A}$, the data analysis task is to decompose the matrix into best estimates $\hat{\mathbf{K}}$ and $\hat{\mathbf{C}}$, which is done without advance knowledge of either factor.

To carry out this task, the pH-dependent distribution curves will be modeled according to equilibrium theory. To factor the model behavior from the data, however, the rows of $\mathbf{C}$ must be linearly independent so that a unique, best fit value of $\hat{\mathbf{K}}$ exists. This can be accomplished by defining the rows of $\mathbf{C}$ as difference composition curves,[25,26] one for each of the n acid–base pairs in the mixture,

$$c_{k_j} = \frac{10^{(\mathrm{p}H_i - \mathrm{p}K_k)} - 1}{1 + 10^{(\mathrm{p}H_i - \mathrm{p}K_k)}} \tag{1.22}$$

which corresponds to a difference absorption spectrum in $\mathbf{K}$ given by

$$k_{ik} = (\varepsilon_{A^-} - \varepsilon_{HA})_{ik} b([HA] + [A^-])_k \tag{1.23}$$

where the difference in the molar absorptivity of the base and acid forms is multiplied by the sample path length b and the total concentration of the particular acid–base pair. To observe total intensity in the data, a final $n + 1$ row is defined in $\mathbf{C}$ in which all of the elements are equal to n. This corresponds to an $n + 1$ row in $\mathbf{K}$, which contains the sum of all absorbing species in solution including non-pH-varying species.

For a given estimate of the pH-dependent composition $\hat{\mathbf{C}}$, which requires estimating the n pK_a's, the least squares set of difference spectra can be found by multiplying the data matrix $\mathbf{A}$ by the right-pseudoinverse of $\hat{\mathbf{C}}$, which is analogous to equation (1.16) above,

$$\hat{\mathbf{K}} = \mathbf{A}\hat{\mathbf{C}}^T(\hat{\mathbf{C}}\hat{\mathbf{C}}^T)^{-1} \tag{1.24}$$

The quality of the fit of the product $\hat{\mathbf{K}}\hat{\mathbf{C}}$ to the data depends on the accuracy of the estimated pK_a's defining $\hat{\mathbf{C}}$. To test the quality of fit, the value of chi-square [equation (1.14) for the unweighted case] is calculated. The estimates of the pK_a's are varied so as to minimize chi-square, usually by a Nelder–Mead simplex algorithm;[27,28] at each step of the nonlinear least squares search for the pK_a's, the linear least squares step of equation (1.24) returns the best estimate of the matrix $\mathbf{K}$ for a given estimate of $\mathbf{C}$.

1.4.3. Acid–Base Mixture Resolution and Error Predictions

To test this method of resolving mixtures of monoprotic acids, data matrices containing absorption spectra of mixtures of two and four acid–base indicators were acquired at intervals of 0.2 pH units over a pH range of 3.0 to 8.4. A plot of the four-component data matrix is shown in Fig. 1.8; the composition of the sample is listed in Table 1.1. From the shape of the data surface, the spectral and pH variation of three acid–base components is apparent, but behavior of a fourth component is not obvious. It is, however, clear that the component spectra and pH dependences are severely overlapped. Despite the severity of the

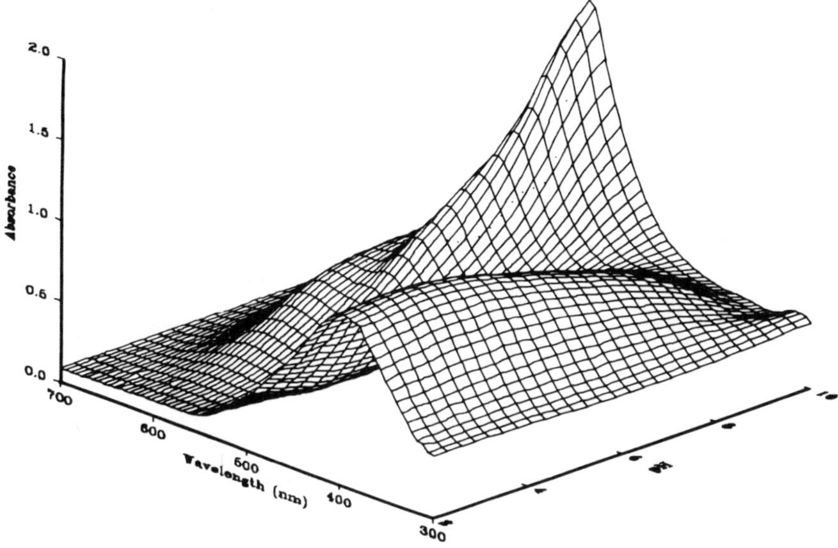

FIGURE 1.8. Spectrophotometric titration of four acid–base indicators. See Table 1.1 for sample composition.

overlap, reiterative application of equation (1.24) to determine the least squares difference spectra as the n pK_a's are varied to minimize chi-square results in an optimum fit to the data matrix, with the results summarized in Table 1.1. The accuracy of the difference spectra extracted from the mixture is illustrated in Fig. 1.9, where the results are compared with spectra of the individual components. The quality of fit is good without any systematic error. The differences between the pK_a's determined from fitting the mixture and those obtained by fitting the

TABLE 1.1. Reiterative Least Squares Resolution of Acid–Base Mixtures

Sample composition	pK_a isolated	pK_a mixture	Observed spectral error, s_k	Predicted spectral error, σ_k[a]
Methyl orange	3.27	3.50	5.4×10^{-3}	5.2×10^{-3}
Bromocresol green	4.77	4.74	3.2×10^{-3}	2.3×10^{-3}
Methyl orange	3.27	3.39	5.5×10^{-3}	5.8×10^{-3}
Bromcresol green	4.77	4.80	3.8×10^{-3}	4.4×10^{-3}
Chlorophenol red	6.07	6.05	3.4×10^{-3}	3.9×10^{-3}
Phenol red	7.71	7.71	4.4×10^{-3}	3.7×10^{-3}

[a] From variance–covariance matrix.

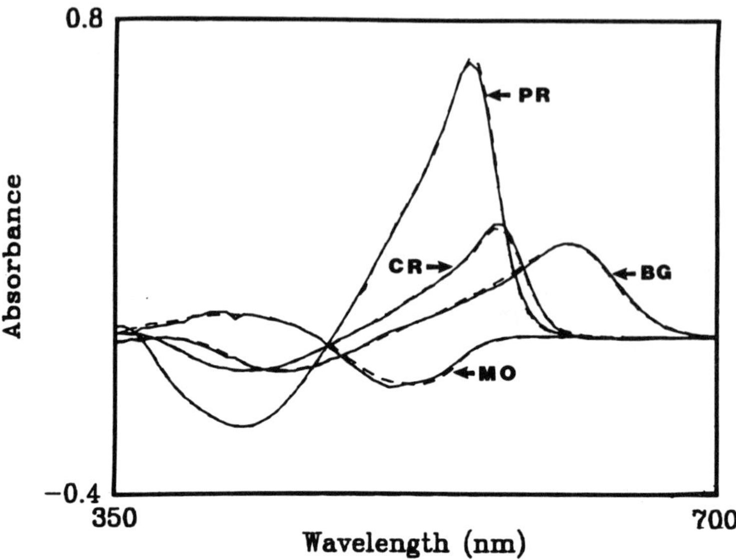

FIGURE 1.9. Difference acid–base spectra (solid lines) resolved from the four-component sample of Fig. 1.8. PR is phenol red; CR is chorophenol red; BG is bromcresol green, and MO is methyl orange. Dashed lines are the spectra of the individual components, plotted for comparison.

isolated indicators average 0.07 pH units, which is much less than the 0.2 pH interval between spectral scans in the data.

The accuracy of the resolved difference spectra extracted from the mixture is again predictable from first principles. The right-pseudoinverse in equation (1.24), $\mathbf{C}^T(\mathbf{CC}^T)^{-1}$, contains $n + 1$ columns corresponding to the $n + 1$ columns of $\hat{\mathbf{K}}$. These column vectors are multiplied by the corresponding rows of the mixture absorbance matrix to extract the estimated spectra in $\hat{\mathbf{K}}$, one row at a time. As a result, the error of this least squares solution depends only on the variance of the mixture absorbance and the degree to which the rows of $\mathbf{C}$ are overlapped—neither of which depend on wavelength. We can, therefore, use the variance–covariance matrix of equation (1.17), which for the right-pseudoinverse is $\mathbf{V} = (\mathbf{CC}^T)^{-1}\sigma^2$, to predict the magnitude of the error in the absorption spectra. Taking a value for the absorbance error from the root-mean-squared residuals, $\sigma = 5.5 \times 10^{-3}$ AU, the standard deviations of estimating the difference spectra are predicted and listed alongside the observed error in Table 1.1. The agreement between the predicted and observed errors is reassuring.

1.5. CONCLUSIONS

Regression methods of spectroscopic data analysis have their theoretical roots in the method of maximum likelihood and least squares analysis. This theoretical basis allows these methods to be derived from simple statistical concepts and applied to multidimensional data. The use of the regression method with models takes advantage of correlations in data in order to reduce the effect of noise and to resolve overlapped component responses. Without exploiting the prior information available in such correlations, they would only reduce the information content of the measurement without providing any benefit. A particular advantage of modeling is realized with spectroscopic methods that combine correlated and uncorrelated measurement dimensions. Modeling the response in the correlated dimension resolves the overlap of multicomponent samples and provides a pure component response in the uncorrelated, more informational, dimension. Using the theoretical basis of regression allows one to predict, *a priori,* the errors of a data analysis procedure. This analysis of errors can be used to properly weight observations contributing different uncertainty to a result, to optimally design an experiment in terms of what observations are made, and to know in advance what errors to expect in the result.

ACKNOWLEDGMENTS

This work was supported in part by the National Science Foundation through grant CHE85-06667 and by the Office of Naval Research. Fellowship support to J.M.H. from the Alfred P. Sloan Foundation is also acknowledged.

REFERENCES

1. H. Kaiser, *Anal. Chem.* **42**(2), 24A (1970).
2. T. Hirschfeld, *Anal. Chem.* **48**(1), 16A (1976).
3. T. Hirschfeld, *Anal. Chem.* **52,** 297A (1980).
4. N. R. Draper and H. Smith, *Applied Regression Analysis,* Wiley, New York (1981).
5. P. R. Bevington, *Data Reduction and Error Analysis for the Physical Sciences,* McGraw-Hill, New York (1969).
6. P. E. Poston and J. M. Harris, *Anal. Chem.* **59,** 1620 (1987).
7. R. B. Lam, *Appl. Spectrosc.* **37,** 567 (1983).
8. A. C. Tam and C. K. N. Patel, *Appl. Opt.* **18,** 3348 (1979).
9. L. A. Currie, *Anal. Chem.* **40,** 586 (1968).
10. G. Strang, *Applied Linear Algebra,* Academic, New York (1976).

11. S. N. Deming and S. L. Morgan, *Experimental Design: A Chemometric Approach,* Elsevier, New York (1987).
12. F. P. Zscheile, H. C. Murray, G. A. Baker, and R. G. Peddicord, *Anal. Chem.* **34,** 1776 (1962).
13. A. Przybylski, *Chem. Anal.* (*Warsaw*) **14,** 1047 (1969).
14. J. Sustek, *Anal. Chem.* **46,** 1676 (1974).
15. S. D. Frans and J. M. Harris, *Anal. Chem.* **57,** 2680 (1985).
16. A. L. Wong and J. M. Harris, *Anal. Chem.* **61,** 0000 (1989).
17. D. V. O'Connor and D. Phillips, *Time-Correlated Single-Photon Counting,* Academic, New York (1984); Chapter 6.
18. J. R. Lakowicz, *Principles of Fluorescence Spectroscopy,* Plenum, New York (1983).
19. F. J. Knorr and J. M. Harris, *Anal. Chem.* **53,** 272 (1981).
20. F. J. Knorr, H. R. Thorsheim, and J. M. Harris, *Anal. Chem.* **53,** 821 (1981).
21. S. D. Frans, M. L. McConnell, and J. M. Harris, *Anal. Chem.* **57,** 1552 (1985).
22. W. H. Lawton and E. A. Sylvestre, *Technometrics* **13,** 617 (1971).
23. M. A. Sharaf and B. R. Kowalski, *Anal. Chem.* **53,** 518 (1981).
24. P. J. Gemperline and J. Craig Hamilton, "Factor Analysis of Spectro-Chromatographic Data", Chapter 2 in this volume.
25. R. I. Shrager and R. W. Hendler, *Anal. Chem.* **54,** 1147 (1982).
26. S. D. Frans and J. M. Harris, *Anal. Chem.* **56,** 466 (1984).
27. J. A. Nelder and R. Mead, *Comput. J.* **7,** 308 (1965).
28. S. N. Deming and S. L. Morgan, *Anal. Chem.* **45,** 278A (1974).

Chapter 2

Factor Analysis of Spectro-Chromatographic Data

Paul J. Gemperline and
J. Craig Hamilton

2.1. INTRODUCTION

The purpose of this chapter is to describe applications of factor analysis to detect and resolve overlapped HPLC (high-performance liquid chromatography) peaks from diode array spectroscopic data. We will begin with a brief introduction to the technique of factor analysis, and discuss its application for detecting overlapped peaks, including the effects of spectral dissimilarity and chromatographic resolution on the limit of detection for severely overlapped peaks. From studies using simulated data matrices, we have developed a technique for investigating both of these effects independently and for estimating the net signal due to a minor component when overlapped with a major component. The technique for performing self-modeling curve resolution of overlapped peaks, called iterative target transformation factor analysis (ITTFA), will also be described. Real and simulated data will be presented throughout to illustrate the various topics.

2.2. BACKGROUND

Consider a hypothetical data matrix **A** whose rows represent the UV spectra of a flowing HPLC effluent over a period of time where a peak is

Paul J. Gemperline and J. Craig Hamilton • Department of Chemistry, East Carolina University, Greenville, North Carolina 27858.

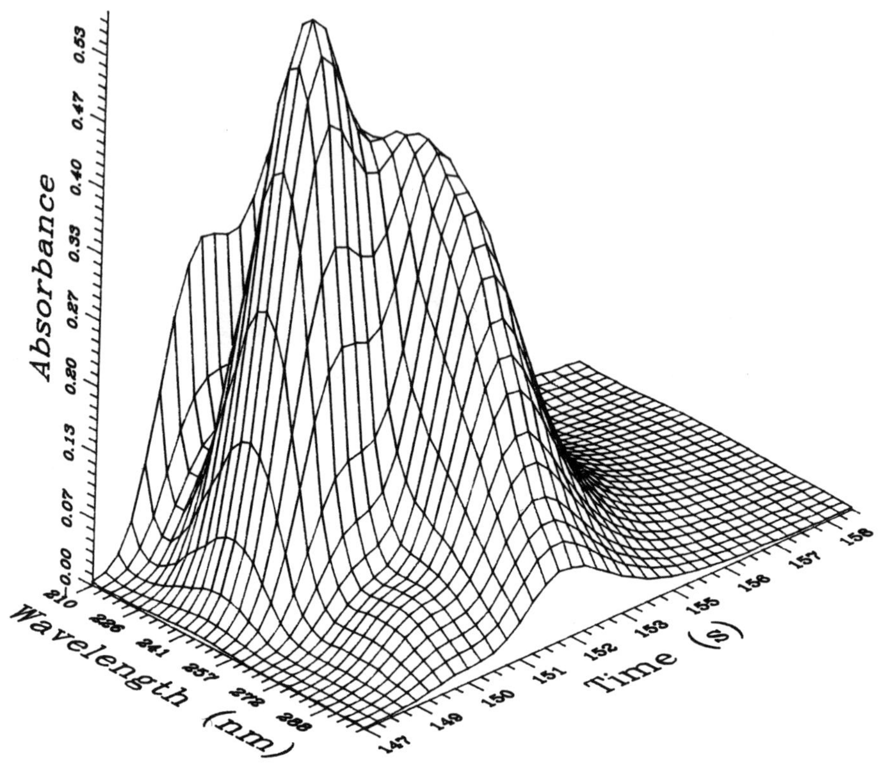

FIGURE 2.1. HPLC absorbance matrix **A** for a two-component mixture of acetophenone and nitrobenzene.

observed to elute. Furthermore, suppose the peak is actually composed of two overlapping components whose UV spectra are somewhat different from each other. A plot of such a data matrix is shown in Fig. 2.1. Two overlapped components, acetophenone and nitrobenzene, are present. Each spectrum in the plot represents a two-component mixture according to Beer's law,

$$A_{i,j} = \sum_{k=1}^{n} c_{i,k} e_{k,j} \tag{2.1}$$

where $A_{i,j}$ is the absorbance at time i and wavelength j; n is the number of absorbing components; $c_{i,k}$ is the concentration of the kth component at time i; and $e_{k,j}$ is the molar absorptivity of the kth component at wavelength j. In matrix form, equation (2.1) can be written

$$\mathbf{A} = \mathbf{CE} \tag{2.2}$$

where the columns of C represent the concentration of the pure components (elution profiles), and the rows of E represent the molar absorptivity spectra of the pure components.

The goal of factor analysis is to decompose the raw matrix A into a matrix of abstract factors, C_{abst}, and E_{abst}, without standards or other prior knowledge:

$$\mathsf{A} = \mathsf{C}_{\text{abst}}\mathsf{E}_{\text{abst}} \tag{2.3}$$

The goal of self-modeling curve resolution, described later, is to transform C_{abst} and E_{abst} to give estimates of the physically meaningful matrices, C_{real} and E_{real}, also without standards or other prior knowledge.

Each column, $\mathbf{c}_i$, of C_{abst} is an eigenvector satisfying the relation

$$\mathbf{c}_i^T(\mathsf{AA}^T)\mathbf{c}_i = \lambda_i \tag{2.4}$$

where λ_i is the eigenvalue associated with $\mathbf{c}_i$, $\mathbf{c}_i^T$ is the transpose of the eigenvector, and A^T is the transpose of A. Several well-known methods exist for calculating the eigenvectors and eigenvalues.[1,2] Because the columns of C_{abst} are mutually orthogonal and normalized, the inverse of C_{abst} is simply its transpose. Using this relationship, E_{abst} is easily calculated according to equation (2.5):

$$\mathsf{C}_{\text{abst}}^T\mathsf{A} = \mathsf{E}_{\text{abst}} \tag{2.5}$$

If the data were free of experimental error, factor analysis would yield exactly n unique eigenvectors and nonzero eigenvalues, where n is the number of components present in the peak. Because of experimental error, factor analysis invariably gives c eigenvectors, where c is the number of columns (wavelengths) in A, assuming there are more rows than columns. Factors can be ranked according to the size of their associated eigenvalue, which is directly related to their ability to account for variation in the original matrix A. The first factor is associated with the largest, most important eigenvalue, and the last factor is associated with the smallest, least significant eigenvalue. Only n of the eigenvectors have physical meaning—those associated with the n largest eigenvalues. The set of factors associated with the n largest eigenvalues can be used to reproduce the original data matrix within experimental error. The remaining $c - n$ eigenvalues are called error eigenvalues, denoted by the superscript zero, λ_j^0, and account only for experimental error.

2.3. DETECTION OF OVERLAPPED PEAKS

Determining n, the number of significant eigenvectors, is equivalent to determining the minimum number of components present in the data matrix, provided the spectra of the associated components are different and the peaks of the associated components are not perfectly overlapped. Three techniques for determining the number of factors will be described here. The first is based on Malinowski's RE (real error) function,[3] and requires prior knowledge of the experimental error, and the other two are based on Malinowski's IND (indicator) function[3] and REV (reduced eigenvalue)[4] functions, both of which require no prior knowledge of experimental error. All three techniques assume the experimental error is homogeneous throughout the data matrix; that is, the standard deviation of the error is the same from column to column and row to row in the data matrix. Malinowski made the unusual assumption that the error is uniformly distributed, rather than normally distributed, in his development of REV.

2.3.1. Estimating the Number of Factors Using RE

The number of significant eigenvectors may be deduced according to the empirical rules described below. The known experimental error is compared with the estimate of the root-mean-square (rms) error in the data obtained during factor analysis. The rms error, or real error[3] (RE), is estimated according to equation (2.6), where r and c are the number of rows and columns in $\mathbf{A}$ (we assume r is greater than c), n is the number of suspected factors, and λ_j are the eigenvalues

$$\mathrm{RE} = \left(\sum_{j=n+1}^{c} \lambda_j/[r(c-n)] \right)^{1/2} \tag{2.6}$$

If after n factors are used, the estimated noise level, RE, is less than 0.0008 absorbance units (AU), then we estimate the number of components to be n, the same as the number of factors. If the change in the estimated noise level going from $n-1$ to n factors is less than 0.00045, then the estimated noise level at $n-1$ factors is considered so slight that only $n-1$ components (factors) are suspected instead of n.

For the results reported in this chapter, the LKB (LKB-Produkter AB) model 2140A diode array detector was used. According to the manufacturer's specifications, the detector has an experimental error of ±0.0005 AU. The noise level in a typical HPLC run may be higher than the specified level, and depends largely on the magnitude of absorption due to the solvent. In this case more precise estimates of the experimental error may be obtained from averages of spectra in a flat baseline region of the chromatogram.

2.3.2. Estimating the Number of Factors Using IND

Malinowski's empirical IND function may also be used to deduce the number of factors in a data matrix. Earlier studies suggested the function should reach a minimum when the proper number of factors is included[(3)]

$$\text{IND} = \text{RE}/(c - n)^2 \tag{2.7}$$

Although the test is not as reliable as the RE test described above, its advantage is that no prior knowledge of the experimental error is required.

2.3.3. Estimating the Number of Factors Using REV

In 1987 Malinowski proposed calculation of reduced eigenvalues, REV.[(4)] The reduced eigenvalues of the error eigenvalues, REV^0, are constant and may be predicted according to equation (2.8):

$$\text{REV}^0 = l\sigma^2 \tag{2.8}$$

where

$$l = r(c - n)\Big/\sum_{j=n+1}^{c} (r - j + 1)(c - j + 1) \tag{2.9}$$

According to equations (2.8) and (2.9), the expected values of the reduced eigenvalues depend only on the experimental error σ and the degrees of freedom. The actual reduced eigenvalues are calculated according to equation (2.10):

$$\text{REV}_j = \lambda_j/(r - j + 1)(c - j + 1) \tag{2.10}$$

Any eigenvalue λ_j that is not an error eigenvalue will have REV_j values significantly larger than those predicted by equation (2.8).

The following test makes use of the REV theory and requires no knowledge of experimental error to estimate the number of significant factors in a data matrix. Starting with the least significant eigenvalue, $(j = c)$, REV_j is calculated. REV_{j-1} is calculated, and the ratio r_{j-1} is obtained according to equation (2.11):

$$r_{j-1} = \text{REV}_{j-1}/\text{REV}_j \tag{2.11}$$

The value of j is decremented and the process is repeated going from least significant to most significant eigenvalues until all $c - 1$ ratios are tabulated. When λ_j and λ_{j-1} are error eigenvalues, the ratio r_{j-1} will be

TABLE 2.1. Analysis of Acetophenone (0.114 mg/ml) and Nitrobenzene (0.0011 mg/ml) Overlapped Peak at $R_s = 0.54$. (Data matrix dimensions: 22×30)

Factor	Eigenvalue	Percent variance	Real error	IND	REV	REV ratio
1	8.19905879	99.98141	0.0015556	3.53×10^{-6}	1.2422×10^{-2}	5807.66
2	0.00130263	0.01588	0.0006082	1.52×10^{-6}	2.1389×10^{-6}	14.84
3	0.00008072	0.00098	0.0004978	1.38×10^{-6}	1.4414×10^{-7}	1.12
4	0.00006617	0.00081	0.0003729	1.15×10^{-6}	1.2897×10^{-7}	1.46
5	0.00004128	0.00050	0.0002575	0.89×10^{-6}	8.8202×10^{-8}	3.19
6	0.00001176	0.00014	0.0002144	0.84×10^{-6}	2.7660×10^{-8}	1.73
7	0.00000614	0.00007	0.0001881	0.84×10^{-6}	1.5985×10^{-8}	2.00
8	0.00000276	0.00003	0.0001770	0.90×10^{-6}	8.0003×10^{-9}	

approximately equal to one. When the first significant eigenvalue λ_n is encountered, the ratio r_{n-1} will be significantly greater than one.

Table 2.1 shows the factor analysis results for a typical two-component overlapped peak of acetophenone (0.114 mg/ml) and nitrobenzene (0.0011 mg/ml) at a chromatographic resolution of $R_s = 0.54$. In this example, nitrobenzene is easily detected even though it is only present at the 1.0% level and is severely overlapped with the acetophenone peak. In Table 2.1, the estimated error is 0.0006 AU after two factors are used, indicating the presence of at least two components. Inclusion of an additional factor only slightly improves the estimated error, further substantiating the presence of two components.

The ratio of reduced eigenvalues is shown in the column labeled "REV ratio." The REV results corroborate the conclusion obtained above. Working backward from the bottom of Table 2.1 shows the reduced eigenvalues 3 through 8 to be nearly equivalent to one another, indicating they are actually error eigenvalues. Not until the second eigenvalue is encountered is a significant increase observed, indicating the second eigenvalue is not an error eigenvalue; that is, it accounts for significant variation in the data.

In Table 2.1, the IND function levels off after four or five factors and does not reach a minimum until seven factors are used. In this application the IND criteria is not as reliable as the RE and REV criteria. It has been shown in earlier work that the IND function is especially sensitive to rows or columns in the data matrix that exhibit slightly unique behavior.[5]

2.3.4. Detection of Too Many Components

Under certain circumstances, too many components may be detected in an overlapped peak. These type of errors are attributed to baseline drift, uncompensated background absorption, and instrumental and chemical factors that give rise to deviations from Beer's law. Baseline drift and uncompensated background absorption are modeled with extra factors during factor analysis, leading to the detection of too many components. Adequate baseline correction is necessary to obtain the expected results. Short-term drift arising from gradient elution experiments also results in extra factors.

Peaks that absorb strongly (absorbance > 0.8 AU) frequently exhibit a nonlinear response with the instrumentation used in this work. Two categories of nonlinear response may be suspected. The first is instrumental; the diode array detector may be operating outside its linear range or strong sample fluorescence may be present. The second is chemical; chemical equilibria or nonideal solution behavior may cause shifts in absorption maxima. If present, these nonlinear effects are modeled with extra factors, leading to the detection of too many components. The simplest remedy for the second class of problems is dilution.

2.3.5. Net Signal and Detection of Minor Components

Under certain circumstances, too few components may be detected in an overlapped peak. Consider a minor component Y overlapped with a major component X. If the concentration of Y is reduced while the concentration of X remains constant, errors will occur when the concentration of Y drops below the detection limit when overlapped with X. The degree of chromatographic resolution between X and Y, the dissimilarity between the UV spectra of X and Y, and the signal-to-noise ratio all work in concert to determine the detection limit. Theoretical calculations to predict the detection limit are hampered by the complicated nature of the interaction between chromatographic resolution, spectral dissimilarity, and peak height.

When standards are available, it is a simple matter to determine experimentally the limit of detection of Y when overlapped with X by holding the chromatographic resolution constant, holding the concentration of the major component X constant, and varying the concentration of Y. In this case, the concentration of Y is reduced until λ_2 becomes so small that Y is no longer detectable.

In the absence of standards, it is only possible to estimate the limit of detection of a minor component Y when overlapped with X. In a series of simulated studies, we have discovered a relationship shown in equation (2.12) that can be used to estimate, within an order of magnitude, the net signal of component Y when overlapped with X:

$$\lambda_2 = h^2 \beta_1 \beta_2 \lambda^* + \lambda_2^0 \tag{2.12}$$

Equation (2.12) takes into account the effect of chromatographic resolution, peak height, spectral dissimilarity, and experimental error on the magnitude of the net signal. The term λ_2 is the net signal (eigenvalue number 2) for minor component Y when overlapped with X; λ^* is the signal for component Y (eigenvalue number 1) when it is not overlapped; and λ_2^0 is the contribution of experimental error to λ_2, represented by the second error eigenvalue. The term h is the height of Y relative to X, β_1 is the fraction of λ^* remaining when Y is overlapped with X at a specific chromatographic resolution and peak height ratio, and β_2 is the fraction of λ^* remaining as a function of the dissimilarity between the UV spectra of X and Y.

2.3.6. Effect of Measurement Error on Net Signal

For two components, the error eigenvalue λ_2^0 is the minimum value of λ_2 to be expected, and depends on the experimental error and the degrees of freedom for the data matrix at hand. Two methods for estimating the expected error eigenvalue may be used. The first is to solve equations (2.8) through (2.10) for the expected error eigenvalue λ_2^0, using $j = 2$ and an estimated value for σ, giving equation (2.13):

$$\lambda_j^0 = (r - j + 1)(c - j + 1)l\sigma^2 \tag{2.13}$$

The second method is to use values of λ_3 from actual data matrices, where λ_3 is assumed to be an error eigenvalue. Equation (2.10) leads directly to equation (2.14). With $j = 2$ one may solve for λ_2^0 by multiplying REV_3 by the proper degrees of freedom:

$$\lambda_j^0 = \mathrm{REV}_{j+1}(r - j + 1)(c - j + 1) \tag{2.14}$$

2.3.7. Effect of Chromatographic Resolution on Net Signal

The fraction of λ^* remaining after considering chromatographic overlap β_1 is calculated according to equation (2.15),

$$\beta_1 = \lambda_2/\mathbf{v}_y^T\mathbf{v}_y \tag{2.15}$$

where λ_2 is the second eigenvalue obtained from the decomposition of a matrix of two vectors, $\mathbf{v}_x$ and $\mathbf{v}_y$, each representing the elution profiles of the pure components. The dot product in the denominator $\mathbf{v}_y^T\mathbf{v}_y$ is the total signal due to Y when not overlapped at all.

Calculations using simulated overlapped chromatographic peaks (Gaussian) were used to investigate the relationship between chromato-

TABLE 2.2. Value of β_1 as a Function of Chromatographic Resolution and Peak Height

Chromatographic resolution	Peak height ratios 1:1	1:1/2	1:1/16	1:1/64
1.00000	0.98174	0.99955	0.99966	0.99966
0.75000	0.89461	0.98526	0.98885	0.98889
0.50000	0.63212	0.82926	0.86420	0.86463
0.25000	0.22119	0.33755	0.39252	0.39340
0.12500	0.06058	0.09583	0.11708	0.11746
0.06250	0.01550	0.02473	0.03064	0.03075
0.03125	0.00390	0.00623	0.00775	0.00778
0.01563	0.00001	0.00002	0.00775	0.00778

graphic resolution and β_1. These results are summarized in Table 2.2 and Fig. 2.2. Peak separation was adjusted to give pairs of overlapped peaks having precisely known resolution. Equation (2.15) was used to calculate the values of β_1. The curves for peak height ratios of 1:1, 1:1/2, and 1:1/64 are shown in Fig. 2.2. Rapid convergence to the curve for 1:1/64 is observed as the height of the minor peak is decreased. Thereafter, no significant changes in the shape of the curve are observed. Since actual peaks are usually skewed, not Gaussian, slightly different results are expected for experimental data.

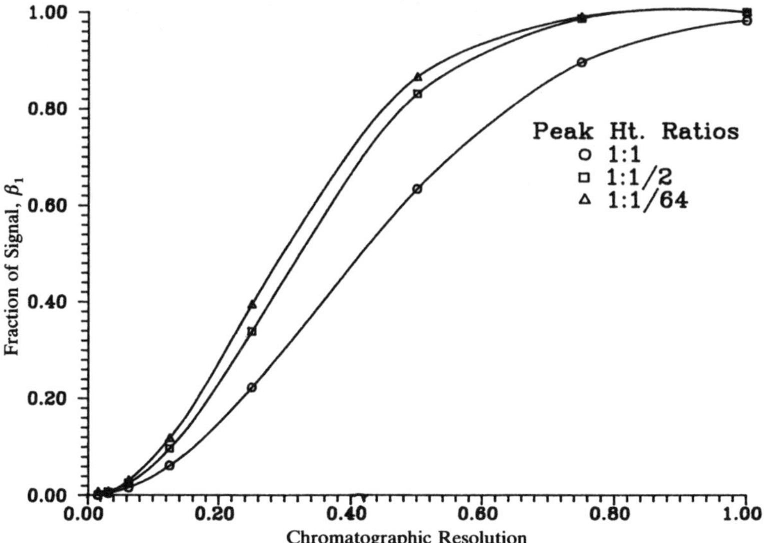

FIGURE 2.2. Effect of chromatographic resolution on net signal.

2.3.8. Effect of Spectral Dissimilarity on Net Signal

The fraction of λ^* remaining after considering spectral dissimilarity, β_2, is calculated in a fashion similar to β_1 according to equation (2.15). Here the vectors $\mathbf{v}_x$ and $\mathbf{v}_y$ represent the spectra of X and Y normalized to unit area. The dissimilarity between pairs of spectra is defined as $1 - \cos(\theta)^2$, where θ is the angle between the vectors, calculated according to equation (2.16),

$$\cos(\theta) = \mathbf{v}_x^T\mathbf{v}_y/\|\mathbf{v}_x\|\,\|\mathbf{v}_y\| \tag{2.16}$$

where $\|\mathbf{v}_x\|$ is the Euclidean norm of the vector

$$\|\mathbf{v}_x\| = (\mathbf{v}_x^T\mathbf{v}_x)^{1/2} \tag{2.17}$$

For superimpossible spectra, the angle between the two vectors will be zero, giving a dissimilarity of zero. Spectra that have no overlapped area at all are orthogonal and will have a dissimilarity of one. This definition of spectral dissimilarity is identical to Lorber's proposed selectivity, ζ_j.[6] His method of calculation may be used to extend the calculations to more than two components.

Table 2.3 shows the dissimilarity between several pairs of UV spectra and the corresponding values for β_2. Plots of these spectra may be found in Figs. 2.3 and 2.4. To illustrate the relationship between spectral dissimilarity and β_2, calculations using simulated spectra (Gaussian-shaped absorption bands) were used to generate a plot of β_2 as a function of spectral dissimilarity (see Fig. 2.5). For reference, the location of some of the pairs of spectra in Table 2.3 is shown on the curve. The plot shows that the fraction of net signal for component Y decreases sharply below a spectral dissimilarity of 0.80.

TABLE 2.3. Spectral Dissimilarity of Example Spectra

Spectral dissimilarity	β_2	Mixture
0.7107	0.2965	Acetophenone–nitrobenzene
0.5443	0.1611	Acetophenone–benzophenone
0.4055	0.0859	Acetophenone–biphenyl
0.2854	0.0416	Nitrobenzene–benzophenone
0.4626	0.1134	Nitrobenzene–biphenyl
0.2339	0.0277	Benzophenone–biphenyl
0.0263	0.0003	Toluene–cumene
0.1065	0.0057	Toluene–phenylcyclohexane
0.0836	0.0035	Cumene–phenylcyclohexane

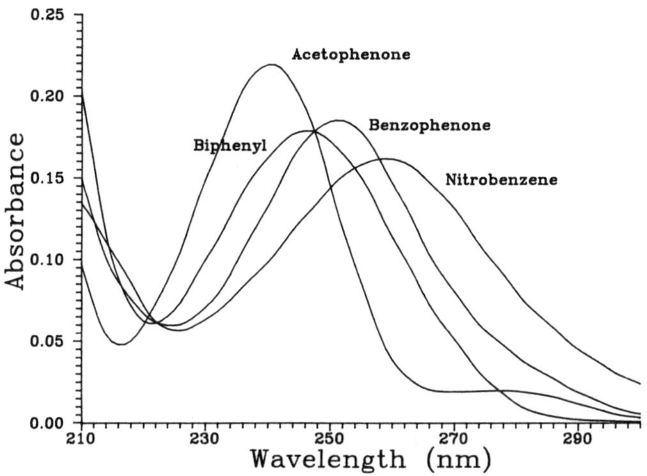

FIGURE 2.3. UV spectra of acetophenone, nitrobenzene, biphenyl, and benzophenone.

2.3.9. Analysis of Simulated Data

The results from Tables 2.2 and 2.3 can be used to estimate the net signal for a minor component Y when overlapped with X. Suppose a certain set of chromatographic conditions gives a resolution of 0.25 between nitrobenzene (major component) and acetophenone (minor component). Suppose the signal due to a certain pure acetophenone peak and pure nitrobenzene peak is $\lambda^* = 7.09$ and 7.09, respectively (the two

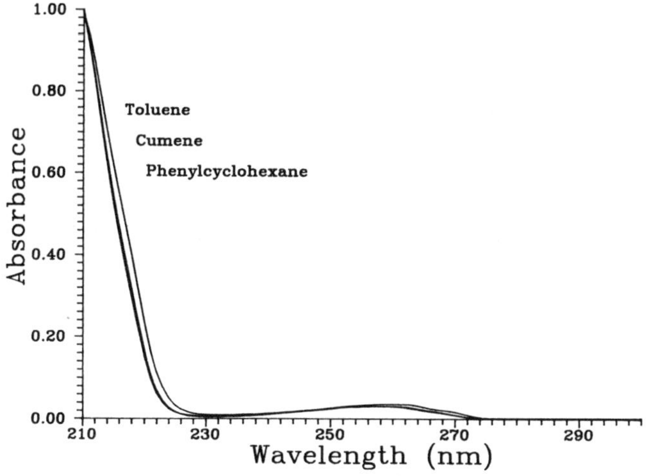

FIGURE 2.4. UV spectra of toluene, cumene, and phenylcyclohexane.

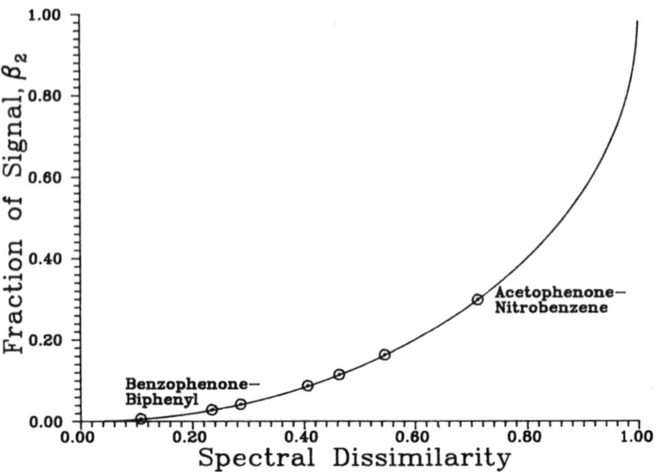

FIGURE 2.5. Effect of spectral resolution on net signal.

pure peaks have equal heights). Finally, suppose the expected measurement error is 5×10^{-4} AU for a 32×30 data matrix, giving a value of 2.2×10^{-5} for λ_2^0 according to equation (2.14). When nitrobenzene is overlapped with acetophenone at a height ratio of 1:1/2, an estimate of the net signal due to acetophenone may be calculated using $\beta_1 = 0.338$ and $\beta_2 = 0.297$:

$$\begin{aligned}\lambda_2 &= [(1/2)^2 \times 0.338 \times 0.297 \times 7.09] + 2.2 \times 10^{-5} \\ &= 0.178\end{aligned}$$

Factor analysis of a simulated data matrix under the above conditions gives an actual value of $\lambda_2 = 0.199$, corresponding to a 40-fold reduction in the original acetophenone signal. The estimated value is low by about 11%, but is within an order of magnitude of the actual value. At a peak height ratio of 1:1/64, the estimated value of λ_2 is 0.00022 compared to the actual value of 0.00034. In this case the net signal has been reduced by a factor of 2×10^4 and the estimate is about 41% low, still within the correct order of magnitude. Simulations involving other pairs of spectra under varying degrees of chromatographic resolution give estimates having the correct order of magnitude, with errors ranging from 0% to 50%.

2.3.10. Analysis of Experimental Chromatographic Data

Two-component overlap studies were performed using mixtures of acetophenone and nitrobenzene in a wide range of ratios to investigate the limit of detection for overlapped peaks. The standard solutions were run on a 25 cm × 4.6 mm C18 analytical column using a mobile phase of 95:5 ratio of methanol to water, giving a chromatographic resolution of about $R_s = 0.5$. Thirty-seven mixed and pure standards were prepared in the ratios shown in Table 2.4. The RE criterion most frequently gave the expected number of components, 30 out of 37 trials, corresponding to a component recognition rate of 81%. In seven trials the RE criterion overestimated the number of components by one. In four of these seven trials, an old bottle of nitrobenzene was used to prepare the standards.

TABLE 2.4. Analysis of Overlapped Peaks at Various Ratios of Acetophenone to Nitrobenzene

Std. no.	Concentration ACP	Concentration NIT	No. suspected components	No. components found RE	No. components found IND	No. components found REV
1	1	1	2	$2(2^+)$	$7(7^+)$	$2(3^+)$
2	1	1/2	2	2	6	2
3	1	1/5	2	2	7	2
4	1	1/10	2	2(2)	6(5)	2(4)
5	1	1/20	2	2	6	2
6	1	1/50	2	2(2)	7(7)	2(4)
7	1	1/100	2	2(2)	7(6)	1(2)
8	1	1/200	2	2	7	1
9	1	1/400	2	2	6	1
10	1/2	1	2	3^+	6^+	3^+
11	1/5	1	2	3^+	7^+	3^+
12	1/10	1	2	$3^+(2)$	$6^+(5)$	$3^+(4)$
13	1/50	1	2	2	6	1
14	1/100	1	2	3	6	3
15	1/200	1	2	2	7	3
16	0	1	1	$2^+(2)$	$4^+(6)$	$2^+(2)$
17	0	1/10	1	1	5	2
18	0	1/20	1	1	3	2
19	0	1/40	1	1	5	2
20	0	1/80	1	1	2	2
21	0	1/160	1	1	2	2
22	1	0	1	1(2)	6(6)	1(2)
23	1/2	0	1	1	7	2
24	1/5	0	1	1	4	2
25	1/10	0	1	1(1)	3(5)	2(1)
26	1/20	0	1	1	4	1
27	1/40	0	1	1	4	2
28	1/80	0	1	1	4	2
29	1/160	0	1	1	2	1

NOTE: (1) Separate trials; $^+$ suspected contaminant in nitrobenzene; see text. Highest acetophenone (ACP) concentration 1 = 0.114 mg/ml. Lowest acetophenone (ACP) concentration 1/200 = 0.0006 mg/ml. Highest nitrobenzene (NIT) concentration 1 = 0.107 mg/ml. Lowest nitrobenzene (NIT) concentration = 0.0003 mg/ml.

Excessive tailing was observed in the nonoverlapped nitrobenzene peak prepared from this bottle. In these four cases we suspect the extra component detected by the RE criterion may be attributed to an unknown contaminant in the nitrobenzene.

The performance of the IND criterion for determining the number of components is unsatisfactory in this application because it consistently overestimates the number of components by four or five. The performance of the REV criterion compared to the IND criterion is much better. In cases where there is no prior knowledge of experimental error, the REV criterion may provide an acceptable starting point for determining the number of factors. Compared to the RE criterion, the performance of the REV criterion is unsatisfactory because it gives the expected result in only 11 out of 37 trials, corresponding to a 30% component recognition rate. The REV criterion overestimates the number of components 21 out of 37 trials and underestimates the number of components 4 times.

The eigenvalues from factor analysis of the pure standards and mixtures in Table 2.4 were used to test the relationship expressed in equation (2.12) with experimental data. The resolution between the acetophenone and nitrobenzene peaks was about $R_s = 0.5$. The results of the calculations are shown in Table 2.5. In part A of Table 2.5, the values $\lambda^* = 14.52$, $\beta_1 = 0.86$, and $\beta_2 = 0.297$ were used. In part B of Table 2.5, the values $\lambda^* = 16.76$, $\beta_1 = 0.86$, and $\beta_2 = 0.297$ were used. An average value of $\text{REV}_3^0 = 2.83 \times 10^{-7}$ was calculated from experimental

TABLE 2.5. Observed and Expected Values of λ_2 for Experimental Data

Concentration (mg/ml)						
Acetophenone	Nitrobenzene	Expected height	Matrix size	Observed λ_1	Observed λ_2	Expected λ_2
A: Varying Nitrobenzene Concentration						
0.00000	0.10036	1	29 × 30	14.52	0.0004	0.0002$^+$
0.10258	0.01004	1/10	36 × 30	17.75	0.0536	0.0374
0.10258	0.00201	1/50	38 × 30	16.82	0.0029	0.0018
0.10258	0.00100	1/100	38 × 30	16.88	0.0015	0.0007
0.10258	0.00005	1/200	36 × 30	17.48	0.0009	0.0004
0.10258	0.00002	1/400	36 × 30	16.91	0.0007	0.0003
B: Varying Acetophenone Concentration						
0.10258	0.00000	1	27 × 30	16.76	0.0010	0.0002$^+$
0.01026	0.10036	1/10	32 × 30	16.35	0.0617	0.0431
0.00205	0.10036	1/50	31 × 30	15.04	0.0021	0.0019
0.00103	0.10036	1/100	38 × 30	15.88	0.0010	0.0008
0.00051	0.10036	1/200	36 × 30	14.78	0.0007	0.0004

NOTE: $^+$ Estimated error eigenvalue.

values of λ_3. This value was used to estimate values of λ_2^0 in Table 2.5 according to equation (2.14).

Fair agreement is observed between the observed values of λ_2 and the expected values of λ_2. At ratios lower than 1/50 the observed values of λ_2 level off, indicating a greater contribution from random experimental error.

2.3.11. Conclusions

The techniques reported here can be used to give fair estimates for the net signal due to a minor component in an overlapped peak. The estimated values are accurate to within an order of magnitude. The results of this study indicate that minor components will be detected reliably when their net signals are at least three times greater than the expected error eigenvalues. More extensive studies will be necessary to establish definitive rules for predicting the limit of detection of minor components in overlapped peaks.

2.4. ITTFA SELF-MODELING CURVE RESOLUTION

Curve resolution of overlapped peaks can be performed using the technique called iterative target transformation factor analysis (ITTFA) as first described by Gemperline,[7] later refined,[8] and also reported by Vandeginste.[9] This technique and other similar techniques[10–13] all owe their inception to the pioneering work of Lawton and Sylvestre in 1971.[14] Unlike Lawton and Sylvestre's technique, and refinements made by other workers, ITTFA is uniquely able to perform curve resolution of more than two components. Examples have been given that have showed curve resolution of three and four overlapped peaks.[7,9]

The primary advantage of the ITTFA curve resolution technique is that no prior knowledge of chromatographic conditions or the overlapped components is required. This makes ITTFA curve resolution ideally suited for troubleshooting difficult separations with unknown components or poorly characterized mixtures. In ITTFA, no assumptions regarding the shape of elution profiles are required, allowing various "real-world" skewed peak shapes to be modeled. For this reason, ITTFA is usually referred to as a *self-modeling* curve resolution technique. ITTFA does not require standards or spectra of the pure components. If such spectra were available, straightforward linear algebra techniques could be used to solve for the elution profiles.

The ITTFA technique starts with the $r \times k$ abstract concentration matrix $\mathbf{C}_{abst}$ described in equation (2.3), where r is the number of rows in

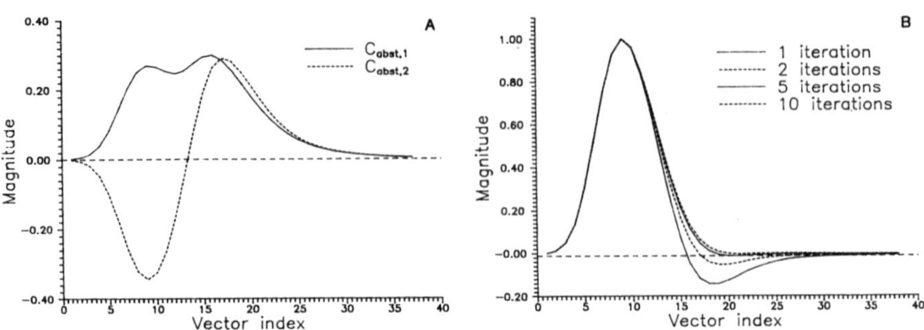

FIGURE 2.6. (A) Abstract concentration vectors for a two-component mixture of acetophenone and nitrobenzene, and (B) iterative refinement of the acetophenone test vector.

the original data matrix and k is the number of components determined from factor analysis. Figure 2.6(a) shows a typical plot of $\mathbf{C}_{\text{abst}}$ for a two-component overlapped peak, in this case acetophenone and nitrobenzene at a resolution of about 0.5. A "needle search" is performed using the target-testing technique to locate the retention times of the overlapped components. To this end, $i = 1$ to r test vectors, $\mathbf{c}_{\text{test},i}(r \times 1)$, are generated to approximate very narrow peaks at all retention times sampled under the peak cluster. Test vectors for the needle search are vectors of zeros except for 1.0 at the index of the retention time i to be tested. These test vectors are fit to the eigenvectors in the chromatographic dimension using the target transformation least-squares technique to give a $k \times 1$ vector of coefficients $\mathbf{t}_i$ according to equation (2.18):

$$\mathbf{t}_i = \mathbf{C}_{\text{abst}}^T \mathbf{c}_{\text{test},i} \tag{2.18}$$

where $\mathbf{t}_i$ minimizes the sum of the squares of the difference between $\mathbf{c}_{\text{test},i}$ and the predicted vector $\mathbf{c}_{\text{pred},i}$:

$$\mathbf{c}_{\text{pred},i} = \mathbf{C}_{\text{abst}} \mathbf{t}_i \tag{2.19}$$

Local minima appear in the residuals of the fit where a test vector coincides with an actual component's retention time. The predicted elution profiles $\mathbf{c}_{\text{pred},i}$ from the best test vectors are next refined in an iterative fashion to eliminate the physically meaningless negative portions of the elution profiles. Figure 2.6(b) shows a plot of the predicted concentration vector for the first component from Fig. 2.6(a) after 1, 2, 5, and 10 iterations. The iterative process is terminated when the error in the test vector is less than the measurement error or when the change in the test vector is too slight to allow further improvement.

2.4.1. Recent Refinements

The vectors from the iterative process have the proper shape but have arbitrary heights and must be scaled to properly model the overall chromatographic response function. To this end, the refined test vectors are assembled into columns of a concentration matrix $\mathbf{C}_{\text{test}}$ with a vector of ones added to model baseline offsets:

$$\mathbf{C}_{\text{test}} = [1 \mid \mathbf{c}_{\text{test},1} \mid \mathbf{c}_{\text{test},2} \mid \cdots \mid \mathbf{c}_{\text{test},k}] \tag{2.20}$$

The pseudoinverse of the resulting matrix, $\mathbf{C}^{+}_{\text{test}}$, is used to calculate the vector of least-squares scalars $\mathbf{s}$ that best fit the test vectors to the overall chromatographic response vector $\mathbf{c}_r$,

$$\mathbf{c}_r \mathbf{C}^{+}_{\text{test}} = \mathbf{s} \tag{2.21}$$

where

$$\mathbf{C}^{+}_{\text{test}} = \mathbf{C}^{T}_{\text{test}}(\mathbf{C}_{\text{test}}\mathbf{C}^{T}_{\text{test}})^{-1} \tag{2.22}$$

These scalars are then used to adjust the peak heights to fit the chromatogram, where s_1 is the coefficient for the vector of ones and is included to adjust for baseline offsets:

$$\mathbf{c}_{\text{ad}\,j,i} = s_i \mathbf{c}_{\text{test},i} + s_1 \tag{2.23}$$

The resulting adjusted vectors are assembled into columns of a matrix, forming $\mathbf{C}_{\text{ad}\,j}$ in the same fashion that $\mathbf{C}_{\text{test}}$ was formed, except that a column of ones is no longer necessary for baseline offsets. The pseudoinverse of $\mathbf{C}_{\text{adj}\,j}$ is then used to estimate the spectra of the pure components

$$\mathbf{A}\mathbf{C}^{+}_{\text{ad}\,j} = \mathbf{E} \tag{2.24}$$

where $\mathbf{A}$ is the original data matrix, and rows in $\mathbf{E}$ are estimates of the spectra.

2.4.2. Sample ITTFA Curve Resolution Results

The standards labeled 1–7 in Table 2.4 were chromatographed on a 25 m × 4.6 mm C18 column using three different mobile phases consisting of methanol and water in the ratios 90% : 10%, 95% : 5%, and 100% : 0% to control the degree of overlap and investigate its effects on the curve resolution results. A replicate of standard 1 and the standards labeled 10–12 were run on a different date using a different 25 cm × 4.6 mm C18 column and mobile phases of 90% : 10%, 95% : 5%, and 97% : 3% ratios of methanol to water. Figures 2.7–2.13 show plots of the chromatograms before (left side) and after (right side) curve resolution.

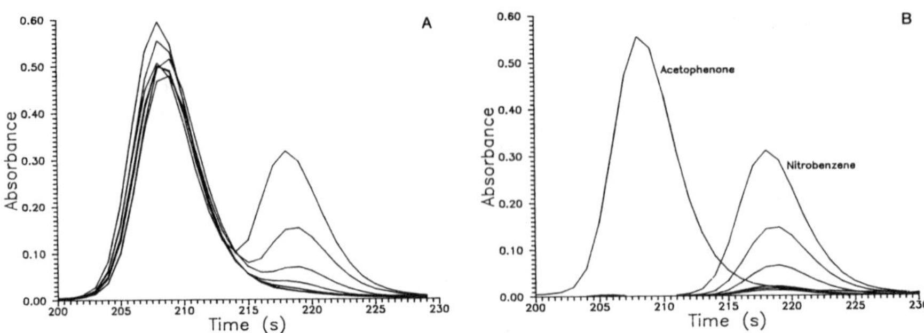

FIGURE 2.7. Chromatograms of standards 1–7 (A) before and (B) after ITTFA curve resolution using 90% methanol/10% water mobile phase.

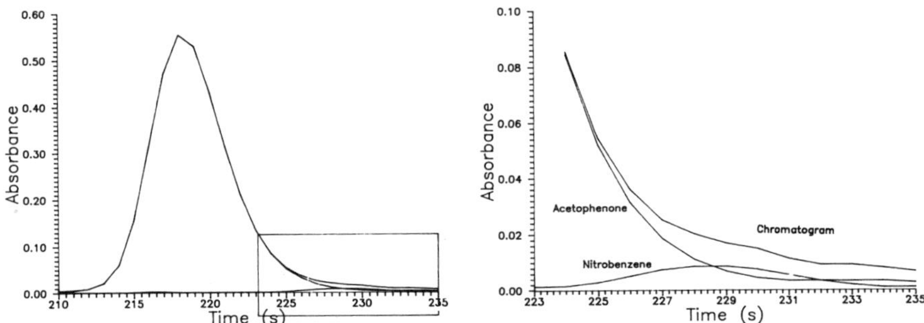

FIGURE 2.8. Curve resolution of nitrobenzene overlapped with a 100-fold excess of acetophenone, standard 7, using 90% methanol/10% water mobile phase.

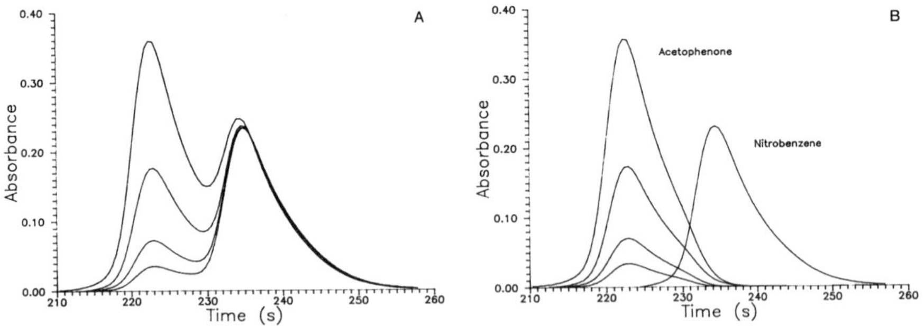

FIGURE 2.9. Chromatograms of standards 1 and 10–12 (A) before and (B) after ITTFA curve resolution using 90% methanol/10% water mobile phase.

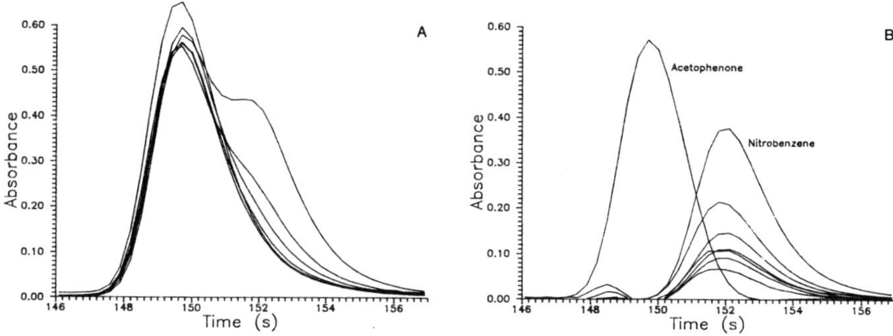

FIGURE 2.10. Chromatograms of standards 1–7 (A) before and (B) after ITTFA curve resolution using 95% methanol/5% water mobile phase.

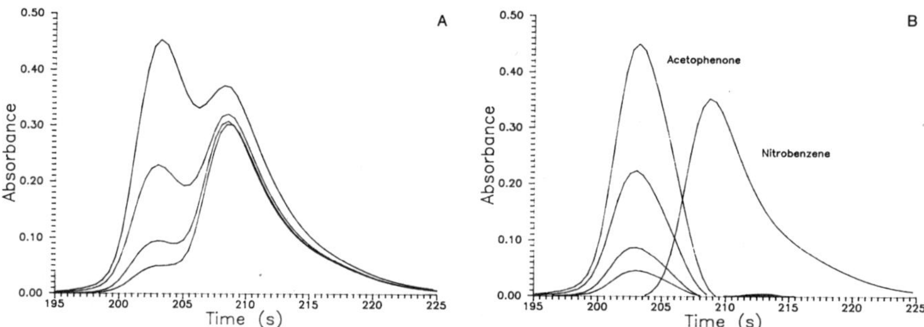

FIGURE 2.11. Chromatograms of standards 1 and 10–12 (A) before and (B) after ITTFA curve resolution using 95% methanol/5% water mobile phase.

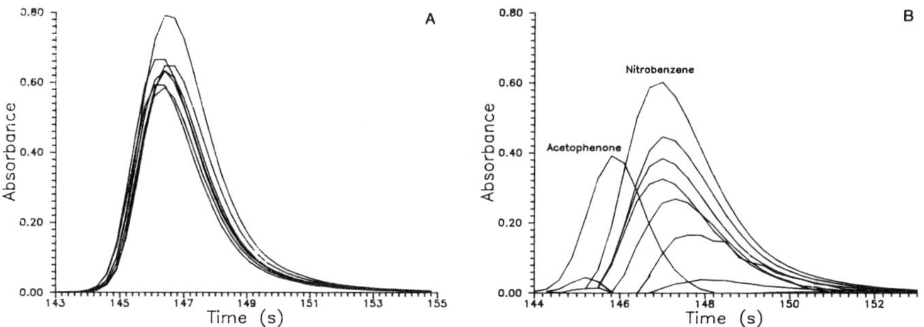

FIGURE 2.12. Chromatograms of standards 1–7 (A) before and (B) after ITTFA curve resolution using 100% methanol/0% water mobile phase.

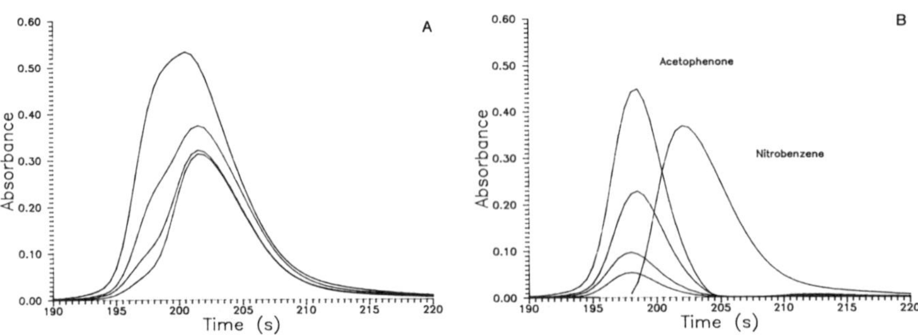

FIGURE 2.13. Chromatograms of standards 1 and 10–12 (A) before and (B) after ITTFA curve resolution using 97% methanol/3% water mobile phase.

To assess the accuracy of the curve resolution results, the ratio of peak heights calculated by the curve resolution technique for standards 1–7 and 10–12 was compared to the ratios obtained when the two peaks were baseline resolved (see Table 2.6). Inspection of Figs. 2.7 to 2.13 and Table 2.6 shows that the curve resolution results for many of the chromatograms are not accurate. Best accuracy is obtained when the overlap between the two peaks is not severe and the concentration of the minor component is not too small. See standards 1–4 and 10–12 using a

TABLE 2.6. Expected and Observed Peak Height Ratios for the Curve Resolution of Overlapped Nitrobenzene (NIT) and Acetophenone (ACP) Peaks

			90% methanol, 10% water		95% methanol, 5% water		100% methanol, 0% water	
Std. no.	NIT:ACP conc. ratio	Expected peak ht. ratio	Observed peak ht. ratio	Error (%)	Observed peak ht. ratio	Error (%)	Observed peak ht. ratio	Error (%)
1	1:1	0.581	0.613	6	0.660	14	1.536	164
2	1/2:1	0.292	0.306	5	0.390	34	1.152	>200
3	1/5:1	0.123	0.129	5	0.247	101	0.901	>200
4	1/10:1	0.065	0.072	11	0.196	203	0.719	>200
5	1/20:1	0.032	0.041	29	0.165	>200	0.526	>200
6	1/50:1	0.013	0.027	108	0.109	>200	0.239	>200
7	1/100:1	0.007	0.022	202	0.106	>200	0.063	>200
10	1:1	0.670	0.624	−7	0.785	17	0.807^{+}	21
11	1:1/2	1.319	1.273	−3	1.404	7	1.407^{+}	7
12	1:1/5	3.280	3.244	−1	3.511	7	3.332^{+}	2
13	1:1/10	6.674	6.868	3	6.878	3	6.491^{+}	−3

NOTE: $^{+}$ Mobile phase of 95% methanol to 3% water used for these standards.

90%:10% ratio of methanol to water mobile phase, for example. Although not shown here, the estimated spectra for standards 1–4 and 10–12 faithfully reproduce the spectra of the pure components of standards. There is a corresponding decrease in the accuracy of the estimated spectra as the accuracy in the estimated peaks decreased.

In all cases, the results for standards 10–12 are consistently better than the results for standards 1–7. One of the difficulties with the ITTFA curve resolution technique is that the height and width of the second peak is frequently overestimated, especially when the height of the second peak is small relative to the overlapping tail of the first peak. This effect is especially pronounced in Figs. 2.8 and 2.9. This effect does not introduce as significant an error in standards 10–12 because the minor component elutes first. This observation suggests a refinement for the technique: Adjust the peak widths of test vectors to an average value during iterative refinement.

2.5. CONCLUSIONS

Factor analysis and ITTFA curve resolution are powerful techniques that can often provide a chemist with useful information while trouble-shooting difficult separations. Factor analysis is a very sensitive technique for determining the presence of overlapped peaks. Its superior sensitivity arises because it makes use of all available data points in a data matrix.

In some cases the resolved peaks and estimated spectra produced by ITTFA curve resolution are faithful reproductions of the underlying curves. Unfortunately, it is not possible to identify the instances where the peak shapes and spectra estimated by any curve resolution technique are inaccurate unless standards are available.

REFERENCES

1. E. R. Malinowski and D. G. Howery, *Factor Analysis in Chemistry,* Wiley, New York (1980), pp. 36–45.
2. W. H. Press, B. P. Flannery, S. A. Teukolsky, and W. T. Vertterling, *Numerical Recipes: The Art of Scientific Computing,* Cambridge University Press, New York (1986), pp. 349–356.
3. E. R. Malinowski, "Determination of the Number of Factors and the Experimental Error in a Data Matrix," *Anal. Chem.* **49,** 612–617 (1977).
4. E. R. Malinowski, "Theory of the Distribution of Error Eigenvalues Resulting from Principal Component Analysis with Applications to Spectroscopic Data," *J. Chemometrics* **1,** 33–40 (1987).
5. M. McCue, and E. R. Malinowski, "Target Factor Analysis of the Ultraviolet Spectra of Unresolved Liquid Chromatographic Fractions," *Appl. Spectrosc.* **37,** 463–469 (1983).

6. A. Lorber, "Error Propagation and Figures of Merit for Quantification by Solving Matrix Equations," *Anal. Chem.* **58,** 1167–1172 (1986).
7. P. J. Gemperline, "A Priori Estimates of the Elution Profiles of the Pure Components in Overlapped Liquid Chromatography Peaks Using Target Factor Analysis," *J. Chem. Inf. Comput. Sci.* **24,** 206–212 (1984).
8. P. J. Gemperline, "Target Transformation Factor Analysis with Linear Inequality Constraints Applied to Spectroscopic–Chromatographic Data," *Anal. Chem.* **58,** 2656–2663 (1986).
9. B. Vandeginste, R. Essers, T. Bosman, J. Reijnene, and G. Kateman, "Multicomponent Self-Modeling Curve Resolution in High-Performance Liquid Chromatography by Iterative Target Transformation Analysis," *Anal. Chem.* **57,** 253–264 (1985).
10. M. H. Sharaf, and B. R. Kowalski, "Quantitative Resolution of Fused Chromatographic Peaks in Gas Chromatography/Mass Spectrometry," *Anal. Chem.* **54,** 1291–1296 (1982).
11. A. Meister, "Estimation of Component Spectra by the Principal Components Method," *Anal. Chim. Acta* **161,** 149–161 (1984).
12. B. R. Kowalski, and O. S. Borgen, "An Extension of the Multivariate Component Resolution Method to Three Components," *Anal. Chim. Acta* **174,** 1–26 (1985).
13. R. F. Lacey, "Deconvolution of Overlapping Chromatographic Peaks," *Anal. Chem.* **58,** 1404–1410 (1986).
14. W. H. Lawton, and E. A. Sylvestre, "Self-Modeling Curve Resolution," *Technometrics* **12,** 617–633 (1971).

Chapter 3

Isolation of Pure Spectra in GC/MS by Mathematical Chromatography: Entropy Considerations

Erkki J. Karjalainen

3.1. INTRODUCTION

The chromatographer can use the computer as a means to separate compounds that do not fully separate on the column. This "mathematical chromatography" reconstructs the pure spectra and concentrations on the basis of the extra information on the spectral axis. Even the best separation methods are not powerful enough to deal with the complexity of biological samples; there are overlaps that can usually only be overcome by further chromatographic manipulations. The number of components that can be measured in a single chromatographic run is much smaller than the number of components present and producing a detectable signal.[1,2] We faced this situation in gas chromatography/mass spectrometry (GC/MS) and developed the alternating regression (AR) method to deal with it.

In modern chemistry the analyst is often faced with the problem of spectrum reconstruction. We encounter this situation with all hyphenated techniques. Several spectroscopies are two-dimensional. We have a

Erkki J. Karjalainen • Department of Clinical Chemistry, University of Helsinki, SF-00290 Helsinki, Finland.

two-dimensional table of observations. One of the axes can describe time in a separation process like chromatography, while the second axis consists of the elements of the spectrum (e.g., intensities corresponding to wavelengths).

The goal of the mathematical chromatographer is to reconstruct the three-dimensional cube of elementary concentrations on the basis of observations in two dimensions (Fig. 3.1). The horizontal layers in the spectral cube correspond to two-dimensional tables containing the concentrations for each spectral species. When these layers are vertically summed together they form our observations. The solution to the reconstruction problem is not possible without introducing some additional information to the solution process besides our two-dimensional table of observed intensities. This additional information is introduced as constraints. The constraints are either clearly recognized as constraints or present in the computer program that is the final realization of the abstract algorithm. All forms of constraint may not be apparent even to the programmer.

The problem of spectrum reconstruction has two aspects. The first partial problem is to find the algorithms for the reconstruction. The second problem is to define the additional information, namely, the constraints, to make the solution possible. The two problems are closely intertwined. Some aspects of the algorithm can act as constraints even though they are not always recognized as such. What is needed is a

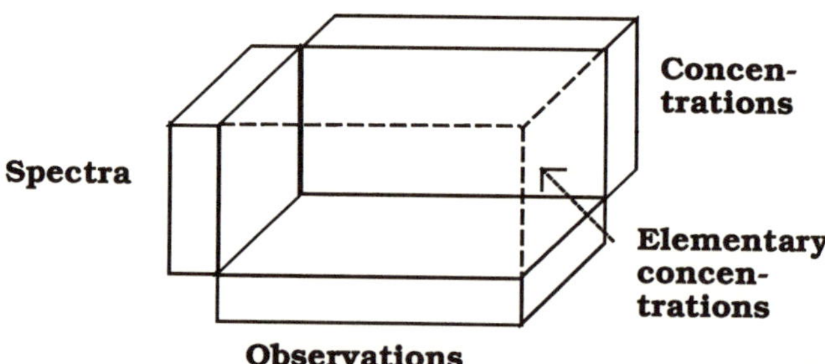

FIGURE 3.1. The spectral "cube" of elementary concentrations can be projected to the neighboring matrices in three directions. Vertically the elementary concentrations sum up to the observation matrix. From right to left the elements sum up to the spectrum matrix (unscaled). Finally, from front to back the elements sum up to the concentration matrix.

The matrices containing the spectra and concentrations can also be considered to form the elementary concentrations in the spectral cube as a product. The two points in each matrix that project into the element in the cube are multiplied. For the multiplication the spectrum vectors must be scaled to a constant length.

In later diagrams the spectrum and concentration matrices have been turned on their side to lie flat in the same plane as the observation matrix. We should not forget that even in the flat representation the full cube is implicitly present.

well-defined set of assumptions necessary for the reconstruction. This set should contain just the minimum needed to find a solution. Over the last ten years we have used spectrum reconstruction algorithms, called the AR method, in drug metabolic analysis. The last incarnation of the programs works in a Macintosh II, a modern 32-bit micro that has an arithmetic processor and 5 MB of fast memory. With time, methods like the AR method will probably be a direct part of the analytical instruments. Today, too much effort is spent in interfacing the different systems.

3.2. ALTERNATING REGRESSION

3.2.1. The Overall Algorithm

The alternating regression, or AR algorithm (Fig. 3.2), is an approach to the reconstruction problem we have used for several years.[3,4] The process is reasonably fast on the Macintosh; the main bottleneck has been the inability of the compilers to take full advantage of the arithmetic processor. The time needed for the analysis corresponds roughly to the run time of the GC/MS. The run times are minutes, not hours.

The AR algorithm starts with a set of assumed spectra that initially can be just random numbers. Using the observation matrix from GC/MS we then solve for the concentrations using regression. The concentration matrix is constrained to be positive and to have just one local maximum for each spectral species. Then we change our view of the problem by transposing all matrices involved. This change in view requires changing the order of the indexes in the computer program. Now we consider the concentrations to be known, and solve for the spectra. After the spectra are constrained to be positive the cycle can go on to the next iteration. After reaching a stable error we stop, typically in less than ten iterations. An effective implementation of the algorithm is important for run-time speed.

We have earlier described the mathematical background to the AR problem.[4] In that presentation the spectral cube was present only in an implicit fashion. The two matrices for spectra and concentrations were shown to be multiplied directly. When the computer storage is large enough to permit the storage of elementary concentrations, it is advantageous to keep all the intermediate terms available.

There are different ways to look at the reconstruction problem as a cube. The first viewpoint sees the spectral cube as a product of two matrices (Fig. 3.1). These two matrices contain the unit spectra and the

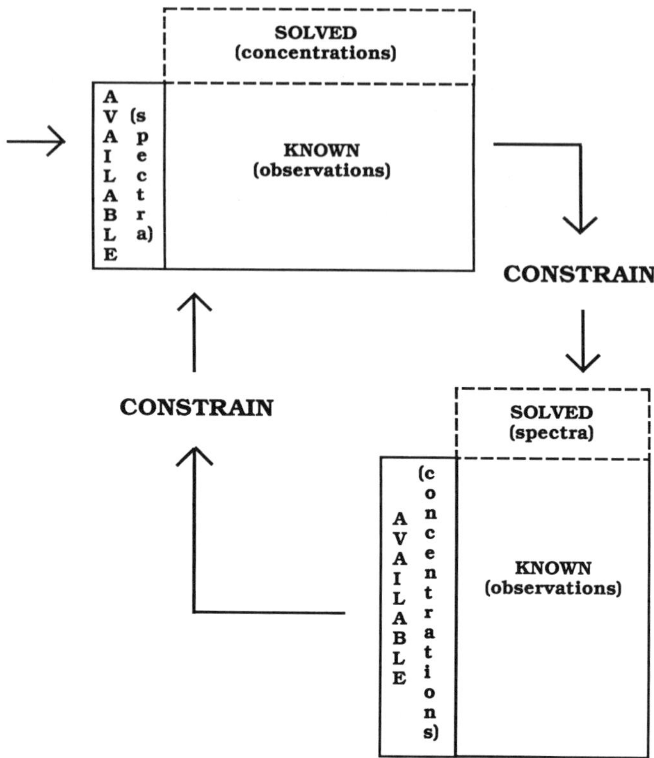

FIGURE 3.2. The basic AR (alternating regression) process is started by choosing a starting basis. The basis is modified by the latter steps. The simplest choice for the starting basis is to use random numbers as elements in spectrum matrix.

Regression is then used to find the concentrations. The concentrations are constrained to be positive and to have a single maximum. The problem is then transposed and the concentrations are used as a basis. The spectra are again solved by regression and constrained to be positive. This cycle goes on until a stable solution emerges. The number of spectral species is a parameter that is input manually by the experimenter. Conceivably, even the choice of this parameter could be automated.

corresponding concentration profiles. If we scale the unit spectra to a constant vector length, we see that the cube is a product of these two matrices. For practical reasons the contents of the cube are normally not stored during the computation, and occur just as temporary variables during the matrix calculations.

Another way of looking at the cube is to think about the spectrum and concentration matrices as summations of the elements in the cube. The cube containing the elementary concentrations can be projected in three different directions, and the elements can be summed for each projection. The three views correspond to observations, spectra, and concentrations.

If we look at the problem in this way, the spectrum matrix is now not scaled to vectors of constant length. The linear length of the vectors corresponding to one spectral species is identical in both spectrum and concentration views. If we sum the components in a spectral vector this sum is the same as the sum of the corresponding concentration vector.

It is customary to scale the spectrum vector to unit length. Mathematically, we could scale the concentration vectors equally well. The concentration vectors could be throught of as "concentration spectra." In practice the spectra are more stable between experiments, so it is more natural to scale them.

The situation we have to deal with in practical analysis however, is different. We have to identify or reconstruct a cube filled with elementary concentrations of spectral elements. This type of problem belongs to the general category of "inverse problems."[(5)] In inverse problems we try to identify the model structure and the values of the parameters from observations in an indirect way.

In inverse problems the input to the problem has often fewer variables than the output desired. The solution we try to find must sum up to our observations. The solution should contain physical components that can be verified by separate experimentation.

The solution we obtain should be stable. A small perturbation in the observations should be reflected as a small perturbation in the solution. The stability and sparseness of the solution are linked to each other. The solution should be sparse in some sense of the word. It should contain the least number of spectral species. Each spectral species should have the minimum number of spectral lines. There is always a limit to the complexity of the solution. This limit is dictated by the true information present in our data. If we try to include too many spectral species in our solution, the solution behaves in an unstable fashion.

The alternating regression (AR) algorithm is a special-purpose algorithm developed for the identification of spectra in chromatographic analysis. It utilizes the extra information present in mass spectra to identify the spectral species and their concentrations. The only parameter that needs to be controlled manually by the experimenter is the number of components needed. This number is gradually increased until the number of components used exceeds the information present in the observations. Then, a smaller number of components is chosen for the final solution. Even then, the existence of the smaller components should be taken with some caution. They should be considered to be tentative until verified by more experiments in different samples.

3.2.2. Selecting the Starting Basis

There are several ways to choose the starting basis for the AR algorithm. The choice of the starting basis is not crucial for the result we get. The result we obtain is always the same. The choice, however, is important for the computer time needed. We should give the matter some thought to find the best starting basis for our purposes.

We can start from scratch and simply use random numbers. Alternatively, we can use existing "library" information. This can be in the form of either known spectra or known retention times. A third alternative is to pick the starting information from the observation matrix itself.

The simplest way to choose the spectra for starting the AR algorithm is to use random numbers. "White noise" can be produced by random number generators. It has been found beneficial to the numerical behavior of the AR algorithm to produce random numbers that also contain a suitable number of "gaps" or zero intensities. The proportion of gaps should be chosen to correspond roughly to the number of empty mass numbers in the observation matrix.

We can graphically show the progress of the AR process for a synthetic set of observations. The data set contains five components. The original situation—the synthetic specta and concentrations—is shown in Fig. 3.3. Figure 3.4 shows the starting situation of AR with random guesses for the spectra. The concentrations are then solved using these spectra as known. Then again, the spectra are solved using the concentrations as "known." The solution after the tenth iteration is shown in Fig. 3.5.

The classical way to start a spectrum identification process is to use a library of known components. This can be used for AR as well. If some component is present in the sample that is absent from the library, this does not prevent it from being found. If enough candidates have been chosen for the starting basis, one of the library elements will be transformed to the new component.

Another starting basis is to use some of the observed mass spectra as a starting point. The observations are mixtures of several spectral species, but they approximate the final solution spectra better than random guesses. The problem is to select the best subset of observations for a starting base. They should span the observations in an optimal manner. The chosen set can be thought of as a sort of "vector basis", which should contain vectors as "orthogonal" to one another as possible.

One way to choose a starting basis for observed spectra is to use stepwise regression to identify a good spanning subset of the spectra. First, we form a sum spectrum of all observations. For this, the

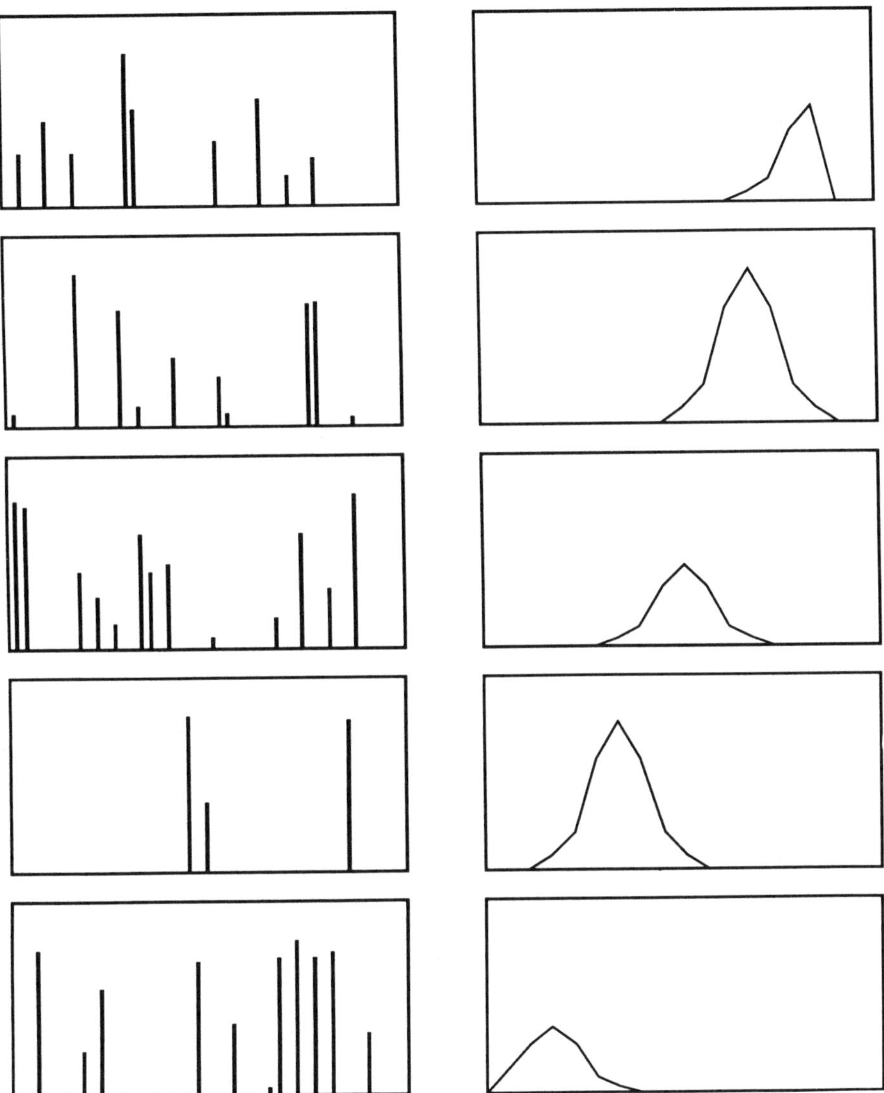

FIGURE 3.3. The original synthetic spectra and concentration profiles used to demonstrate the AR algorithm (reprinted with permission from Ref. 4).

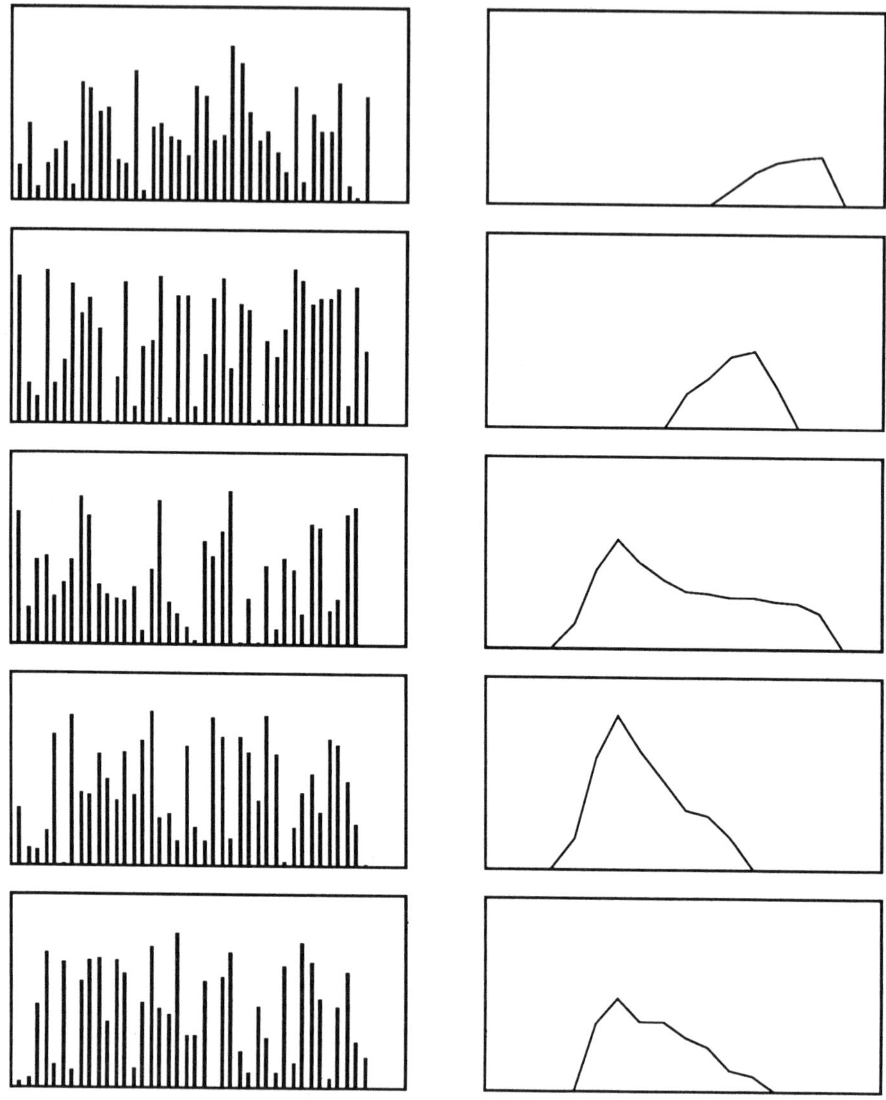

FIGURE 3.4. The starting situation of AR. To the left are spectra filled with random numbers. To the right are concentration profiles calculated on the basis of these (reprinted with permission from Ref. 4).

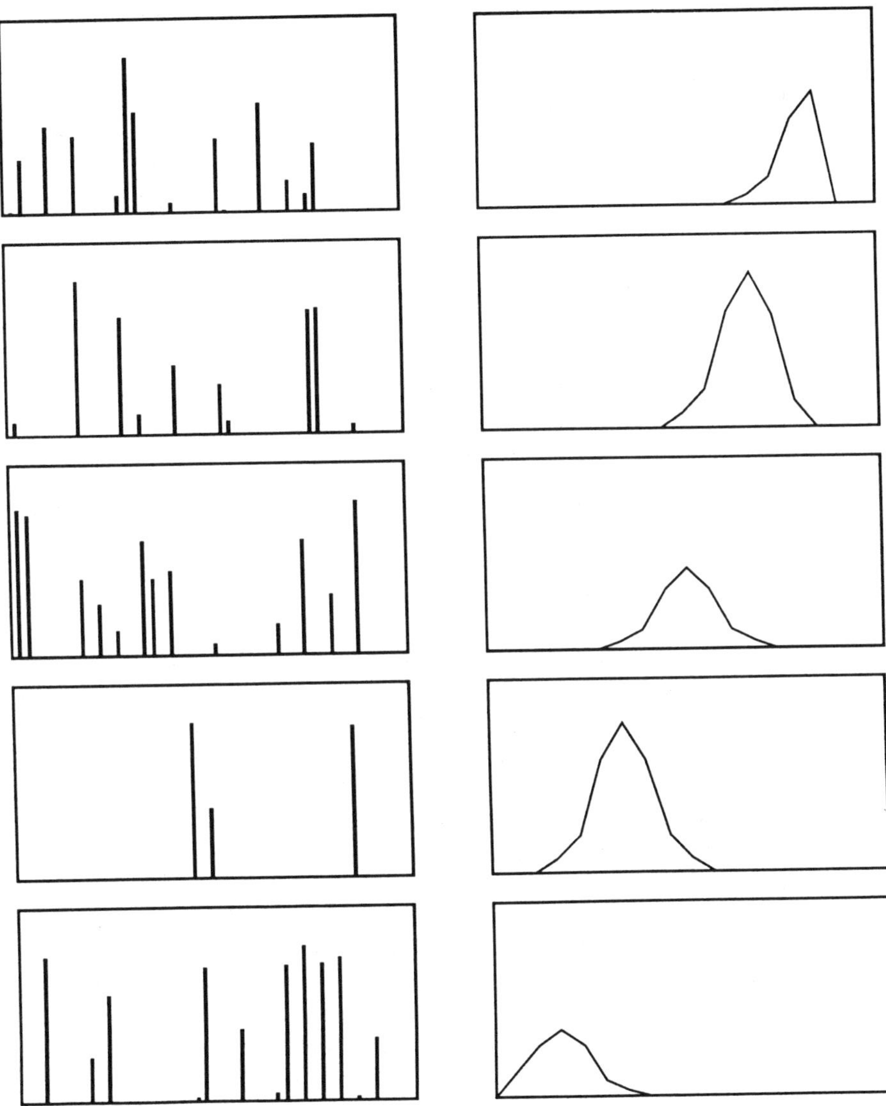

FIGURE 3.5. The solved spectra and concentration profiles after the tenth iteration of AR. The figure shows no differences when compared to Fig. 3.3 (reprinted with permission from Ref. 4).

observation spectra are simply averaged. Next, we try to explain the average by a weighted combination of observation spectra. The stepwise regression process brings explanatory spectra into the basis until a satisfactory total correlation has been obtained. During the process, some spectra may be thrown out of the basis and replaced by others. With regression it is possible to get a feel for the total number of components present in the observations. When the final subset of observed spectra has been chosen, they can be used as a starting point for the later AR steps.

Often it is a good idea to use several alternative starting bases for a given situation. If there are differences in the solutions we obtain, we should check a number of things. First, we should check the number of components we use for the solution. Second, we should check if our data has been properly filtered to eliminate noise.

3.2.3. Constraining the Solution

The only way to solve an inverse problem like spectrum reconstruction is to introduce some additional constraints to the problem. Without these extra constraints the solution is not possible even in principle. The problem is finding the minimum set of constraints needed to get a solution.

There are a number of simplifying constraints in the full-spectrum reconstruction problem. Each horizontal layer in the spectral cube is a product of two vectors. This means that the number of truly independent elements in the problem is much smaller than the number of elements in the layer.

There are a number of constraints based on physical arguments. The spectral intensities and concentrations are generally taken to be positive. The elementary concentrations are assumed to be linearly additive.

For its function, the AR method utilizes some additional constraints. The main constraint introduced by AR is to constrain the overall shape of the chromatographic peaks to a purely monotonic function with a single local maximum. A peak is not permitted to have more than one maximum. Finally, the number of the independent spectrum and concentration elements is constrained to keep the solution sparse.

3.3. OPTIMAL PREPROCESSING OF DATA

3.3.1. The Role of the Apodization Function

We have found it practical to use an apodization, or windowing, function before applying the AR algorithm. As we stated earlier, each

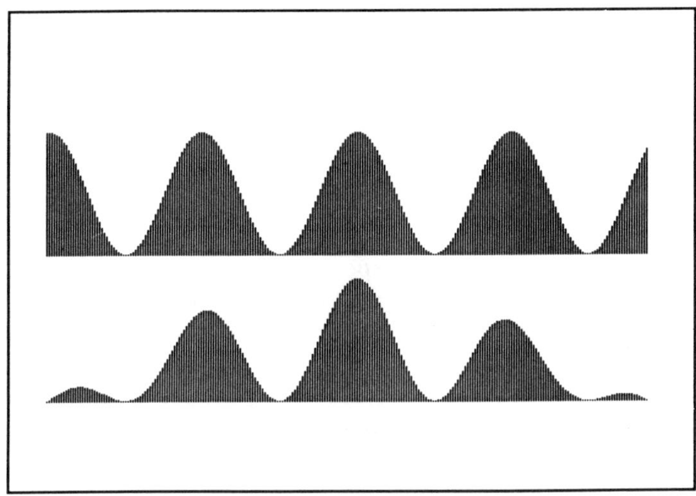

FIGURE 3.6. The windowing function. The upper curve shows the original total abundance for 40 spectra. The lower curve is obtained when the upper curve is weighted by a cosine-shaped windowing function.

mass trace used in AR should have at least one value that has an intensity of zero. If this goal is not met the AR algorithm does not always behave in a stable fashion. The practical remedy for this problem is to utilize a windowing function. A windowing function is a weighting function that emphasizes the middle part of the observations while both ends gradually taper off to reach a value of zero (Fig. 3.6). With a weighted observation series the AR behaves in a stable fashion.

In Fourier transform analysis it is customary to use a windowing function on the observed time series before the Fourier transform is calculated.[6] This is beneficial for several reasons. The endpoints of the time series represent an abrupt change. This rapid change corresponds to high-frequency components in the spectrum. We do not want to get these high-frequency components because they are not actually present in the original signal. The appearance of these high-frequency signals is a sampling artifact.

We have used Fourier smoothing methods for our data. In this work we have found that the observation matrices should always be weighted by a windowing function. If we do not perform this preliminary step the results from the smoothing step remain unsatisfactory. As we stated earlier, the observations for each mass trace should in all cases contain an intensity value of zero.[4] If the observation originally do not contain this zero baseline, the use of the windowing function certainly forces both ends of our observations to this zero value.

The windowing function has an effect on the retention times at both ends of the observation matrix. To be safe, we should measure retention times only for compounds that are completely contained in our observation interval.

3.3.2. Selecting the Optimal Band-Pass

Filtering the observation matrix with a suitable digital filter is critical for the success of any subsequent spectrum processing, AR included. We should remove high-frequency noise from our observations by limiting these components with a low-pass filter. Unnecessary local features can cause the later processing steps to get stuck instead of going on to larger features. If we filter too much we are going to lose essential data. It is very critical how we choose our degree of smoothing.

The choice of the degree of filtering should be done carefully. The filtering should be adjustable for each mass trace. What we shall describe next is a system that has produced good results for us.

The filtering can be done in the Fourier domain. We can limit the number of frequencies we use in reconstructing the waveform. If we omit the higher frequencies, the resulting waveform is smoother. The change from low to high frequencies should not be abrupt but gradual. The amount of error we can expect in a given mass trace can be predicted on the basis of the statistics of the ion counts. Low counts should be filtered more to give satisfactory results. The difference between the original and smoothed readings is used to guide the degree of filtering in a feedback loop.

A second point that concerns the degree of filtering is based on the complexity of the solution for which we are looking. We have found that a system based on the expected number of spectral components functions well in practice. If we expect to find ten spectral components, each mass trace should not contain much more information than that. We can analyze our smoothed mass trace to find the number of components it contains. We can analyze the smoothed mass trace to find the number of local maxima for the curve and its first and second derivatives.

By analyzing the mass trace we can find the maximum number of local chromatographic peaks it can contain. We can gradually increase the degree of smoothing until the maximum number of peaks corresponds to our expectation. This feedback loop is very critical for the success of any later processing. The degree of smoothing should be chosen in an adaptive fashion to produce information in a form that is optimized for the later processing steps.

High-pass filtering is often very useful for the separation of overlapping chromatographic components. It helps to separate the chromatographic peaks in the observation matrix. If we use too much high-pass filtering, the low points reach negative values, which is not the desired effect. Still, some degree of emphasis of the high-frequency components is beneficial for the success of later steps. A somewhat similar situation exists in several other fields besides chemical analysis. The algorithms used in computed tomography often use high-pass filtering of the projections before applying the other steps in the reconstruction.[7] This helps to prevent "smearing" of the high-frequency components in the final reconstructed picture. High-pass filtering, used with care, is a useful tool in the preprocessing of observations before the AR algorithm.

We see that the highest frequencies should be eliminated. At the same time some high frequencies should be emphasized. The proper elimination of the highest and lowest frequencies corresponds to the choice of the optimum band-pass filter. The success of the latter processing steps like AR depends on doing the preprocessing in an optimal way.

3.4. BALANCING THE "FIT" AND ENTROPY

3.4.1. The Object Function in Spectrum Reconstruction

The central problem in any optimization is to find the proper definition for the object function to be minimized. In the spectrum reconstruction problem we try to optimize two conflicting goals. First, we try to get as perfect a fit as possible between the model and observations. Second, we try to get the simplest solution possible, resulting in the simplest, "cleanest" spectra. The only way to satisfy these two constrasting goals is to define a new object function combining the two partial goals into a new object function.

3.4.2. Measuring the Fit

The fit between model and observations can be measured in several ways. The most common measure is the sum of squares of the differences. This is L2-norm for the error. An alternative formulation is to use the sum of the absolute differences, or the L1-norm. If we use the L1-norm for fitting a model, the norm has an advantage in robustness. Occasional outliers do not have as much effect on the final solution as do errors when using a L2-norm.

If we use enough parameters the fit can be made perfect, but the solution found in this fashion is not the optimum solution we are seeking. We should try to minimize the number of parameters. One indicator of the quality of the solution is to have a measure for the length of the spectrum and concentration vectors in our solution. We find that increasing the number of spectral species beyond a certain point also increases the length of the solution vectors.

The process of AR uses error norms rather indirectly. The L2-norm serves as an intermediate tool when solving for spectra or concentrations. We have also tried other ways to solve the regression step. Some of the alternate methods used positivity constraints for the solution. These more sophisticated methods were not worth their cost in increased computing time.

The main advantage of the AR method is perhaps its low computational cost. The method consists of two main steps that are simple linear algebra, solving a small system of linear equations. The number of unknowns in the regression depends on the number of spectral species, not on the number of observations or the number of spectral lines. The progress of the AR iterations can be followed by calculating the fit after each cycle.

If we want to further refine the solution found by AR we are forced to use methods that are more costly in computer time. Nonlinear optimization can be used to find a more precise solution. The quality of the solution depends on the quality of the object function we define for the optimization problem. The simple metric between the observations and values predicted by the model is not sufficient. We need to optimize other factors in addition to explaining the observations.

3.4.3. Measuring the Entropy

The problem of finding a truly optimum solution is not easy to define in precise mathematical terms. The solution should fit the observations well. Additionally, it should be sparse. If we try to find a minimum solution with the least number of elements, we need a measure for the simplicity. A classical measure for this has been the entropy as defined in the information theory by Shannon.[(8)]

We can develop measures that describe the entropy of the spectra and concentrations. The problem is to find a reference distribution for both. The frequency of occurrence of a certain value should be relative to that of the "reference" distribution. For positive physical entities we can assume that the reference distribution is roughly logarithmic.

We should consider quantities other than just spectra and concentrations. What is relevant in forming the observations is not spectra or concentrations, but their product terms. The product terms fill the spectral cube we have already discussed. The best measure for the information content of the solution is the entropy of the elementary concentrations in the spectral cube.

To calculate the entropy we first form the distribution function for the elementary concentrations. The distribution function is a histogram. We divide the interval between zero and the largest single observation into a number of "bins," which are chosen in a logarithmic fashion. The entropy is then calculated as defined by the Shannon formula

$$H = -\sum p_i \log p_i \qquad (i = 1, \ldots, n)$$

where H is entropy, p_i is the probability of finding a given intensity value for the elementary concentration, and n is the number of classes of frequency in the intensity distribution.

The final object function is a combination of the term describing the fit and the term describing the entropy. The combined object function is then minimized by the nonlinear optimization process. To use nonlinear programming with a large number of parameters we need a good initial solution. The AR method produces a good starting solution at a low computational cost.

The entropy can also be used simply as a complement to the AR algorithm. For this purpose we do not need to measure the entropy by the Shannon formula. We can use a simpler method to constrain the solution to be sparse.

We can constrain the solution to be simpler by limiting the number of active elements in the spectral cube. For this purpose we constrain the number of nonzero elements in the cube to a certain number (Fig. 3.7). This constraint results in making the spectra and concentration curves simpler. The proportion of low intensities in the spectra is smaller. For the concentration curves the proportion of low concentrations gets smaller and the shape of the concentration curves gets steeper. If we limit the number of elements to a number that is too small, the concentration curves start to lose their smooth shape.

The number of nonzero elements in the spectral cube gives a measure of the complexity of the solution found. The AR process can be complemented to include a constraining step for the number of active elements in the spectral cube. This produces an effect that is similar to that obtained with the nonlinear optimization but at a lower computational cost.

3.5. APPLICATIONS OF AR

3.5.1. Steroid Mixtures

The analytical problem that catalyzed the development of AR was the identificatioı of forbidden drugs used in sports. The analysis of

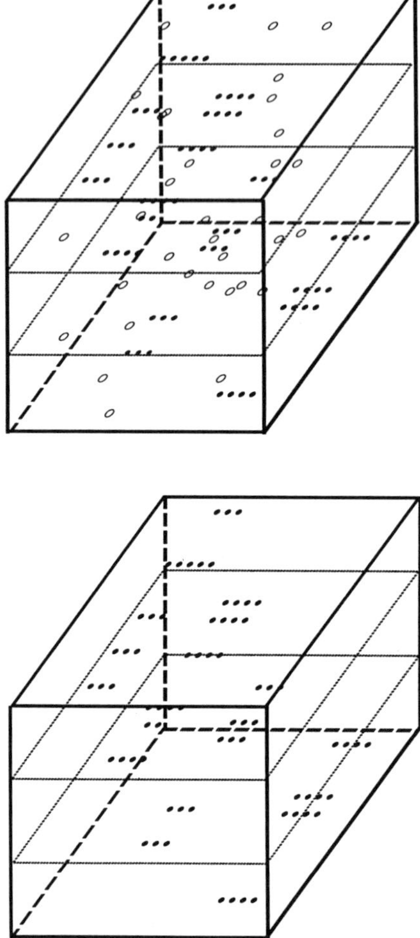

FIGURE 3.7. The entropy of the spectral cube can be measured by the number of nonzero elements. The original solution in the upper picture can be constrained to produce a simpler solution in the lower picture.

doping control samples in connection with sports is a continuous challenge for the analyst. The list of forbidden drugs is rather long. Additionally, new compounds that are pharmacologically active but not yet present in a spectrum library should be found and identified.

The highest analytical sensitivity can be obtained in those cases where the compound is known already in advance. In those cases we can tune the instruments to known masses using specific monitoring of chosen ions. If we do not know the spectrum of the compound we are looking for, we must use scanning of complete spectra with a loss in sensitivity. The AR method can recover minority components hidden under larger peaks. In this sense AR can improve the analytical sensitivity.

The methods we have used have been described earlier in detail.[3,9] The data were collected with a Hewlett-Packard 5995B gas chromatograph/mass spectrometer using a continuous scanning mode. The data were transferred to a Macintosh II over a serial data line for further analysis by AR.

The practical advantage of AR in doping analysis is the ability of the method to find smaller components. In doping cases it is important to be able to prove that the compound has been metabolized before entering the urine sample. It is useful to identify a number of known metabolites and the parent compound. These secondary compounds are often present in small concentrations, and the amount of the sample available for the analysis is limited. We have used AR in doping cases to obtain supplementary evidence. The series of spectra were dissected by us without knowing the chemical identities of the suspected compound. A pure spectrum of the parent drug was found in a case where the presence of the unmetabolized drug could not be shown by looking at the original spectra.

A second field of potential application of AR is steroid profiling. We can shorten the run time in GC/MS and allow more overlap between peaks. The loss in chromatographic resolution is compensated by AR. What we gain is higher sample throughput.

3.5.2. Aroma Compounds

These data were obtained from an industrial laboratory in the form of a magnetic tape. The data were transferred to a Macintosh II over a serial line from a minicomputer equipped with 9-track magnetic tape unit. The data were then analyzed by AR in the Macintosh. The decomposition of the total ionization into element spectra produced an error of 7.4%, when the error norm was the sum of absolute deviations between the measurements and the model (L1 error norm). If the error was calculated as the vector length of the differences and the original observations, the error was found to be 5.53%.

To calculate the vector length of the error matrix the individual residuals are squared and summed. Finally, a square root is taken from the result. To get the vector length of the original observations, the sum of squares is first calculated. The square root of this sum represents the vector length of the original observations.

The quality of the solution can be seen in two plots. If we plot the sum of the absolute deviations as a function of the scan, we see that the

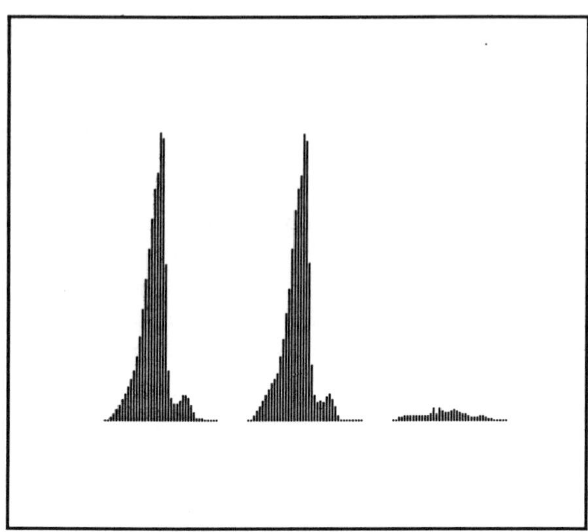

FIGURE 3.8. The difference between the AR results and observations. The leftmost curve shows the original total abundance. The middle curve is the total abundance predicted by AR. The rightmost curve is a summation of the absolute differences. The absolute differences have been calculated for each element of the observation matrix.

majority of the error is found in the middle spectra (Fig. 3.8). This is a consequence of the shape of the abundance curve that has been weighted by a windowing function. A second plot shows the distribution of the absolute errors for different mass numbers in the spectrum (Fig. 3.9). There is a general correspondence to the overall shape of the average spectrum. The distribution of the errors can also be analyzed for the elements of the observation matrix. Figure 3.10 shows an average spectrum on the left. In the two panels of the figure we see the positive and negative error terms. The errors have been magnified by a scaling factor of five to make them clearly visible, while the average spectrum on the left has been given unscaled.

The spectra recovered were reported to the laboratory that had originally collected the spectra. They told us they had found by heartcutting methods four spectra that were similar to those we found by AR. This supports the view that the spectral entities found directly by AR can be isolated by further experimental steps.

Two of the resulting spectra were found to be highly correlated. These compounds are probably isomers, and they were combined for reporting. The concentration curves obtained are shown in Fig. 3.11 and the total abundance is shown in Fig. 3.12.

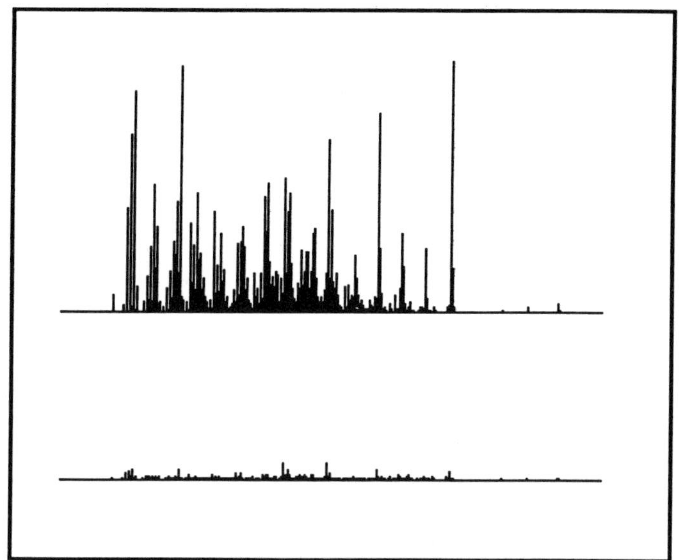

FIGURE 3.9. The difference between the spectra observed and calculated by AR. The upper trace shows the sum of all spectra in the observations. The lower trace is the sum of the absolute differences between the elements of the observation matrix and those predicted by AR.

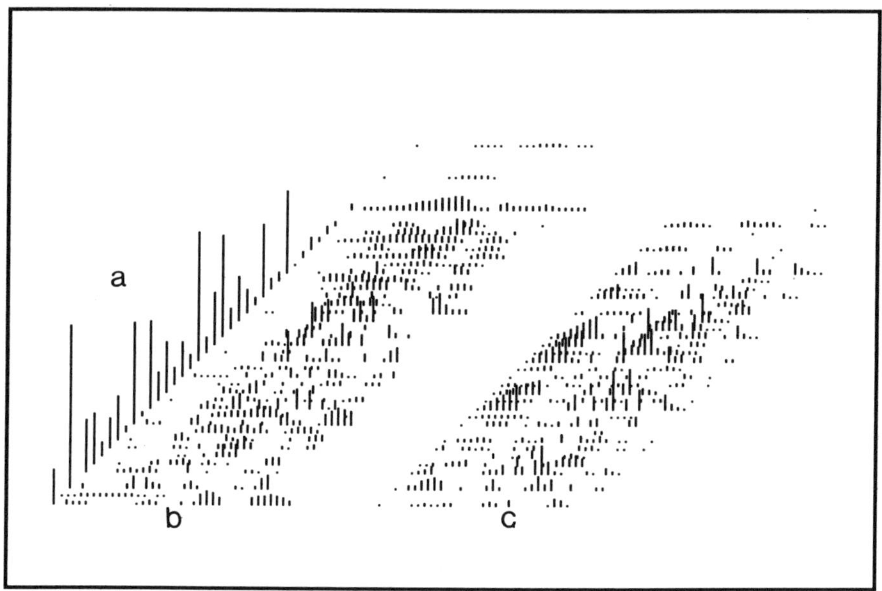

FIGURE 3.10. The distribution of the residuals for the observation matrix. The vector on the left (a) shows the magnitude of the average spectrum that has been scaled down by a factor of five. The two rectangular areas (b) and (c) show the residuals between the original observations and those predicted by AR. The leftmost matrix (b) shows the positive residuals. The righthand matrix (c) shows the negative residuals that have been changed in sign.

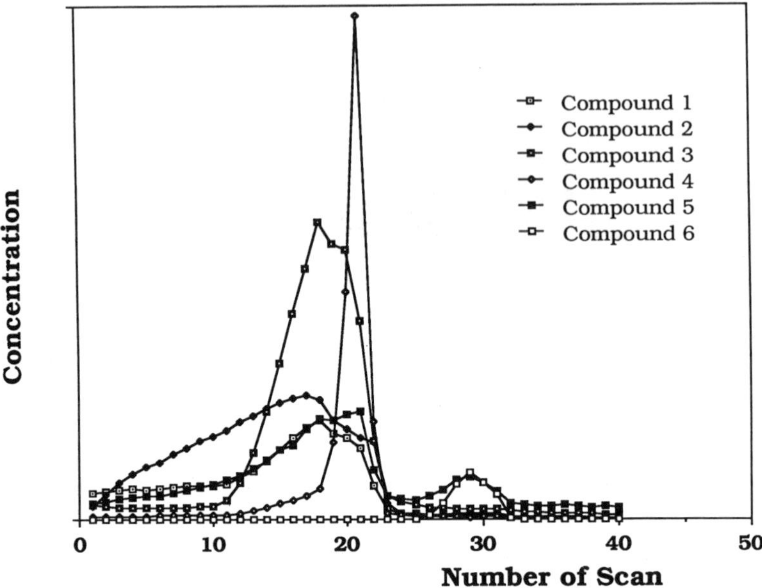

FIGURE 3.11. The resolved concentration profiles of a mixture of aroma compounds.

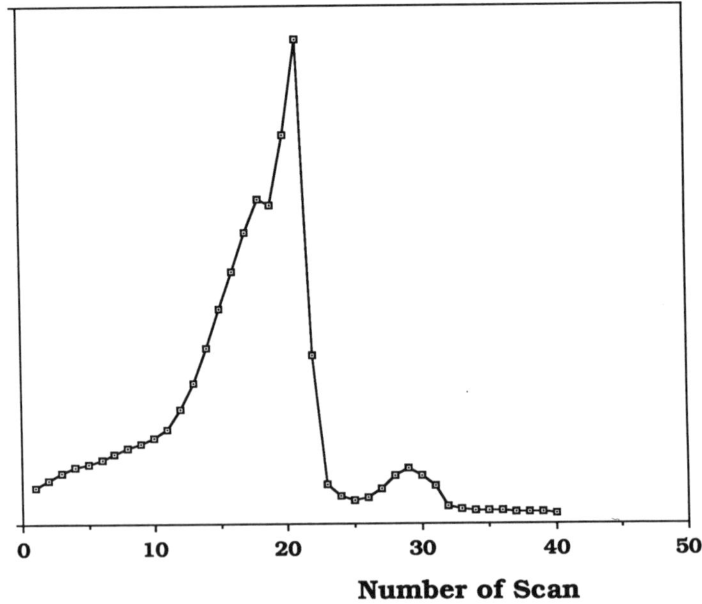

FIGURE 3.12. The total abundance curve of a mixture of aroma compounds.

3.6. CONCLUSIONS

The reconstruction of pure spectra in GC/MS and similar spectroscopies has a very practical motivation as the cost of computing goes down. The alternating regression (AR) method developed by the author is one way of finding the spectra. Drug analysis, as a part of antidoping laboratory activities, provided the experimental setting for developing the computer-based methods. The computer can separate compounds that remain overlapped after physical separations.

Practical experience shows that the AR method can locate novel compounds in those cases where we have no idea in advance about the resulting spectra. The AR method itself is not dependent on having a library of known compounds, although libraries can be used as a starting point.

Another way of looking at the ability of AR to find new spectral species is to consider AR as a way to get more sensitivity from the GC/MS instrument. The improved sensitivity can again bring about savings in materials, time, and cost. In profiling studies, the run times can be made shorter with savings in analysis times.

Finally, the cost of computing is going down at a faster rate than the cost of analytical instruments. Relatively, it is often better to do more calculations on data that cannot be deciphered by just inspection than to strive for data that can be interpreted by manual methods. If the amount of latent information in instrument output is sufficient, the computer can be a most efficient tool in interpreting it. The role of the analyst remains essential; he is merely working at a higher conceptual level. Instead of poring over the instrumental readings the modern analyst constructs computer-based systems to do the spectral analyses for him.

REFERENCES

1. D. Rosenthal, "Theoretical Limitations of Gas Chromatographic/Mass Spectrometric Identification of Multicomponent Mixtures," *Anal. Chem.* **54,** 63–66 (1982).
2. L. J. Nagels, W. L. Creton, and P. M. Vanpeperstraete, "Determination Limits and Distribution Function of Ultraviolet Absorbing Substances in Liquid Chromatographic Analysis of Plant Extracts," *Anal. Chem.* **55,** 216–220 (1983).
3. E. J. Karjalainen and U. P. Karjalainen, *"Mathematical Chromatography"—Resolution of Overlapping Spectra in GC/MS, Medical Informatics Europe 85, Proceedings,* Helsinki, Finland 1985, pp. 572–578 Springer-Verlag, Berlin (1985).
4. E. J. Karjalainen and U. P. Karjalainen, "Mathematical Chromatography in GC/MS. Finding Pure Mass Spectra", in: *Clinical Chemistry Research Foundation Library,* Vol. 2, pp. 1–48, Helsinki (1987).
5. A. Tarantola, *Inverse Problem Theory,* Elsevier, Amsterdam (1987).
6. G. D. Bergland, "A Guided Tour of the Fast Fourier Transform," *IEEE Spect.* **6,** 41–52 (1969).

7. H.-W. Ridder, "Ein neuer Algrithmus zur gefilterten Rückprojektion in der Computer-Tomografie," *Medizin. Phys.* pp. 379–384 (1979).
8. C. E. Shannon, "A Mathematical Theory of Communications," *Bell Syst. Tech. J.* **27,** 379–423 (1948).
9. K. Kuoppasalmi and U. Karjalainen, "Doping Analysis in Helsinki 1983. The First IAAF World Championships," in: *Clinical Chemistry Research Foundation Library,* Vol 1, pp. 1–40, Helsinki (1984).

Chapter 4

Signal Processing Techniques for Remote Infrared Chemical Sensing

Robert T. Kroutil, John T. Ditillo, and Gary W. Small

4.1. INTRODUCTION

Passive remote infrared (IR) sensing can provide a valuable method for the detection and identification of gaseous pollutants.[1–4] Remote monitoring chemical sensors are capable of detecting absorptions and emissions of low-concentration chemical vapor clouds using an ambient temperature atmospheric background. For many pollution-monitoring problems IR spectroscopy represents the only viable approach.

An IR remote chemical sensor consists of a sensor and a signal processor that operate in parallel to give an indication of the presence of a pollutant. The sensor detects the signatures of all chemical vapors and backgrounds, while the signal processing algorithms discriminate between the pollutant signatures and the background emissions. Typical instrumentation required for chemical remote IR sensing usually consists of a Michelson-based Fourier transform infrared spectrometer (FTIR) that has a wave number resolution of at least four inverse centimeters. The FTIR used is modified in comparison to laboratory FTIR spectrometers

---

Robert T. Kroutil and John T. Ditillo • U.S. Army Chemical Research, Development, and Engineering Center, Aberdeen Proving Ground, Maryland 21010. Gary W. Small • Department of Chemistry, University of Iowa, Iowa City, Iowa 52242.

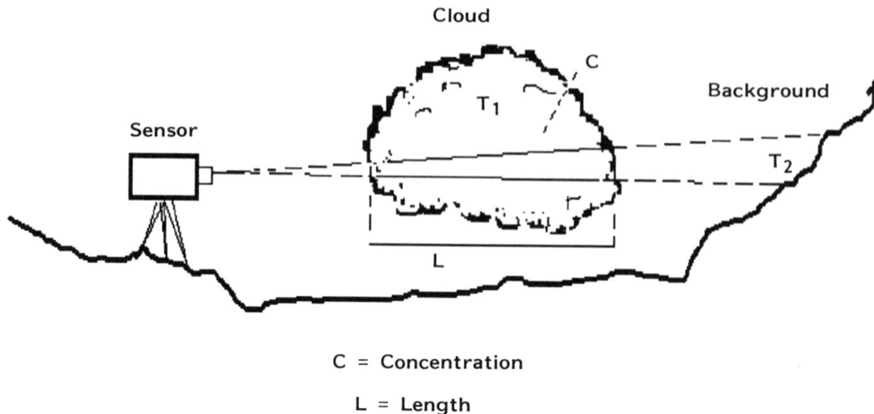

FIGURE 4.1. Principle of operation of a passive remote IR chemical sensor. The spectral emission band intensity is governed by three parameters: the cloud length, the cloud concentration, and the relative temperature difference between the cloud and its background. The spectral features can be either absorption ($T_2 > T_1$) or emission ($T_1 > T_2$).

in that the source is removed and a specialized set of collimating optics is added. Current sensors used in stationary fixed-site applications measure the background radiant emission profile as a function of time. When a cloud moves into the field of view of the spectrometer, a change in the radiant energy profile is detected. This stationary fixed-site application is shown in Fig. 4.1.

Remote IR sensors must operate in relatively complex chemical environments. For this reason, an advanced spectral pattern recognition algorithm is needed to process the many spectral components of the signal. The U.S. Army Chemical Research, Development, and Engineering Center (CRDEC) has been interested in IR spectral pattern recognition since the late 1960s. The first efforts were with Bendix (ERIM) and General Electric.[5–8] These efforts were of limited success to the chemical detection problem, although they did contribute to other applications, such as pattern recognition of earth resources. The Army continued to work on the problem throughout the mid-1970s, culminating with a successful demonstration of pattern recognition techniques for remote detection.[3,9,10] Based on these results, the Honeywell Corporation (now Brunswick) has continued to develop frequency-domain spectral pattern recognition methods that use background subtraction and linear discriminant analysis techniques.

An addition to these efforts was the development of a time-domain algorithm that sensed a change in the emission profile of the instrument.[11] This effort to develop an "event indicator" applied the Gram–Schmidt method to time-domain data for automatic analysis. The Gram–Schmidt method has been used widely for the analysis of GC/FTIR laboratory interferograms. Continued research by CRDEC and the University of Iowa has focused on digital signal processing concepts for the detection of chemical vapor signatures for cases when an interferometer was mounted on a mobile platform.[12]

Recently, the FTIR literature has indicated that remote chemical sensors have been used in mobile environments including helicopters, aircraft, and earth-orbiting satellites.[13–18] When FTIR interferograms are collected from these platforms, the background emission profile changes rapidly with respect to time. In this case, the change in emission profile versus time cannot be used for the detection of the presence of a vapor absorption or emission signature. Clearly, a signal processing technique is needed to remove the background spectral features in order to distinguish the chemical vapor IR absorption or emissions. In addition, the detection problem is complicated by the large dynamic range of the background radiant emission versus that of a typical vapor cloud. Reduction of the dynamic range of all background features could enhance the discrimination of the chemical vapor signatures.

The collection of data from mobile FTIR applications is further complicated by the extreme data-processing requirements in which an inteferometer may collect up to 30 two-wave number scans per second. For many pollution-monitoring applications, size, power, cost, and weight limitations require that a relatively low-powered single-board computer be used for real-time data analysis. If the application requires an automatic alarm at high data rates, then the Fourier transformation may be infeasible for single-board computers because of the length of computation times. If the transform is determined to be a luxury that cannot be afforded, then a computationally effective method should be developed for the detection of vapor signatures.

This chapter describes several mathematical techniques developed for the detection of remote IR vapor signatures. The techniques include frequency-domain digital filtering, time-domain digital filtering, and the computationally expensive maximum entropy method (MEM) transformation. In addition, a description of several recent developments in advanced digital signal processors (DSPs) is included. The DSP chips have the potential to revolutionize signal processing concepts for remote IR sensors because processing rates are extremely high. Current DSPs are capable of up to 25 million instructions per second. Future remote sensing processing schemes will be developed around the advanced DSP technology.

4.2. THE PASSIVE REMOTE SENSING PROBLEM

To illustrate the method of detection of a target chemical species located at some distance from a sensor, one can consider the following theoretical description. If one totally fills the field of view of the sensor, then the radiance incident on the sensor is the result of the radiances from (1) the background (which is assumed to begin immediately behind the target cloud), (2) the target cloud, and (3) the intervening atmospheric gases. The radiance from an infinitesimal layer is

$$N = \int_0^x \{k_T(\lambda)N_x(\lambda) - k_T(\lambda)N'(\lambda) + k_A N_x(\lambda) - k_A N'(\lambda)\}\, dx \quad (4.1)$$

where k_T and k_A are the extinction coefficients of the target gas and the atmosphere, respectively, with units of reciprocal length. N_x is the radiance of a blackbody at the temperature on the infinitesimal layer. (The target gas is assumed to be at the temperature of the infinitesimal layer.) N' is the radiance incident on the infinitesimal slab, traveling to the sensor, and x is a length variable that is integrated independently for range and target cloud length. All of these variables are functions of the wavelength as a function of the resolution of the instrument.

Assuming homogeneous atmospheric and target cloud species, the integration of equation (4.1) gives the power incident on a passive sensor[(19)]

$$P = [T_A T_T N_{BG} + (1 - T_A T_T)N_T]A_c\Omega_s \quad (4.2)$$

where T_A is the atmospheric transmittance, T_T is the target cloud transmittance, N_{BG} is the radiance of the background, N_T is the radiance of a blackbody at the ambient temperature, A_c is the collector area, Ω_s is the solid angle of acceptance of the sensor. The atmospheric cloud transmittance is

$$T_A = e^{-k_A R} \quad (4.3)$$

where R is the distance to the target cloud. The target cloud transmittance is

$$T_T = e^{-k_T L} = e^{-\alpha_T CL} \quad (4.4)$$

where α_T (m^2/mg) is the absorptivity of the target cloud, C is the concentration of the target cloud (mg/m^3), and L is the length of the target cloud (m). It is assumed that $R \gg L$.

Two factors are implicit in equation (4.2). First, under normal conditions, N_{BG} and N_T differ at most by a few percent. If $N_{BG} = N_T$, then $P = N_{BC}A_c T_0$, and the incoming radiation contains no information about the spectral properties of gases in the intervening atmosphere, although the magnitude of P remains about the same. Therefore, if detection is to occur, it is essential, that $|N_{BG} - N_T| = \Delta N > 0$. The temperature associated with a given radiance level and optical wavelength can be determined from Planck's function. Therefore, the required difference can be interpreted in terms of the temperature. There is an occasional question as to whether the necessary difference should be interpreted in terms of air temperatures or equivalent radiometric temperatures of the target cloud and the background. The interpretation to some degree depends on the constraints of the model. In this case, the target cloud is assumed to be sufficiently small with the temperature being the primary crieria. That is, the air temperature of the target and the effective radiometric temperature of the background, regardless of the emissivity of the cloud, are primary considerations. The temperature of the background is more difficult to define under conditions of good atmospheric transmittance. The measured ratiometric temperature results from integration over a long path, which is nonhomogeneous with respect to temperature, species, and concentration. The effective temperature generally varies with the wavelength of the radiation. When atmospheric transmittance becomes low, the range becomes short and the background temperature approaches the local temperature and becomes more easily identified with a particular air mass. This lack of precise knowledge of temperatures does not cause particularly difficult detection problems, but it places severe limits on the measurement capabilities of passive IR systems. The second factor implicit in equation (4.2) is that the signal due to the target is not $1/R^2$ dependent as long as the target cloud fills the field of view. The signal does depend on atmospheric transmittance, which is range dependent.

The range-dependent signal discrimination problem, as developed in the preceding discussion, requires that an extremely small temperature difference be measured at a remote distance. Of all techniques currently used, only the Michelson interferometer has the minimum detectable limits and optical bandwidth necessary for detection. The sensitivity and minimum detectable limits are based on two advantages, the optical throughput (Jacquinot's advantage) and the multiplex advantage (Fellgett's advantage).

The current hardware used by our group to obtain Jacquinot's and Fellgett's advantages consists of a right-angle Michelson flex-pivot interferometer, a cooled Hg(CdTe) infrared detector, and 16-bit analog-to-digital converter electronics. A single-axis rotating mirror is used for positioning the instantaneous field of view (IFOV) for a number of collection positions. The Michelson interferometer used by our group consists of a rugged porch-swing flex-pivot design. The original design was developed by Walker and Rex of the U.S. Air Force Geophysics Laboratory in Cambridge, Massachusetts.[20] The detector is a single element which must be cooled to 85 K by a split-cycle Stirling cryogenic cooler. The signal collected from the detector is analyzed and digitized by a 16-bit analog-to-digital converter and provided to the signal processor for data analysis.

4.3. FREQUENCY-DOMAIN ALGORITHMS

In stationary applications, automatic operation of a passive interferometer begins with the collection of a reference spectrum of a background scene, presumably without a target gas present. Subsequent spectra can then be checked for the presence of target gases. The detection process begins with a subtraction of a background and a scene spectrum to eliminate all background spectral features. In addition, many spectral features caused by various optical elements in the sensor can also be eliminated.[21] The reference subtraction has been particularly effective in removing these components. A second purpose to subtract a reference spectrum is to reduce the dynamic range of the signal, much of which does not contain useful information. It has also been determined that further improvements in performance result from the use of a boxcar filter. This operation reduces problems associated with long-term temperature drift.

4.3.1. The Linear Discriminant

The final step in the stationary operation of a passive remote sensor is to use a linear discriminant for the identification of a pollutant spectrum. A linear discriminant can be used by passive IR sensors to classify the pollutant spectra into one of two distinct classes: (1) those that contain the pollutant spectral features and (2) those that do not. Linear discriminant functions can be described in the following equation as[22]

$$g(x) = w_0 + \sum_{i=1}^{N} w_i X_i \tag{4.5}$$

where w is the weight vector, X^i is the response vector, w_0 is the threshold vector, and $g(x)$ is the frequency response function.

The linear discriminant divides the feature space by a hyperplane. The weight vector used has a number of components or in this case a number of optical frequencies. The values of these weights are determined by a process called "training" which produces an optimum set of weights for the detection of a specified pollutant gas. The process gives a response function that is a minimum response to the spectral interferents. In general, the weights are chosen to produce the maximum possible response to the pollutant gas while constraining the interferant response to a value much less than an automatic alarm threshold.

A linear discriminant process assumes that the two classes are linearly separable into two distinct groups. The weight vector w is obtained by using a convergence technique known as the perceptron convergence theorem. This theorem uses a weight vector to find all examples in a given set that are misclassified on the positive side of the hyperplane. The misclassified samples are used to estimate a new weight vector and a revised hyperplane surface. The samples are classified a second time with the misclassified samples, forming the new error function. This error classification of the weight vector is done until all of the samples are classified correctly or it is assumed that the set is not linearly separable.

While the process of reference subtract, boxcar filtering, and operation with linear discriminants has proven satisfactory for operation of a passive interferometer in stationary applications, it is too limiting a requirement if that sensor is to be operated from a moving vehicle. For this application, the initial approach to the elimination of the background features was to substitute another reference for the background. A frequency-domain filter was then developed to be used with a linear discriminant operation. The best substitute reference was determined to be a local blackbody reference spectrum.

4.3.2. Improved Frequency-Domain Filtering Techniques

To determine whether frequency-domain digital filters are useful in the separation of target gas spectral features from all other features, an analysis was performed of the audio frequencies present in a transformed spectrum. This analysis was used as a guide for the spectral filtering operation. Using this method, the separation of target gas spectral features may be performed only if the target gas features have much different bandwidths from those of the background features. A Fourier transformation was used to convert the optical spectrum to an audio frequency spectrum. Several analyses were performed on the frequencies present in the audio spectrum. Figure 4.2 is typical of audio frequencies encountered when using passive IR field sensors in that it shows that the signature contains a large low-frequency component. A spectral filter can be used to eliminate the low-frequency components while retaining the high-frequency components.

The conclusion from the preliminary analysis of both laboratory and

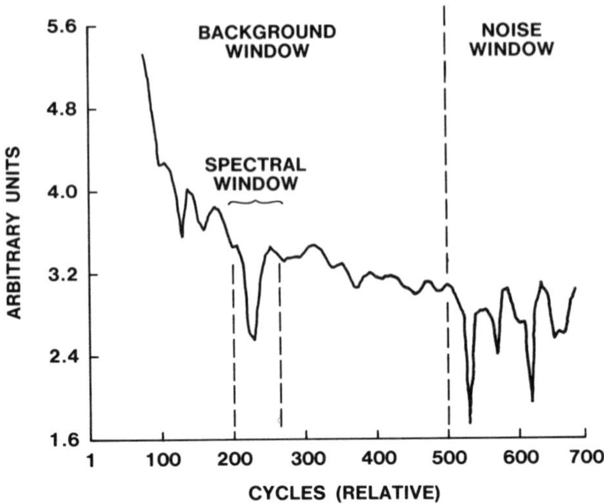

FIGURE 4.2. The relative audio frequencies of a Fourier transformation of the spectral features.

field test data sets showed that a considerable percentage of the bands due to the background might be eliminated by simple digital filters. More specialized digital filters can be used to eliminate even more of the unwanted background signals. These filters typically eliminate all but a narrow range of audio frequencies. In addition, some of the filters do not reduce the signal-to-noise ratio of the spectrum by an appreciable amount. An example of this type of filter is a simple Butterworth filter. Figure 4.3 shows a difference spectrum between a normalized target gas spectrum and a normalized blackbody reference spectrum. Because there is a large temperature difference between the blackbody spectrum and the target gas spectrum, a shift in the spectral response is induced into the difference spectrum. Low-pass coupled with high-pass digital filtering can be used to eliminate this shift.[23,24] In addition, it was found that the filtering step improves the selectivity of the linear discriminant for interferents because most of the common spectral atmospheric interferents have a radically different band-pass than those of the target gases.

In 1983 the Honeywell Corporation, under contract to CRDEC, developed the first optimized frequency-domain digital filter to eliminate broad background features.[23] Performance of the initial filter was good, although it was determined to not be adequate for all cases of remote sensing data because the technique filtered a large portion of the vapor phase spectral features. To become more selective, other concepts such as the median filter with replacement, the hybrid median average filter,

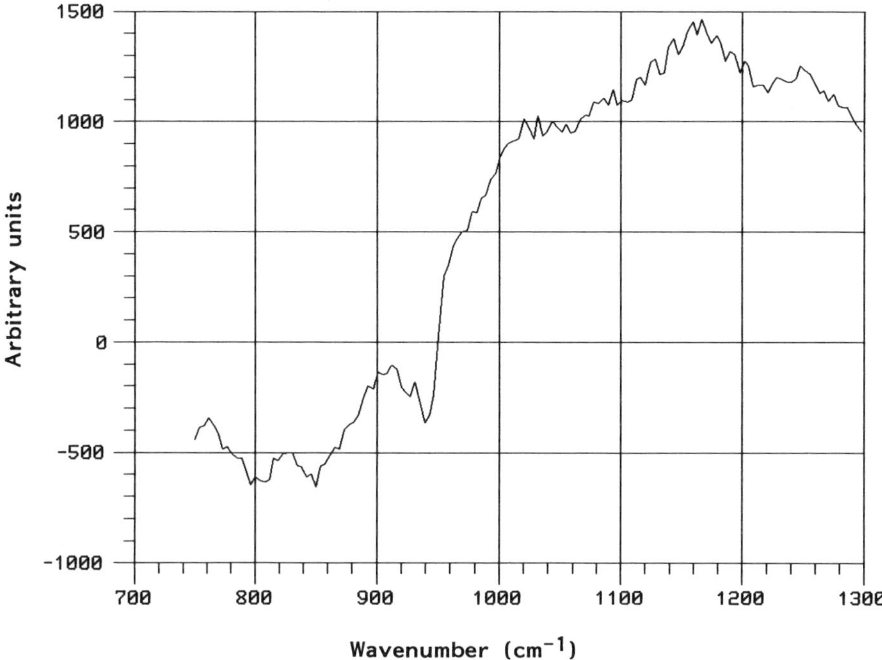

FIGURE 4.3. Difference spectrum of a normalized target gas spectrum subtracted from a normalized blackbody reference spectrum. The target spectral simulant was sulfur hexafluoride (SF_6).

the maximum likelihood filter, and the expanded central difference filter were used.[24] The performance increased relative to that of the baseline approach; however, the signal-to-noise output of the filters was still not adequate for the discrimination of all sets of training data.

The Honeywell Corporation reached two significant conclusions from the filtering experiments. First, frequency-domain digital filters could be used to aid the discrimination between background spectral features and those of chemical vapor signatures. Second, it was determined that many of the specialized digital filters designed to enhance the spectral bands worked well at the edges of the band, but they could not conserve the peak energy level in the bands. Later, Honeywell also determined that the absorption and emission cases were effected differently by the filter operation. Because of this filter effect an additional class on the two-class threshold concept was thought to be needed.[25] Although the initial research by Honeywell was very successful in the elimination of spectral features, some broad spectral features still remained after use of the frequency-domain filter. These features degraded the performance of the linear classifier.

A problem that can occur in any filtering approach is that the nonseparated features are usually treated as a linear sum of the radiometric signals from individual contributors. This is a necessary assumption if the pattern recognition problem is to be solved. The assumption, however, will become invalid at some point with increasing demands of performance. This effect can be seen if equation (4.1) is rewritten with terms describing the contribution from the optical train,

$$N = \int_0^x \{k_A(N_x - N') + k_T(N_x - N') + k_{OT}(N_{OT} - N')\}\, dx \quad (4.6)$$

This equation can be solved in a closed form for special cases. Although equation (4.6) was not integrated, it can be seen from equation (4.2) that, in general, the individual contributors are not linearly separable. The spectral filtering results show clearly the presence of spectral contributors at stages in the processing where they should not appear. Clearly, the supposition of linearity has been applied too early in the processing for the digital filters. Further efforts must be made to separate the signatures of the various contributors. This can be done by either physical and/or signal processing techniques. Physical techniques generally involve making better optical components; for example, lower absorbing components. For this effort, we did not want to use this option; therefore, we took the approach of trying to improve the filtering.

Enhancement of useful spectral features may be done in either the frequency domain (the spectrum) or the time domain (the interferogram). Although filtering in the frequency domain is usually considered lower risk, it was felt that for an interferometer in which the background is changing rapidly, the attention should be given to the problem of correlation in the interferogram space. Originally, it was thought that this would have two advantages: (1) reduced computation time and (2) less loss in useful information. The application of filtering techniques in the time domain is very similar to that in the spectral domain.

4.3.3. Apodization Methods for Digital Filtering

In the time domain, the central fringe of an interferogram is at zero retardation of the moving mirror of the interferometer. Although the central fringe contains information from all spectral wavelengths, the fringe contains a disproportionate amount of information concerned with the general-scene blackbody radiation curve.[26,27] Separation of these spectral features can be performed by apodizing, clipping, or deconvolving the central portion of the interferogram fringe. If the interferogram is reasonably blanked or apodized, then degradation of the spectrum might not be too severe.

Apodization and deconvolution schemes used in the Fourier transformation can be considered as weighting functions that can be applied to the interferograms. Normally, they are used to reduce the amplitude of the side lobes caused by the transformation process.[26] The basic assumption of this work was that apodization functions might be used to eliminate spectral signatures of the background as well as signatures from the optical train of the system. If a proper apodization function could be derived, then background and optical train features could be separated from the pollutant features and the supposition of linearity restored. Several special-purpose apodization functions were developed to remove broad background spectral components.

Because the apodization function reduces side lobes present in the spectrum, one can think of it as a form of spectral filtering. The effect of using a triangular apodization, as currently used in many commercial FTIR software packages, can be seen by comparing spectra with and without sharp spectral features. The interferogram that has only a broad signature in the frequency domain has a large central fringe with modulations that rapidly approach zero intensity as the mirror path length differs. In contrast, the interferogram with the narrow spectral features present has a significant modulation even at large mirror retardations.

Codding and Horlick[27] have studied several novel apodization functions that give unique modifications to the spectral transform. One such apodization is the notch filter, which eliminates broad spectral features while retaining the sharp, narrow features. The notch filter worked well with laboratory data obtained with conventional IR spectrometers; however, it was found to be of marginal use on remote sensing data that typically have low signal-to-noise ratios. The difficulties of using a notch filter lie in the weighting given to the wings of the inteferogram where the contribution of the narrow features is greatest. The wings of the interferogram are regions that have a decreasing signal-to-noise ratio.[28] The notch filter, while reducing the low-frequency information, increases the high-frequency noise signal. To eliminate the high- and low-frequency spectral components, a new apodization processing concept would have to be developed that would consider spectral features of moderate width nearly equal to that of the pollutant gas signatures.

Interferograms were taken from a passive interferometer in a remote sensing field test to provide a basis for the development of an apodization function. A blackbody reference interferogram was collected from all sets of interferograms. The blackbody source was actually internal to the optical train of the instrument. The reference interferogram and the collected field interferograms were computed twice using different apodization functions. The first apodization function was triangular with a width of 100 points and having a peak maximum at the same point of the zero retardation fringe. The second apodization function weighted the points in the wings of the interferogram, but had a wider base of 300 points. Each of the two apodization functions is shown in Fig. 4.4, plotted with a reference interferogram. The computed spectrum, using the first apodization function, contained mainly the broad spectral envelope. The computed spectrum from the second apodization function also contained the broad features along with simulant features and features from the optical components of the sensor. The difference of these two spectra was taken to eliminate the broad features of the background. The resulting difference spectrum contains the spectral features of the simulant as well as the features from the optical train. Finally, the elimination of the signatures due to the system optical train can be made by subtracting the difference reference spectrum from the difference field spectrum. The projected result is a signature that contains

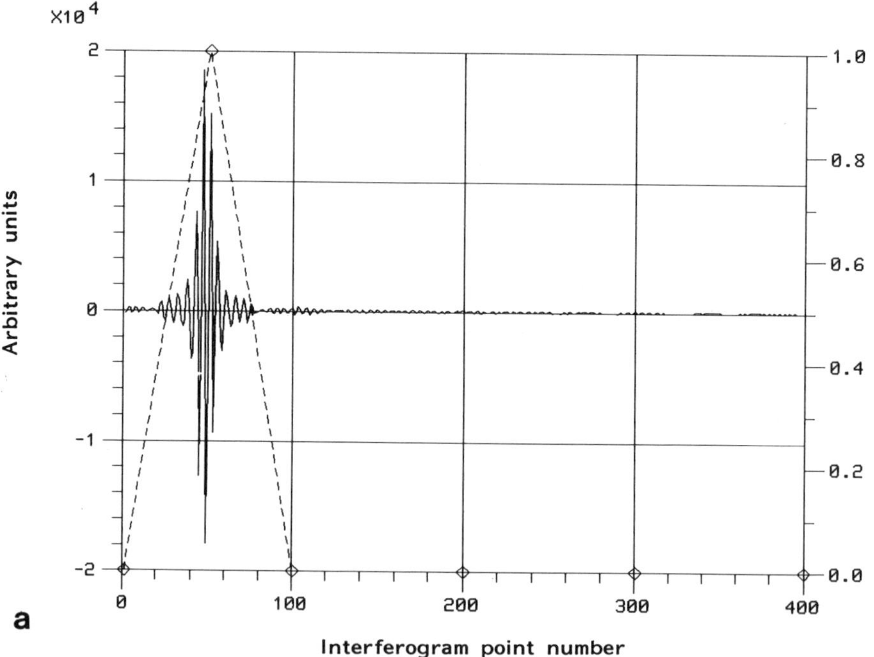

FIGURE 4.4. Two apodization functions are plotted along with an interferogram collected from a reference blackbody source. The apodization function in the upper figure (a) has a base of 100 points. The apodization function in the lower figure (b) has a base width of 300 points.

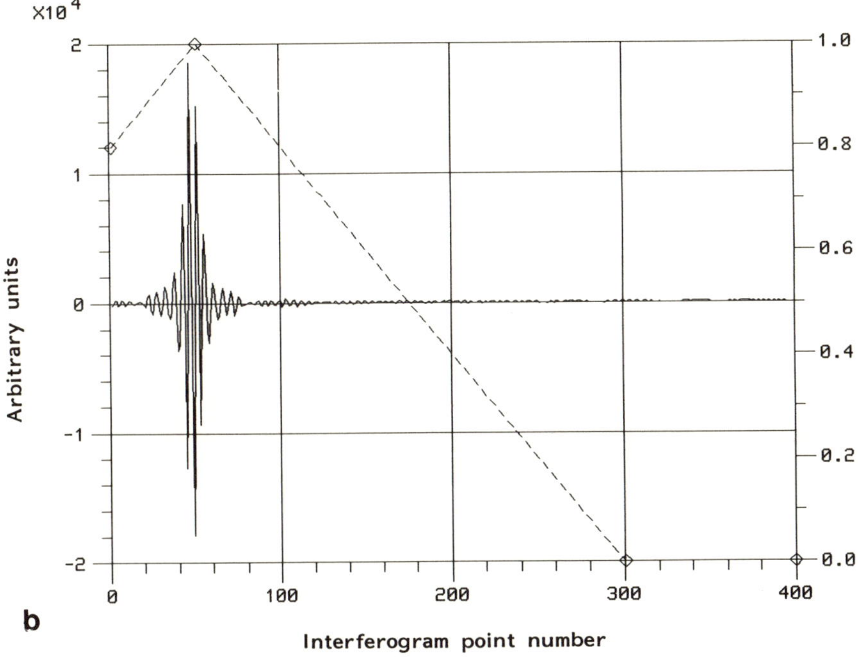

FIGURE 4.4. *(continued)*

only features from residual atmospheric gases and simulants in the sensor's field of view. Figure 4.5 shows that this objective can be achieved using the simulant sulfur hexafluoride (SF_6) when the signal-to-noise ratio is high. This method was successfully applied to approximately 50 sets of interferograms.

The apodization process has two disadvantages and two advantages over filtering techniques used in the spectral domain. The first disadvantage is that the apodization process currently uses two Fourier transformations for each interferogram. This is because the interferogram must be phase corrected for each apodization. This is not considered a major disadvantage because the phase correction processing time is considered negligible. The second disadvantage is that the response may not be fully separated, and thus could cause a loss in the processed spectral features.

The first advantage of using an apodization process is that it may produce better separation of optical train features than the previous filtering efforts. This improved filtering should improve performance of the pattern recognition methods that depend on linear suppositions of all radiometric contributors. A second advantage of the apodization process is that it leads to a generalized processing scheme.

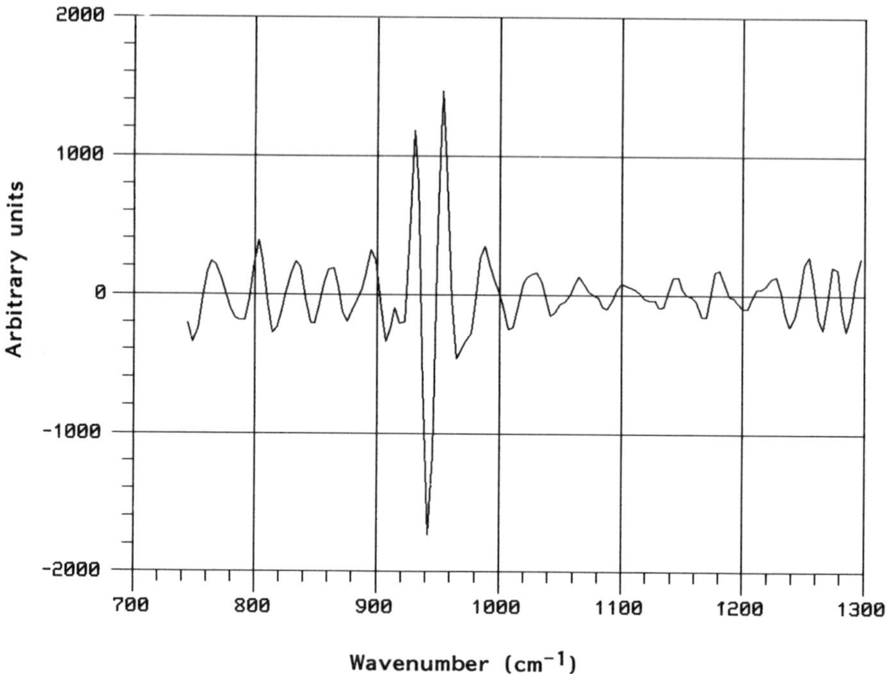

FIGURE 4.5. A difference spectrum taken of a Fourier transformation of a narrow apodization function subtracted from a Fourier transformation of a broad apodization function.

4.4. TIME-DOMAIN DIGITAL FILTERING

Because the apodization function is essentially a form of a digital filtering function, it can be argued that the Fourier transform using apodization functions as digital filters is an inefficient processing strategy for time-domain interferograms. The problem with an interferogram-based analysis, however, centers on the composite nature of the waveform. The periodic signals corresponding to the individual frequencies co-add to form the interferogram. Information regarding the specific frequencies of interest is therefore obscured by frequency information that has no bearing on determining the presence of the target compound.

This problem is illustrated by an inspection of Fig. 4.6. Segments (points 150–350 in the 1,024-point interferogram) from six interferograms are plotted. Based on an inspection of the corresponding spectra, the lower three interferograms indicate the presence of SF_6 in the field of view of the spectrometer, while the upper three interferograms indicate

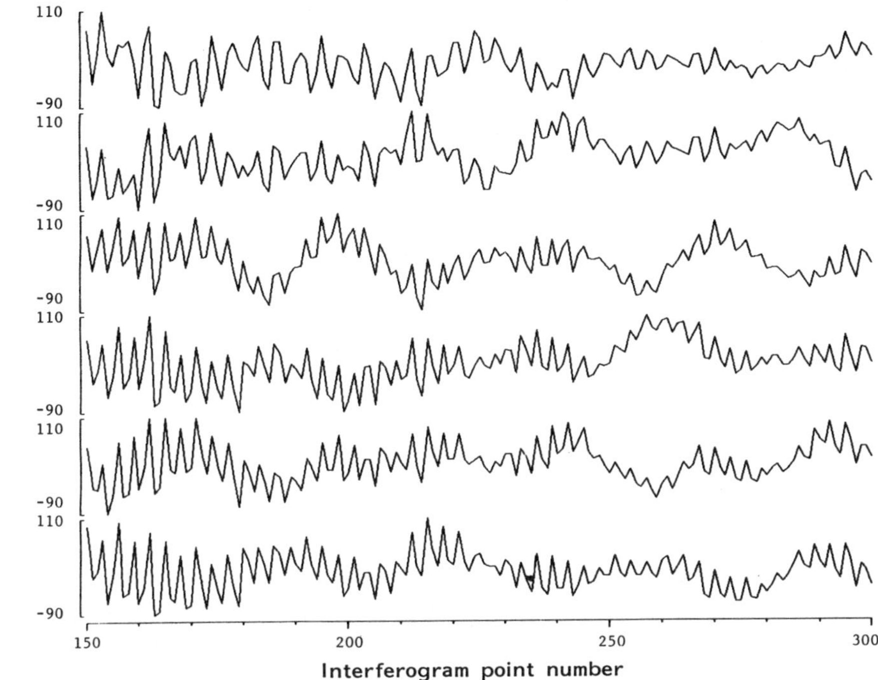

FIGURE 4.6. Segments (points 150–350 in a 1,024-point interferogram) from six interferograms. The top three interferograms contain the background, while the lower three interferograms contain background plus simulant spectral features.

no SF_6 present. It is clear from an inspection of the figure that no reliable detection procedure can be based on the raw interferogram information.

4.4.1. Linear Finite-Impulse Response Filters

Digital filters are mathematical transforms that operate on time-domain data in a frequency-dependent manner. The result of the transformation is that information pertaining to certain frequencies is suppressed. The most common type of digital filter is termed a nonrecursive or finite-impulse response (FIR) filter. The basic form of this filter is

$$Y_i^* = f_1 Y_{i-N} + \cdots + f_{N+1} Y_i \tag{4.7}$$

where Y_i^* is a filtered data point resulting from the application of the filter, Y_i is the corresponding raw data point, and the f-terms are weighting coefficients that determine the frequency dependence of the filter. Each term is weighted by a different coefficient, f_i through f_{N+1}. By way of example, the familiar moving average filter has the above form, with each coefficient equal to $1/(N + 1)$. Effectively, the filtered point is formed from a linear combination of the corresponding raw data point and a set of preceding raw data points.

The basis of the methodology used here is to develop narrow-band-pass digital filters that operate on interferograms, isolating the characteristic bands of target molecules. The resulting filtered interferograms can then be analyzed directly for the presence of the target chemical species.

The generation of a digital filter begins with the definition of the frequency-domain form of the filter. Effectively, a frequency filter is derived that possesses a band-pass centered on a characteristic spectral band of the target species. To generate an SF_6 filter, a Gaussian function was defined, centered at 940 cm^{-1}, and possessing a standard deviation of 20 spectral points. If this Gaussian function is multiplied by a sample spectrum, then the filtered spectrum results. The inverse Fourier transform of this filtered spectrum is a corresponding filtered interferogram. If the desired time-domain filter were available, the action of the filter on the sample interferogram would produce a filtered interferogram identical to that obtained through the Fourier transform, Gaussian multiplication, and inverse transform. The filter described by equation (4.7) is, in reality, a linear model that relates a set of independent variables (e.g., raw interferogram points) to a dependent variable (e.g., filtered interferogram points). The f-terms in the equation are directly analogous to coefficients determined through a multiple linear regression analysis. A regression analysis can be performed to derive the coefficients, producing a filter that can then be applied.

The initial work in developing the SF_6 filter employed consecutive points behind the point being filtered, as defined by equation (4.7). In the regression procedure, consecutive terms were added to the linear model until the correlation coefficient stabilized. As an example, a 21-term filter was developed, characterized by a correlation coefficient of 0.902. The computations were conducted over interferogram points 150–350. Figure 4.7 presents the results of applying this filter to the six raw interferogram segments depicted previously in Fig. 4.6. The filter produces dramatic results. The three lower interferogram segments, known to contain SF_6 information, are now clearly different from the upper segments, known to contain no SF_6 information.

The lower three interferograms depicted in Figs. 4.6 and 4.7 contain strong SF_6 signals. These result from the helicopter data runs in which the spectrometer is observing the SF_6 signal against the earth, a strong IR radiator. Much weaker signals are observed in the ground-based data

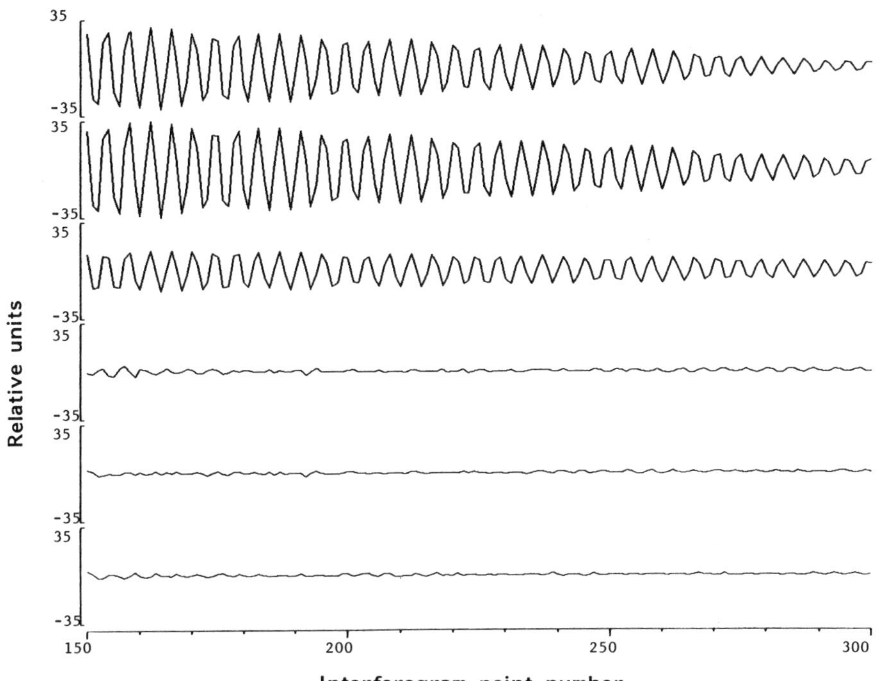

FIGURE 4.7. Interferogram segments plotted in Fig. 4.6 shown after the application of the derived interferogram filter. The lower three segments, known to correspond to SF_6-containing interferograms, are now clearly different from the upper three segments, known not to contain SF_6 spectral features.

collection. The derived SF_6 filter exhibited increased difficulties in distinguishing the active interferograms as the signal level decreased. After an evaluation of the filter derivation procedure, two goals were established: (1) narrowing the filter bandpass, and (2) increasing the correlation coefficient in the regression.

As noted above, equation (4.7) dictates the use in the filter of consecutive points previous to each point being filtered. In the regression analogy, each of these points is a separate independent variable. Some of these variables are not statistically significant in terms of their contribution to the model. There are standard techniques in regression analysis to select the significant independent variables out of a pool of candidates.[29] One such technique, stepwise regression, builds the model in a stepwise manner, adding at each step the most significant independent variable out of the pool of candidate variables. In an evaluation of this procedure, a pool of independent variables was used, consisting of the 50 points prior to each point being filtered. The stepwise procedure was allowed to select the most significant variables to add to the filter. A much narrower Gaussian function was used, specified by a standard deviation of one spectral point. Additionally, a modified interferogram range was used, consisting of points 200–300. This range was found to be slightly superior to the 150–350 range used previously.

Through this procedure, a 27-term filter was found with a correlation coefficient of 0.985. This filter was judged superior to the previous filter in its ability to process low-signal data. In an evaluation of performance, the filter was applied to an entire ground-based data run. To characterize the results of filtering, the squared intensities of the points in each filtered interferogram segment were summed. Figure 4.8 is a plot of these intensities for each interferogram in the data set. The peaks in the plot signal the presence of SF_6. An examination of the corresponding transformed spectra confirmed this result.

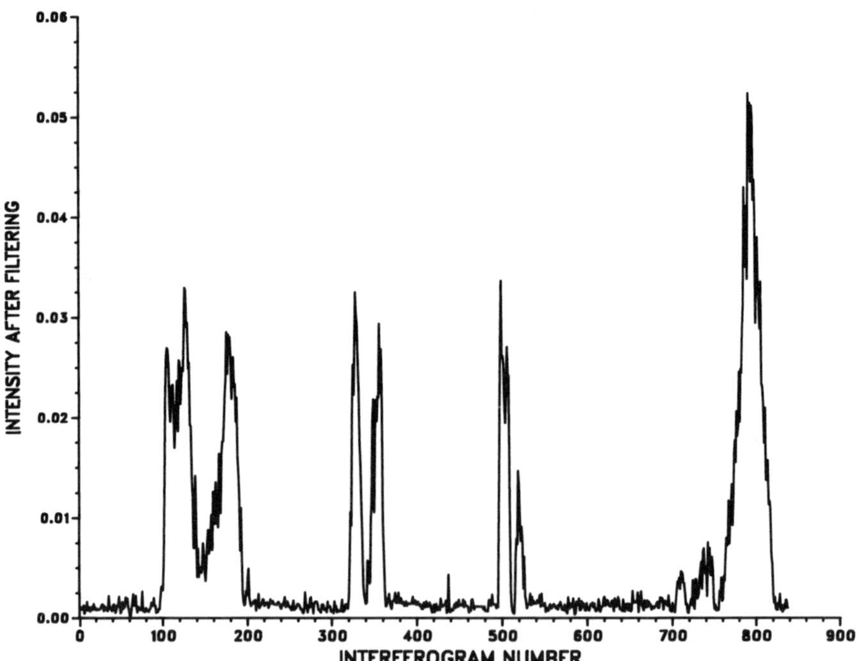

FIGURE 4.8. The results of a sum-of-squares addition of filtered interferogram points. The x-axis represents the number of interferograms collected as a function of time. The intensity of the sum-of-squares indicates the presence of sulfur hexafluoride (SF_6).

As noted previously, a Gaussian with a standard deviation of one spectral point was used in the generation of the filter used in Fig. 4.8. Fourier transformation of numerous filtered interferograms revealed, however, that the derived filter still was not able to achieve the specified width of the Gaussian. Moreover, attempts to derive multiple-band filters for dimethylmethylphosphonate (DMMP) were characterized by low correlation coefficients.

4.4.2. Matrix Filters

The investigation of these effects led to the conclusion that the filter-generation methodology could be improved further by the adoption of a modified strategy. It was decided that the basic limitation of the current methodology is the reliance on one filter to process a relatively broad region of the interferogram. By deriving separate filters to be used on each interferogram point, it was hypothesized that the overall accuracy of the filtering process could be increased. To perform this derivation, however, a set of interferograms must be employed. For example, a filter to be used on interferogram point 155 must be derived through the use of point 155 across a set of interferograms. It is clear that the use of a set of filters may offer increased accuracy. It is also true that such a procedure may be faster computationally. Since each filter is now performing a simpler task, fewer terms in each filter may be needed. The derivation of a set of filters is a direct extension of the methodology discussed previously. Separate multiple regressions are performed, one for each interferogram point treated. The result of this process is a set of filters stored as a matrix of filter coefficients. For this reason, we have termed the resultant filter set a "matrix filter."

This methodology was tested by developing matrix filters for SF_6 and DMMP. It was determined that when a set of interferograms is used, the presence of random low-frequency noise prevents the regression procedure from converging. For this reason, each interferogram was preprocessed with a four-term low-frequency cut-off filter to eliminate the low-frequency noise. Additionally, the increased complexity of the three-band-pass filtered DMMP signal dictated that interferograms be normalized before application of the filter. The resulting matrix filters performed in a manner superior to all filters previously determined. In addition, it was discovered that individual filters with low correlation coefficients could be eliminated from the analysis with improved results. It was verified that the matrix filters accurately produce the desired width of the Gaussian function. For SF_6, a Gaussian with a standard deviation of five spectral points was found to be optimum. The DMMP filter was based on three Gaussians, each with a standard deviation of one spectral point. The individual filters averaged 16 coefficients for SF_6 and 37 coefficients for DMMP. Interferogram processing was based on 53 filters for SF_6 and 56 filters for DMMP.

Figures 4.9 and 4.10 present the sum squared intensities of interferograms in two ground-based data sets after application of the SF_6 and DMMP matrix filters, respectively. Figure 4.9 represents the same data set treated in Fig. 4.8. The results obtained through the application of the matrix filter are clearly superior to those based on the standard filter. In addition, four times fewer arithmetic operations were required to apply the matrix filter. Figure 4.10 represents our first successful DMMP filter. The data set represents very low signal-to-noise data. The broad peak at the right of the plot corresponds to interferograms whose spectra definitely show the presence of DMMP.

The digital filter analysis procedure is an easily automated method

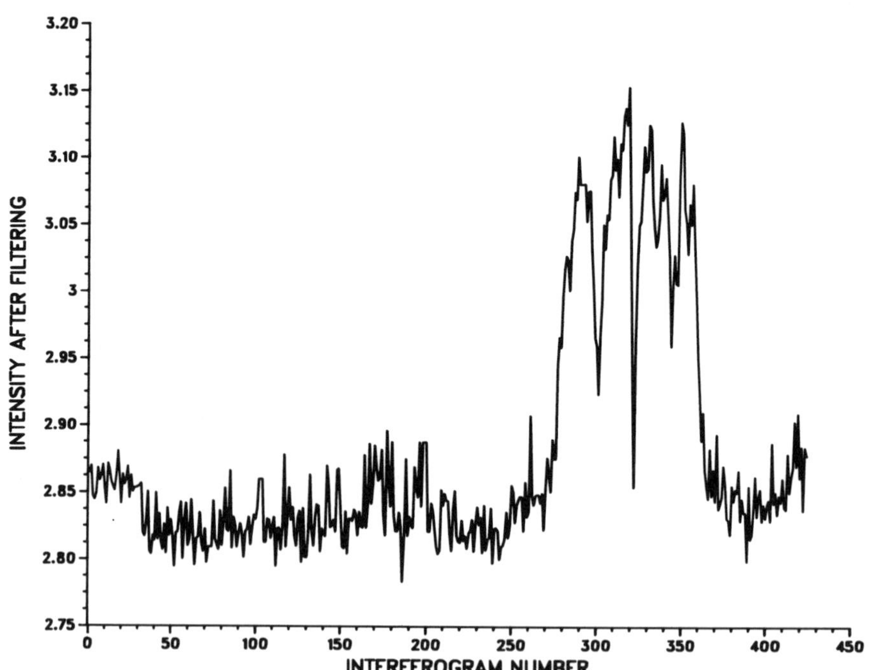

FIGURE 4.9. The results of a sum-of-squares calculation of the results from the matrix filter for DMMP. From a visual inspection of the transformed data DMMP was present from interferogram subfile 270 to 370.

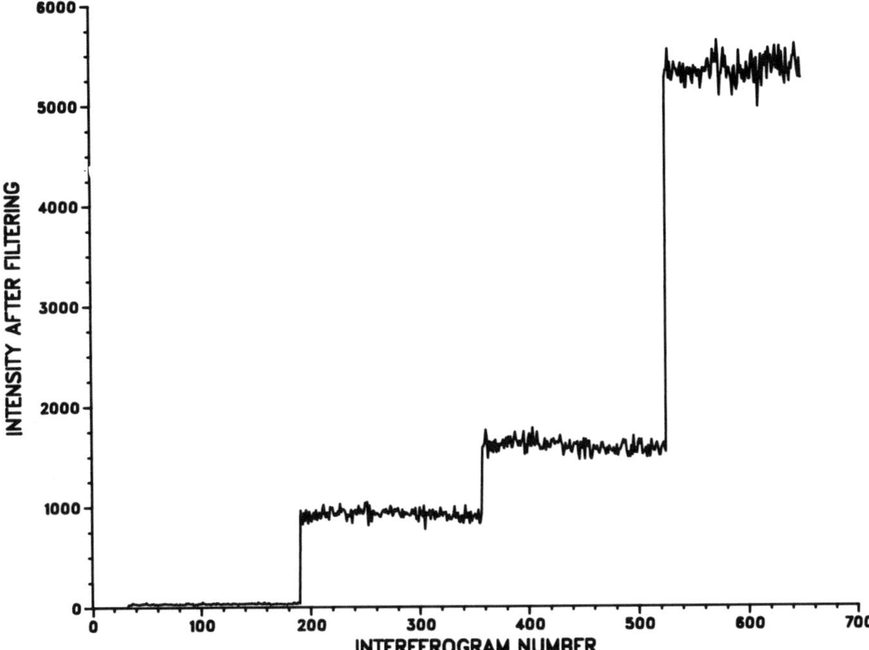

FIGURE 4.10. The figure represents a digital filter and a sum-of-squares calculation of several known concentrations of the simulant SF_6. The figure does suggest that the interferogram filtering approach could be a quantitative method under controlled conditions. Further data is required to determine the limitations of the method for quantitative analysis.

that requires no stable spectral background. The derivation of filter coefficients is a general procedure, amenable to a variety of target chemical species. The selectivity of the time-domain filter is keyed by the frequency of the filter used and the functional form of the filter.

An additional consideration of digital filters concerns the absolute sensitivity of the filters when used with quantitative data. Figure 4.11 shows the sum-of-the-square responses of a set of interferograms run when the concentration in a 1-m gas cell was increased. The arbitrary units on the y-axis roughly scale to the known concentration. This figure does suggest that the time-domain digital filter might have a role as a quantitative indicator. Further data collected in a controlled environment is required to investigate this observation in a rigorous manner.

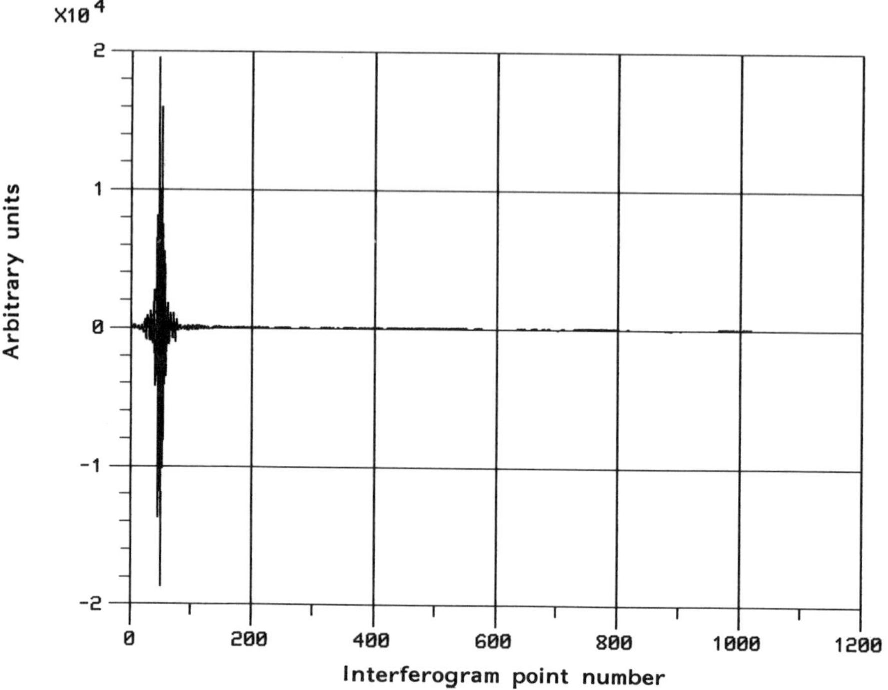

FIGURE 4.11. A 1,024-point interferogram collected from an absorption experiment. Sulfur hexafluoride was used as an absorbing media.

4.5. MAXIMUM ENTROPY TRANSFORMATIONS

The maximum entropy transformation method (or MEM) is a technique that has been used in a variety of disciplines to estimate a time series.[30] The method is actually a linear model that contains coefficients that can describe random and nonrandom events in a time series. The estimation of coefficients is typically derived in the time domain using a simple multiple linear regression. The technique can generally be divided into one of two classes. The first class, which is the method most discussed in the literature, is known by a variety of names: the maximum entropy transformation (MEM), the autoregressive model (AR), the periodogram method, or the all-pole model.[31–36] The second class is popularized with the names of moving average transformation (MA), autoregressive moving average (ARMA), or is simply referred to as the all-zero model.[37–39] The first model type refers to a model that has coefficients in the denominator that estimate the real frequencies in a spectrum. In contrast, the second model approximation has all of its coefficients in the numerator and, thus, must attempt to fit the time series with an equation that is essentially a simple polynomial function.

The all-pole model, with all of its coefficients in the denominator, can fit functions that have very narrow spectral features. In fact, the method has been used previously in several other analytical areas such as FT/NMR, FT/MS, and Raman spectroscopy, where a need exists to extract high-resolution features from short interferogram segments.[40–44] It has, however, not been used to transform short data segments of FTIR interferograms from remote sensing measurements. In each of the FT/NMR and FT/MS references an early version of the Burg MEM transform[45–48] was compared to the AR spectral estimator[31,49] for a variety of different spectra.

The all-pole approximation to the Fourier transformation can be shown in the following equation:

$$\frac{a_0}{1 + \sum_{k=1}^{M} a_k^{Z^{K2}}} \approx \sum_{J=-M}^{M} i^{Z^J} \tag{4.8}$$

where

$$Z \equiv e^{2\pi i f \Delta} \tag{4.9}$$

where the value a_k is the coefficient, f is the frequency, and M is the model order. The unique feature concerning equation (4.8) is that although the value of the right-hand term approximates zero outside the range of $-M$ to M, the left-hand model approximation term will have nonzero values outside the range. Choosing a valid number of terms to properly fit the true model order can give a series expansion that extrapolates the function outside of the sampling window. Thus, the MEM has the capability of being able to fit spectral features that have very sharp features. The basic difference between the Fourier transform and the MEM for short IR interferogram segments lies in the assumption that the data continues in a predictable manner outside of the sampling window. Obviously, the resolution depends on the predictability of the data at the ends of the sample interval. Because the MEM transform uses data in the sampling window to predict the data outside of the interval, one serious problem in the method is created. That is, the model order, or the approximation to the model order, must be known in advance of the determination of the coefficients. For data collected from remote sensing IR experiments, an optimum model order is usually never known in advance. Thus, semiempirical methods to select a proper model order were investigated using the passive remote sensing data.

Because all-pole models have the property of fitting very sharp spectral features collected from short time-domain data segments, they have an application for the modification of the hardware design of a remote IR sensing spectrometer. FTIR remote spectrometers suffer from the fact that the signal is contained in data that spans over 16 bits of dynamic range (i.e., 16-bit analog-to-digital converter electronics are required). If a short scan segment could be transformed, then digitized data could be used that consisted, for example, of only eight bits of resolution in the abscissa. Furthermore, advantages could possibly be gained by processing the same short segment as is processed by the linear digital filter, which is located slightly removed from the center burst, as that of the digital filter. Digital filters may be able to filter out several of the background spectral problems and thus leave an inteferogram containing a more limited number of spectral frequencies in which to fit a set of coefficients. The advantage in using a digital filter prior to the use of a MEM transformation could mean a smaller model order might be used to fit the features of interest.

To compare a digital filter and the MEM transformation versus that of a Fourier transformation and a background subtraction, a set of interferograms were collected on a Michelson flex-pivot interferometer.

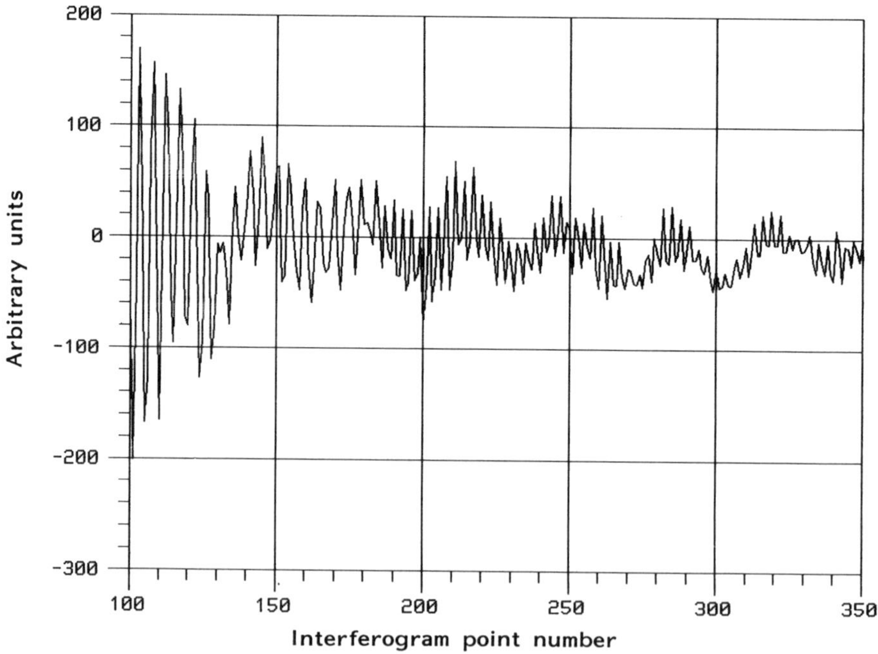

FIGURE 4.12. A short section of the 1,024-point interferogram shown in Fig. 4.11.

Figure 4.11 shows a 1,024-point interferogram collected when the optical path of the interferometer had a 300 K blackbody and a 1-m path length gas cell that contained a moderate concentration of a strong spectral absorber, sulfur hexafluoride. (SF_6 has a strong absorption at 940 cm^{-1}.) The interferogram is characterized by a large dynamic range near the center burst. The center burst region is not used in the filtering and MEM transformation because filtering a linear function is difficult over the center burst region. Instead, the linear digital filter and MEM transformation were used away from the center burst. The short section shown in Fig. 4.12, which starts at interferogram point 100 and ends at point 300, was digitally filtered and transformed by an MEM. The combination of a digital filter and the MEM transform indicated that spectral background features can be eliminated. Figure 4.13 shows the FFT in comparison to the filter and MEM transform. The upper spectrum contains numerous features associated with the background. The spectra shown in this figure were collected on an interferometer mounted on a half-ton truck

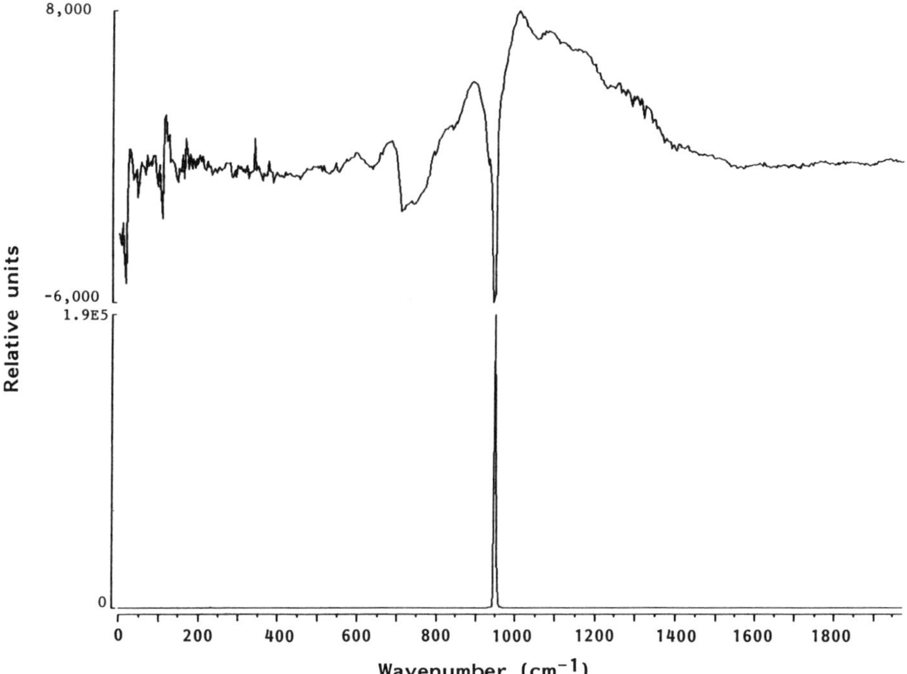

FIGURE 4.13. The lower spectrum shows the power spectrum obtained from a MEM transformation of the short interferogram segment. The top spectrum is the Fourier transformation subtracted from the background spectrum of the entire 1,024-point spectrum.

travelling approximately 20 mph on a rough dirt road. The background features shown in the upper plot are a difference in the total radiative emission of the background. The lower plot was digitally filtered with a 20-term finite-impulse response (FIR) filter which was optimized for a $20\,cm^{-1}$ full-width half-maximum (FWHM) band with a peak maximum at $940\,cm^{-1}$. A 5-term all-pole transformation model was used to generate the power spectrum. Figure 4.13 shows that, at least in the cases where the signal-to-noise ratio is high, digital filtering and MEM transforms can be used successfully on short interferogram data segments.

Model order, if selected incorrectly, can cause an estimation error in the data. As in any regression scheme, if the model order is not chosen high enough, then the fit will average out the narrow spectral features. If the model estimation order is chosen too great, then an over determination of the linear system is made and false spectral features appear. Figure 4.14 shows four plots of the same data as shown in Fig. 4.13,

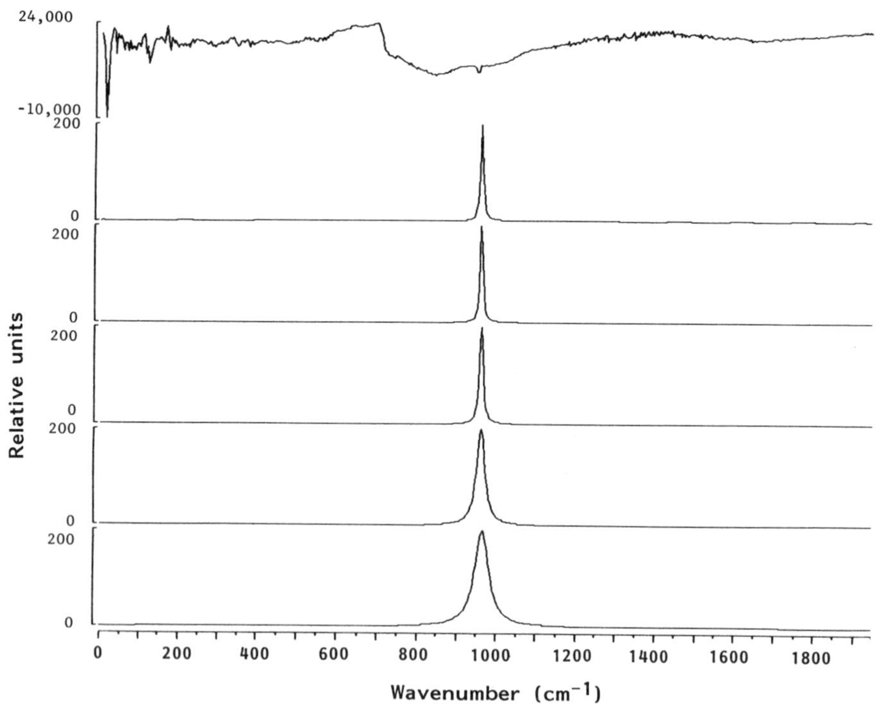

FIGURE 4.14. The five lower spectra indicate the differences in the MEM transformation when different numbers of fitting coefficients are used. The number of coefficients were varied from two coefficients (the lower spectrum) to 20 coefficients (the upper spectrum). The top spectrum is a background subtraction showing the position of the SF_6 absorption.

except that the model order was varied from 2 coefficients to 20 coefficients. The features show that for two coefficients the peaks are wide and the function does not sufficiently fit the data. Twenty coefficients seem to overdetermine the system. The effects of overdetermination result as spurious peaks and band splitting. The net result is that for optimum results a proper model order must be obtained by some method. Also, the model order should be chosen in advance of the actual multiple regression and subsequent power spectrum calculation. In addition, an error function, which could be essentially an R-squared value or a fitting correlation coefficient, needs to be used to infer the transform model "goodness-of-fit."

Several semiempirical methods to identify the statistical model have been described by Akaike,[50] Box and Jenkins,[51] Wang,[52] Jones,[53] and Berryman.[54] In each of these methods, a model that yields a minimal error function of the linear system is developed. The techniques are iterative in the respect that an error function based on the error fit of the regression coefficients is developed. In the regression analysis, the error of fit decreases as the number of fitting coefficients is increased. A simple method of determining the model order is to find an error minimum, or, rather, to determine when the error remains constant for a model order of M when compared to an $M - 1$ model order. The problem remains to select empirically the proper residual error value. In the analysis of real time-domain data, finding the value of the error function when the plot becomes very "flat" can be difficult. Figure 4.15 is a plot of an error function as the number of coefficients is increased. In this example, the error curve continually decreases and does not have a minimum. By inspection of the transformed spectra, one can see that a good spectrum was obtained for 16 coefficients.

Akaike[50] suggested an information criterion that estimates the error by a mean log-likelihood function. The likelihood function provides a means to give a multiple decision rather than a hypothesis-testing procedure. The maximum likelihood estimate was found to be very sensitive to small variations of the error function.

The function given by Akaike[50] can be described as

$$I(p) = -2\log(MLV) + 2P \tag{4.10}$$

where MLV is the maximum likelihood function, P is the optimum number of coefficients, and $I(p)$ is the minimum relative error value. If a Gaussian probability of distribution of the data is assumed, then the Akaike information criterion becomes

$$I(p) = \log(E) + 2P/N_e \tag{4.11}$$

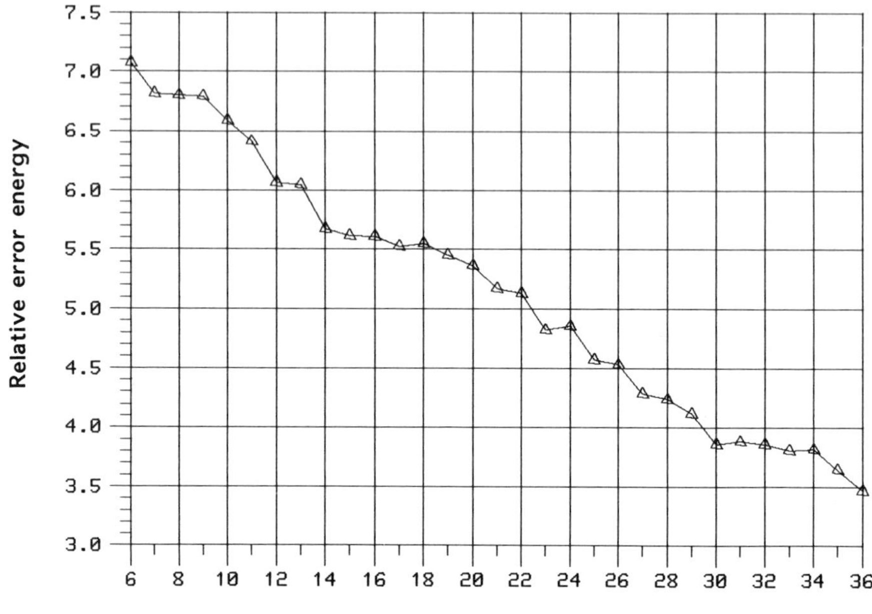

FIGURE 4.15. The figure shows a plot of the residual relative error function from the multiple regression coefficient fit. The fit was repeatedly calculated for an increasing model order.

where E is the relative error function, p is the optimum number of coefficients, $I(p)$ is the minimum relative error function, and N is the number of data points used in the transformation. In our case, an effective value of 150 points was always used. It should be noted that Akaike used an "effective" value of N if the data were windowed in some manner. For example, an empirical value of $N_e = 0.4N$ for use with a Hamming window was suggested. The value was determined by a ratio of the effective energy under the window relative to that of a rectangular window. For all of the IR remote sensing data transformed in this chapter, we used a rectangular window; therefore, our effective value of N and actual number of data points were always the same.

In equation (4.11) the first term decreases logarithmically as a function of model order, while the second term increases. The function produces a minimum when the combination of the two functions is small. Figure 4.16 shows a typical plot of the Akaike error criterion for the MEM transformation. This plot is very typical of all of the data collected

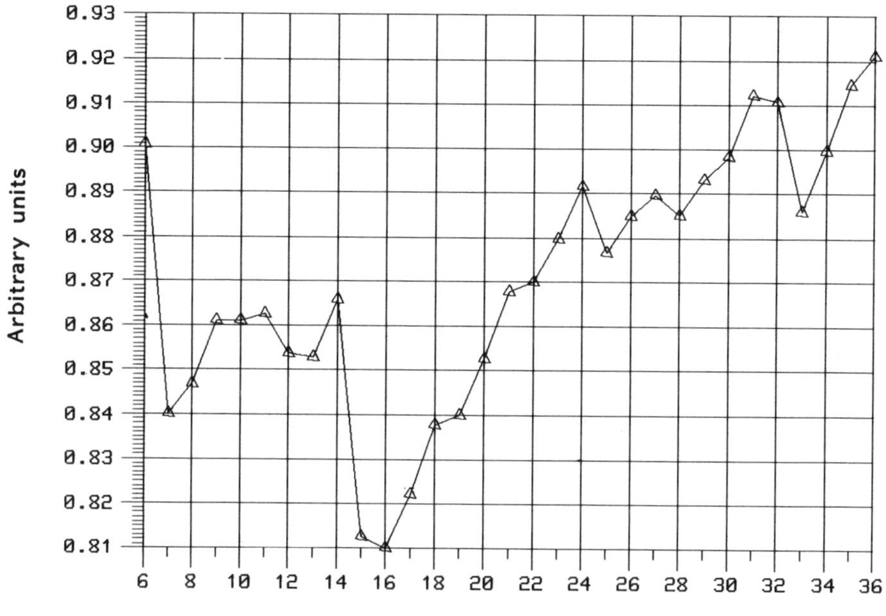

FIGURE 4.16. The figure shows a typical plot of the Akaike error log function for a range of coefficient model orders. The top spectrum is a Fourier transformation and background subtraction.

from our remote sensing experiments. The figure shows a minimum at 16 coefficients. Figure 4.17 shows four plots of the interferogram segment with varying numbers of coefficients. On close inspection, one might claim that 16 coefficients were an optimum value. Interferograms from 20 sets of data were transformed and plotted using the Akaike error log estimator. Inteferograms from all of the data sets show the same general shapes of the log error estimate. All of the interferograms had optimum numbers of coefficients that ranged from 4 to 32 coefficients.

In addition to the number of coefficients affecting the model order, it can be shown that the MEM spectrum is related to the signal-to-noise ratio of the signals present in the interferogram segment. A small signal-to-noise ratio and a large number of bands present in an interferogram segment make it increasingly difficult for the all-pole model to fit increasingly complex data. Various spectral estimates showing the number of coefficients needed for an optimum fit of the data were made with the SF_6 spectra. Figure 4.18 shows five interferogram segments with

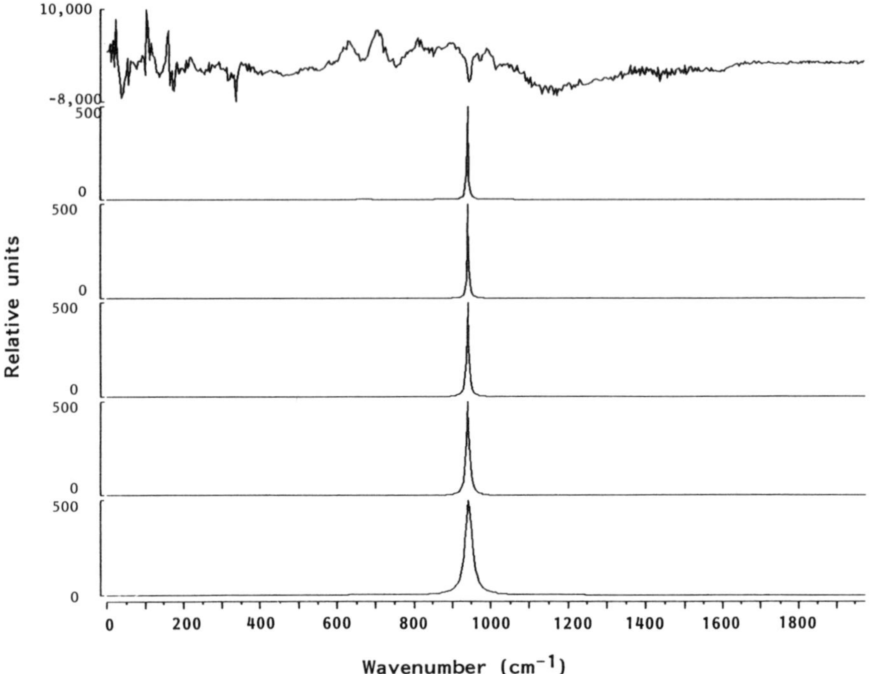

FIGURE 4.17. The five lower plots were obtained when using different model orders for the MEM transformation of the same short section interferogram.

increasing signal-to-noise ratios of the SF_6 feature. This spectral simulant was ideal for the experiment because it has only one strong spectral absorption; therefore, a limited range of sine-wave frequencies would be expected to be present. The interferograms in the figure were digitally filtered with the optimized SF_6 filter and then transformed with a MEM transform. Figure 4.18 shows a plot of the relative intensity ratio of the SF_6 absorption versus the number of coefficients used for an optimum fit using the MEM transform. As can be seen in the figure, a steadily increasing model order is required to fit the spectral features in the time-domain segment. Figure 4.19 shows the effect of the model order on the relative signal-to-noise ratio in an interferogram segment. A digital filter preprocessing step might be used as an advantage when using an MEM transformation because some of the background frequencies are removed and a lower number of coefficients would be required to fit the model. The digitally filtered time-domain segments fit with lower model order because (1) there are fewer frequencies present to fit and, (2) the signal-to-noise ratio of the SF_6 is higher.

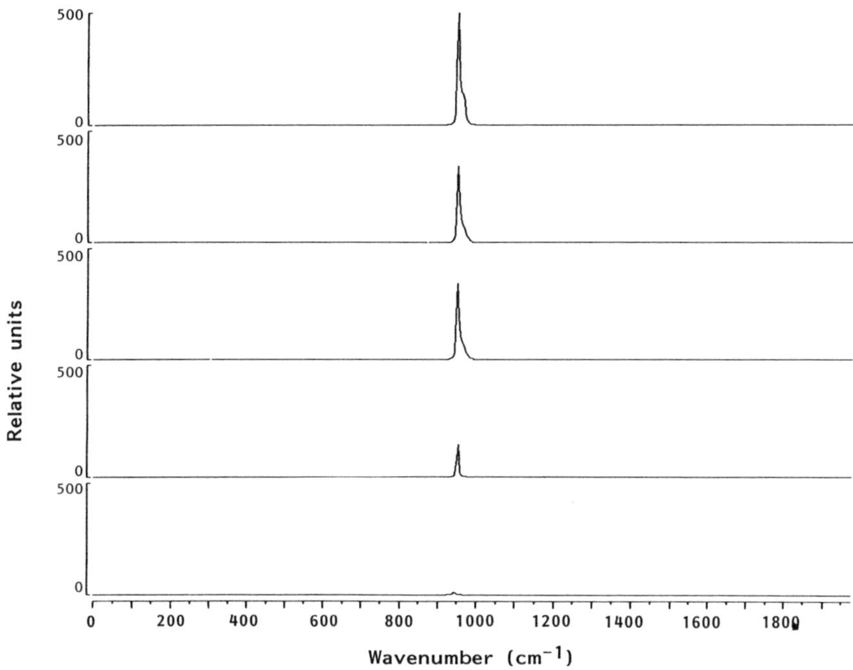

FIGURE 4.18. Five MEM transformed interferogram segments. The spectra contain an increasing contribution in the signal-to-noise ratio for the SF_6 feature.

Despite all of the problems noted, the MEM analysis method provides an extremely valuable tool to evaluate short IR interferogram segments. The method is capable of evaluating very narrow features from data collected using analog-to-digital converters with a limited dynamic range. High-speed remote sensing applications in which the background features are changing very rapidly could benefit from the evaluation of a short segment. Evaluation of short data records could mean that IR interferometer hardware need only scan over a small distance and have only limited dynamic ranges for the analog-to-signal converter electronics. Spectra could be transformed that include features of moderate band-pass width.

Although autoregressive techniques provide advantages in the evaluation of short data segments, there remains several significant drawbacks in the estimation of model order. Robust techniques must be developed to provide useful statistical measures of the error associated with an identified time-series model. The Akaike log-likelihood method provides one statistical method of evaluating a Gaussian probability of distribution. Other methods of hypothesis testing have been described in

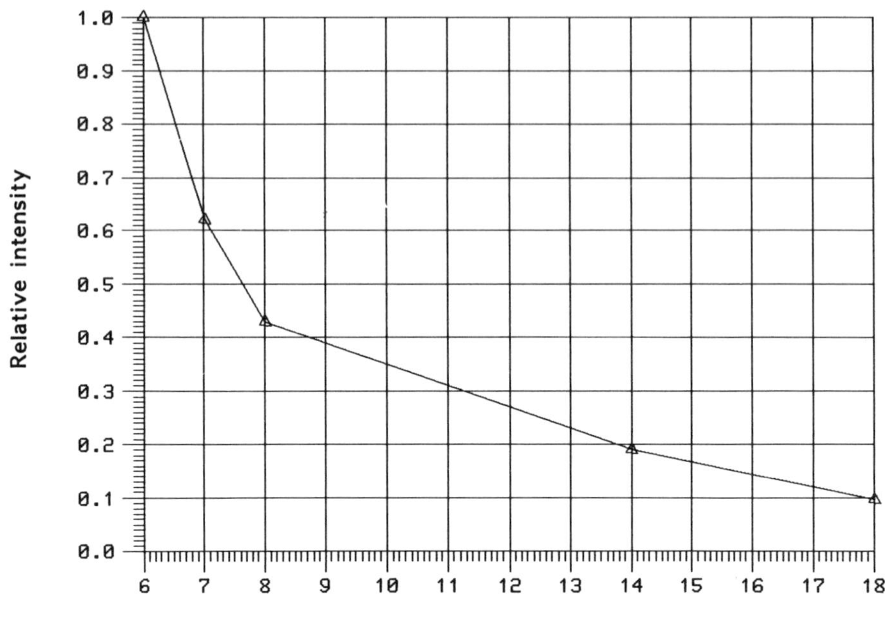

FIGURE 4.19. The figure shows the relative intensity ratio of the sulfur hexafluoride feature versus that of the optimum model order when using an MEM transformation.

the literature which give several advantages in practical applications.[55–61] Akaike's method is moderately successful because the Akaike Information Criteria (AIC) criteria does not use a fixed statistical significance to compare various model orders. As the model order is increased, an increase in the variability of the error function estimate occurs when the number of parameters is increased.[50]

Future research on all-pole time series analysis models must concentrate on developing a reasonable method for selecting and optimizing the model order without the help of any subjective judgement. Successful frequency transformation using time series analysis should be independent of signal-to-noise fluctuation and the fitting efficiency of the model. Fitting efficiency can be complicated by the number of Gaussian bands present. Obviously, more research remains to be performed before time series analysis is a standard mathematical method for use in remote IR sensing.

4.6. DIGITAL SIGNAL PROCESSING HARDWARE

Remote IR chemical sensing demands automatic algorithms (e.g. FFTs, MEMs, and digital filters) that give a real-time analysis of the data.

Real time, in the case of a Michelson interferometer, is considered to be the time between data acquisition phases of the mirror movement. This requirement for immediate feedback places a high demand on the data analysis processor incorporated into the sensor.

The current generation of chemical sensors employs general-purpose processors to perform data analysis on the acquired data as well as to perform other sensor functions. These other functions include data acquisition, control, mirror movement, and system diagnostics. With the requirement to perform all of these tasks, even the fastest of the general-purpose processors are incapable of performing data analysis at rates exceeding 10 kHz. Mobile remote sensors, with scanning rates approaching 30 scans per second, must look elsewhere for the necessary performance to perform real-time data analysis.

In recent years a new class of specialized processors has been developed to perform signal processing calculations.[(62–64)] These digital signal processors (DSPs) have internal architectures optimized to perform the repetitive sum-of-products calculations found in many of the signal processing algorithms. These calculations include finite-impulse response (FIR) and infinite-impulse response (IIR) filters, fast Fourier transformations (FFT), and matrix manipulations.

The popularity of these DSP chips has until recently been confined to the telecommunications industry; however, they are now finding new applications in many seemingly unrelated disciplines. They are now commonly used in radar/sonar, speech analysis, biomedical analysis, and a wide variety of test instrumentation. This popularity has spawned a large number of DSP solutions from several dozen hardware vendors. The solutions and their application to remote sensing will now be discussed.

4.6.1. General-Purpose Processors versus the DSP

The performance requirements to perform real-time data analysis for the mobile chemical sensor can easily surpass that of today's conventional microprocessors. This is particularly true when applications demand several of the time series analytical methods. Conventional processors, like those found in most desktop microcomputers, are designed to perform a wide variety of functions that result in lackluster performance during math operations. To overcome this shortcoming, many microcomputer users purchase an optional numerics coprocessor.

The coprocessors are microcomputers that have been optimized to perform a variety of mathematical functions. These functions include integer and floating point arithmetic as well as some algebraic functions. Coprocessors can increase the performance of a microcomputer dramatically; however, even the slowest of today's DSPs can outperform the

TABLE 4.1. Processing Times for a Variety of Processors Are Compared on Completion of a 32 Tap Filter and a Complex FFT

Processor type	Model no.	32 Tap FIR 2,048 point (μsec)	FFT 1,024 point (μsec)
General-purpose CPU	Intel 8086	645	305
	Intel 80286	265	170
	Motorola 68000	310	165
	Motorola 68020	120	82
General-purpose DSP	TI 320C25	4.2	13.2
	Motorola 56001	4.2	5.0
	AT&T DSP32C	2.5	4.4
	Analog Devices ADSP2100	5.5	7.6
Application-specific DSP	Zoran ZR34325	NA	1.7

processor–coprocessor combination by a factor of ten on common functions used for signal processing applications. Table 4.1 shows *approximate* algorithm execution times for several general-purpose processors as well as several of the DSPs.

4.6.2. DSP Architecture

Why are the DSP processors so much faster than the general-purpose processors? The answer is in the differences in architecture of these two families of processors. To perform the multitude of functions required by the desktop microcomputer, the internal architecture of the general-purpose processor has not been tailored to any particular function. Most microprocessors use a single-bus architecture in which both program instructions and data flow across the same set of data lines. This architecture, known as von Neumann architecture, can result in a bottleneck caused by the flow of data on the data bus.

The size of the general-purpose process registers can also have a serious effect on computational performance. Intel's 8088 and 80286 processors have only 16-bit-wide registers, while the 80386 possesses 32-bit registers. To perform math operations, the microprocessor breaks the numbers into manageable portions and performs a series of software operations to obtain the desired result. This process requires many machine cycles to complete. Numeric coprocessors reduce the number of required cycles by employing larger registers. The Intel 8087 numeric processor has 80-bit-wide internal registers; however, it still requires multiple cycles to perform even the simplest mathematical computation.

DSP chips are distinguished from the general-purpose processor–coprocessor by their ability to perform instructions in a single cycle.

Internal architecture of the DSPs is optimized to perform single-cycle computations that allow faster performance of the sum-of-product calculations required by many digital signal-processing algorithms. The performance is obtained through the use of hardware multipliers–accumulators, Harvard architecture, or pipelining.

Hardware multipliers and adders of the DSP eliminate the software overhead required by conventional processors in mathematical operations. These units allow the DSP to perform operations in a single cycle and insure sufficient register width for accurate results. These multipliers and accumulators are arranged to optimize the multiplication followed by addition type operations.

To take advantage of the multiplier's high-speed capabilities, the DSP must insure a steady flow of data into it. To achieve this, many different techniques are employed; however, most manufacturers use some variation of the Harvard architecture. Unlike the von Neumann approach, the Harvard architecture uses separate program and data memories, each having its own bus or buses. In a digital filtering operation, this architecture allows the data and a corresponding coefficient to be fetched from memory along separate buses and loaded into the multiplier simultaneously while an instruction is fetched on the program bus. This results in a very high throughput device.

Analog Device's ADSP2100 is an example of the Harvard internal architecture. The block diagram for this device in Fig. 4.20 shows the separation of address and data buses for both program and data memories. The ADSP2100 also uses a separate "result" bus for interconnection between computational units.

Pipelining is another scheme used to insure an adequate flow of data to the multiplier–accumulator. In a pipelined architecture, each instruction is composed of several steps such as FETCH, DECODE, MULTIPLY, and ADD. Each subsequent instruction is likewise divided; however, it always follows one step behind the previous instruction in the pipeline. In other words, while the first instruction is decoding the instruction fetched one cycle before, the second instruction is being fetched from memory. While requiring multiple cycles to complete an entire instruction, once the pipeline is filled, a result is obtained every cycle.

4.6.3. Classes of DSPs

Although all DSPs share in common the architectural features mentioned above, DSPs are divided into three distinct classes. These classes are (1) general-purpose DSPs, (2) application-specific DSPs, and (3) custom DSPs. Each class has its own advantages and disadvantages.

The general-purpose DSP is a single-chip integrated circuit designed

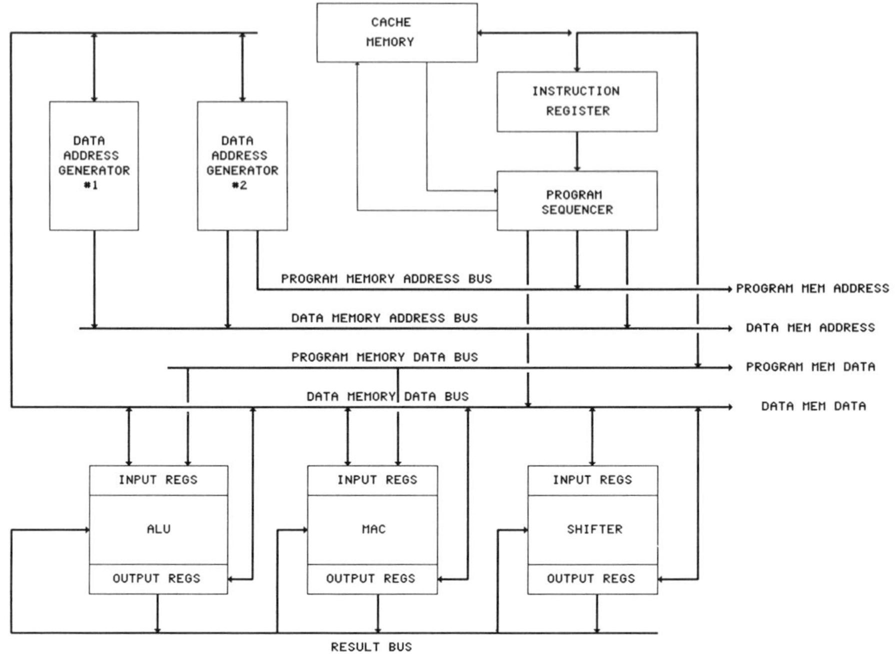

FIGURE 4.20. Block diagram of the ADSP2100.

to allow the greatest flexibility as well as provide good overall throughput. These integrated circuits come in configurations ranging from 16-bit fixed-point to 32-bit floating-point architectures and are capable of performing signal processing on signals up to 200 kHz. Commonly offered features include zero overhead looping, bit reversed addressing, and external interfaces to serial and parallel devices. Numerous combinations of on-board and off-board memory are also available from several manufacturers. The general-purpose DSP architectures make them ideal for a wide range of applications. They can be easily implemented along with conventional microprocessors in many system designs.

The general-purpose DSP can be programmed to perform any of the digital signal processing algorithms much like a conventional microprocessor. They are generally programmed in their native assembly code; however, many of today's DSPs are also capable of being programmed in the "C" programming language. This high-level programming capability makes software porting from microprocessor-based systems much easier, and results in shorter design times.

Processors of this class include Texas Instruments's 320 family

(32010, 32020, 320C25, and the 32030), AT&T's DSP32C, Motorola's 56001, and Analog Devices's ADSP2100. Development boards for the popular IBM PC/XT/AT/compatibles are often available from the manufacturer or from third-party vendors. These boards are either used for the development of stand-alone designs or as a high-speed digital signal processing coprocessor for the host system. Assemblers, compilers, simulators, and debuggers are often bundled with these development boards.

The second class of DSPs are the application-specific, or hardwired, DSPs. These single-chip processors are "preprogrammed" to perform a limited number of digital signal processing functions such as FFTs and FIRs. This specificity results in a loss of flexibility. It does not, however, significantly reduce software development. The Zoran ZR34325 is one such processor in this class. This chip is capable of performing a fast Fourier transform in as few as three lines of code using commands such as LD (load data from internal memory), FFT (fast Fourier transformation), and ST (store data to external memory). This hard-wired approach results in a modest performance improvement over the general-purpose processors. The real benefit of using these chips, however, is the ease of software development.

The last class of DSPs are the custom, or building block, DSPs. This "design your own" approach to digital signal-processing hardware is reserved for those applications in which performance is paramount. Custom-designed systems are capable of processing signals in the megahertz range. These designs are time intensive, however, and lack the support required by many users. The building block approach involves "gluing" a set of functionally specific chips together to obtain the desired result. While less complicated than designing from scratch, this technique is not recommended for the hardware novice.

4.6.4. DSP Hardware for Remote Chemical Sensors

The benefits of using DSPs for remote sensing chemical sensors are easily seen. The desire to detect chemical clouds from a ground or air platform requires fast scanning sensors. These designs demand signal processing capabilities that greatly exceed the performance of the conventional microprocessor. The U.S. Army Chemical Research, Development, and Engineering Center of Aberdeen Proving Ground is currently developing a passive IR sensor for these mobile applications that employs digital signal processing hardware. This sensor, based on a ruggedized Michelson interferometer, is targeted for scan rates of up to 30 scans per second while performing real-time data analysis. Based on the desired scanning rate, the large dynamic range requirement, the design-time deadlines and the in-house technical capabilities, the AT&T DSP32C has been selected as the target processor for this project.

The DSP32C is a CMOS 32-bit floating-point processor based on a pipelined, von Neumann architecture. This 25-MIPS (million instructions per second) device contains twenty-one 16-bit fixed-point registers for use in control, address, and logic functions and, in addition, four 40-bit accumulators to perform 32-bit floating-point mathematical operations. On-chip memory includes 2 K of read-only memory and 4 K of random access memory. The DSP has an off-chip memory capability of 16 MB. The DSP32C also supports serial I/O and a parallel I/O channel designed for easy interfacing to either an 8- or a 16-bit microprocessor. Figure 4.21 shows the block diagram of the DSP32C.

To accomplish all of the tasks not related to digital signal processing required by the sensor, an Intel 80C186 microcontroller is used in conjunction with the DSP. The microcontroller's mission is to control the sensor mechanics, data digitization, and data transfer as well as prepare the incoming data for the digital signal processing operations. These sensor tasks are best performed by the multipurpose microcontroller and are not appropriate for the DSP32C. Figure 4.22 shows a block diagram of the circuit currently under development at CRDEC.

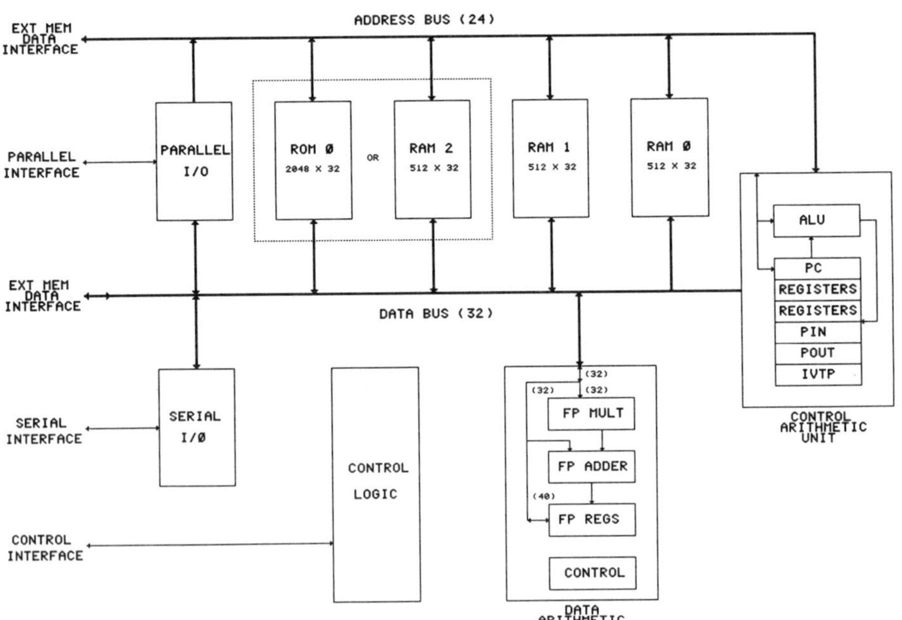

FIGURE 4.21. Block diagram of the DSP32C.

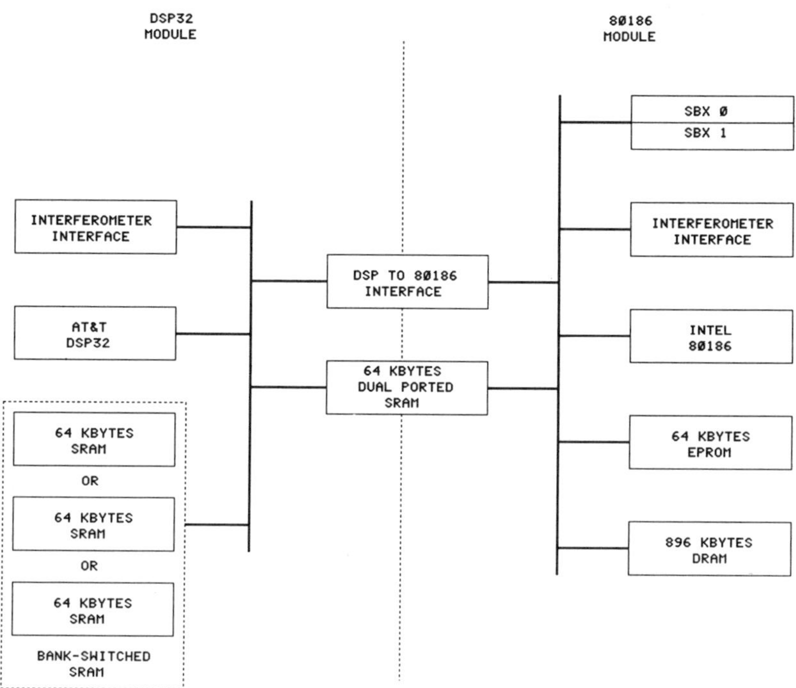

FIGURE 4.22. Block diagram of a multipurpose signal processing board to be used with a DSP32C board.

4.7. CONCLUSIONS

Digital filters, apodization functions, and maximum entropy transformations have been applied to interferograms collected from a remote IR chemical sensor. Maximum entropy transformations allow for the use of a short scanned interferogram data segment for the analysis of moderate-resolution spectra. Digital filtering of interferograms represents the first successful implementation to analyze time-domain data from a mobile passive IR interferometer. The "matrix filter" approach represents one strategy to develop an optimum set of digital filters for interferogram processing.

Many parameters need to be optimized when using digital filters. The functional form (Gaussian, Lorentzian, etc.) of the frequency-domain filter and the optimum bandwidth for the problem will affect the sensitivity and selectivity of the filter. A drawback to the overall procedure is that any interfering species absorbing or emitting frequencies in the filter band-pass will contribute to the resulting filtered signal.

Digital signal processing hardware will play an integral role in the development of the next generation of IR remote chemical sensors. The performance of the DSPs will allow the use of faster scanning hardware while, at the same time, allow for much more computationally intensive transforms. The newly applied MEM transformation functions will enable high-resolution features to be extracted from short interferogram data segments. In the future, advances in signal processing will enable passive remote sensors to be capable of making detections of gaseous pollutants without the need of a reference background subtraction.

REFERENCES

1. P. L. Hanst, in: *Advances in Environmental Science and Technology* (J. N. Pitts, and R. L. Metcalf, eds.), Wiley, New York (1971).
2. S. H. Chan, C. C. Lin and M. J. D. Low, *Environ. Sci. Technol.* **7,** 424–428 (1973).
3. H. Walter and D. Flanigan, *Appl. Opt.* **14,** 1423–1428 (1975).
4. M. J. D. Low and F. K. Claney, *Environ. Sci. Technol.* **1,** 73–74 (1967).
5. R. H. Dye and A. Prostak, CRDEC Interim Report on Contract No. DAAA15-68-C-0521, August 1969.
6. R. A. Dye and A. Prostak, CRDEC Final Report on Contract No. DAAA15-68-C-0521, December 1969.
7. D. R. Morgan and A. Roberts, CRDEC Interim Report (Phase 1) on Contract No. DAAA15-71-C-0186, September 1971.
8. D. R. Morgan and A. Roberts, CRDEC Interim Report (Phase 2) on Contract No. DAAA15-71-C-0186, September 1971.
9. D. F. Flanigan and H. Walter, CRDEC Report No. ED-TR-74070, January 1975.
10. E. N. Webb and H. Walter, CRDEC Report No. ARCSL-TR-77054, October 1977.
11. T. L. Isenhour and J. Marshall, Interim Report on Subcontract to CRDEC Contract No. DAAK11-79-C-0051, December 1980.
12. G. W. Small, R. T. Kroutil, J. T. Ditillo and W. R. Loerop, *Anal. Chem.* **60,** 264–269, (1988).
13. W. F. Herget and J. D. Brasher, *Opt. Eng.* **19,** 508–514 (1980).
14. L. Mertz, U.S. Department of Commerce Report No. AD602936, p. 120.
15. L. C. Block and A. S. Zachor, *Appl. Opt.* **3,** 209 (1964).
16. R. A. Hanel, B. Schachman, F. D. Clark, C. H. Prokesh, J. B. Taylor, W. M. Watson and L. Chaney, *Appl. Opt.* **9,** 1767 (1970).
17. W. F. Herget, Abstracts of Papers, SPIE 1981 International Conference on Fourier Transform Infrared Spectroscopy, Colombia, SC, pp. 449–456.
18. J. W. Kauffman, Abstracts of Papers, SPIE 1981 International Conference on Fourier Transform Infrared Spectroscopy, Colombia, SC, pp. 426–435.
19. S. Chandrasekhar, *Radiative Transfer,* Dover, New York (1960).
20. R. A. Walker and J. D. Rex, *Proc. Soc. Photo-Opt. Instrum. Eng.* **191,** 88 (1979).
21. J. R. Bryson and D. R. Snyder, CRDEC Interim Report DAAK-11-79-C-0051, December 1981.
22. R. O. Duda and P. E. Hart, *Pattern Classification and Scene Analysis,* Wiley-Interscience, New York (1973).
23. F. Ohnsorg, Final Report CRDEC Contract No. DAAK11-82-C-0058, January 1985.
24. M. Schultz and R. Johnston, Software Development Manual on CRDEC Contract No. DAAK11-82-C-0058, July 1985.

25. F. Ohnsorg, personal communication, April 1985.
26. P. R. Griffiths, *Chemical Infrared Fourier Transform Spectroscopy,* Vol. 43, Chemical Analysis Monograph Series, Wiley, New York (1975).
27. E. G. Codding and G. Horlick, *Appl. Spect.* **27,** 85 (1973).
28. H. Mark and M. J. D. Low, *Appl. Spect.* **25,** 605 (1971).
29. N. R. Draper and H. Smith, *Applied Regression Analysis,* 2nd ed., Wiley-Interscience, New York (1981).
30. S. M. Kay and S. L. Marple, *Proc. IEEE* **69,** 1380–1419 (1981).
31. J. Makhoul, *Proc. IEEE* **63,** 561–580 (1975).
32. R. T. Lacross, *Geophys.* **36,** 661–675 (1971).
33. A. H. Nuttall, Naval Underwater Systems Center Technical Document 5419, May 1976.
34. A. H. Nuttall, Naval Underwater Systems Center Technical Document 5501, October 1976.
35. T. J. Ulrych and T. N. Bishop, *Rev. Geophys. Space Phys.* **13,** 183–200 (1975).
36. I. Barrondale and R. E. Erickson, *Geophys.* **45,** 420–446 (1980).
37. J. Capon, *Proc. IEEE* **57,** 1408–1418 (1969).
38. S. L. Marple, *IEEE Trans. Acoust., Speech Signal Process.* **ASSP-28,** 441–454 (1980).
39. E. Parzen, *IEEE Trans. Acoust. Automatic Control* **AC-19,** 723–730 (1974).
40. V. Vita, E. Massaro, L. Guidoni, and P. Barone, *J. Mag. Res.* **70,** 379–393 (1986).
41. A. Rahbee, *Inter. J. Mass Spec. Ion Processes* **72,** 3–13 (1986).
42. F. Ni and H. A. Scheraga, *J. Mag. Res.* **70,** 506–511 (1986).
43. F. Ni and H. A. Scheraga, *J. Raman Spec.* **16,** 337 (1985).
44. O. W. Sorensen, G. W. Eich, M. H. Levitt, G. Bodenhausen and R. R. Ernst, *Prog. Nucl. Mag. Res. Spec.* **16,** 163 (1983).
45. J. P. Burg, *Proc. of the 37th Meeting of the Society of Exploration Geophysicists,* Oklahoma City, OK, October 31, 1967.
46. J. P. Burg, NATO Advanced Study Institute on Signal Processing with Emphasis on Underwater Acoustics (Enschede, Netherlands), August 12–23, 1968.
47. J. P. Burg, *Geophys.* **37,** 375–376 (1972).
48. J. P. Burg, Ph.D. dissertation, Department of Geophysics, Stanford University, Stanford, CA, May 1975.
49. J. S. Lim, *IEEE Trans. Acoust., Speech Signal Processing* **ASSP-26,** 197–209 (1978).
50. H. Akaike, *IEEE Trans. Automatic Control* **AC-19,** 716–723 (1974).
51. G. E. P. Box and G. M. Jenkins, *Time Series Analysis: Forecasting and Control,* Holden-Day, San Francisco (1970).
52. R. J. Wang, *J. Acoust. Soc. Amer.* **52,** 33–38 (1971).
53. R. H. Jones, *Geophys.* **41,** 771–773 (1976).
54. J. G. Berryman, *Geophys.* **43,** 1384–1391 (1978).
55. H. Wold, *J. Roy. Statist. Soc. B* **11,** 297–305 (1949).
56. M. S. Bartlett and P. H. Diananda, *J. Roy. Statist. Soc. B* **12,** 108–115 (1950).
57. M. S. Bartlett and D. V. Rajalakshman, *J. Roy. Statist. Soc. B* **15,** 107–124 (1950).
58. A. M. Walker, *J. Roy. Statist. Soc. B* **12,** 102–107 (1950).
59. A. M. Walker, *Proc. Cambridge Phil. Soc.* **54,** 225–232 (1957).
60. P. Whittle, *Hypothesis Testing in Time Series Analysis,* Almquist and Wiksell, Uppsala, Sweden, (1951).
61. E. J. Hannan, *Time Series Analysis,* Methuen, London (1960).
62. K. E. Marrin, *Computer Design,* pp. 69–87 (September 15, 1985).
63. W. L. Rosch, *PC Week,* pp. 61–70 (May 10, 1988).
64. D. Taylor, A. Genusor, and F. L. Rami, *ESD,* pp. 81–86 (January 1988).

Chapter 5

Imaging Spectrometry of the Earth: A Breakthrough and a Nightmare

Alexander F. H. Goetz

5.1. INTRODUCTION

Imaging spectrometry is a method for surveying, mapping, and understanding the earth based on the simultaneous acquisition of images in hundreds of contiguous spectral bands. It is a technique at the tip of a broad-based pyramid built on the foundation of practical remote sensing through aerial photography that began over 130 years ago. Balloons and pigeons carried the first cameras aloft and now high-flying aircraft, shuttles, and still higher-flying unmanned platforms carry their successors today. Imaging spectrometry is also built on a base of modern optics, detector technology, and computers for data handling and analysis. The rationale for imaging spectrometry lies in the desire to better describe the earth in the form of both pattern and process, to see and measure and characterize in a way that can not be done using our visual senses alone.

The purpose of this chapter is to review the characteristics of earth surface materials that make them amenable to identification through reflectance and emission spectroscopy, to discuss the development of imaging spectrometers, and, finally, to discuss the greatest challenge facing the users of this technology—data available.

Alexander F. H. Goetz • Center for the Study of Earth from Space, Cooperative Institute for Research in Environmental Sciences, Campus Box 449, University of Colorado, Boulder, Colorado 80309.

5.2. BACKGROUND

Remote sensing is any measurement made from afar, and that includes seeing with our eyes. Until the 1970s, the primary data for remote sensing was collected by aerial cameras using black and white film, and, beginning in the 1960s, color and color infrared film played an increasing role. Although the disciplines of photogrammetry and geomorphology dominated the field of interpretation, gross spectral reflectance differences among materials, such as vegetation and soils, in color infrared photography alerted photo interpreters to the possibility of the use of spectral remote sensing to provide more definitive identification. In 1972 the first Landsat, then called ERTS-1, carries on board a multispectral scanner (MSS) that collected images in four spectral channels and relayed data to earth pixel-by-pixel in digital form. These were the first multispectral data sets available to a wide variety of discipline users. The newly evolving technology of digital image processing made it possible to analyze four-channel spectra by statistical and other means and draw thematic maps of the earth's surface cover based on spectral reflectance characteristics.

Geologists are in the forefront of attempting to use multispectral data for mineral identification and mapping. Geologists have a number of techniques at hand to help solve problems such as optical mineralogy, x-ray diffraction, x-ray fluorescence, microprobe analysis, and age dating to name a few. These are laboratory techniques, however, and the results still must be put into the proper context, because field relationships are very important. Remote sensing data are an important aid in the field because the field relationships are integral to the data set.

Aerial photography, in particular stereo air photos, have been used to compile and interpret information for reconnaissance-level geologic mapping, rock-type discrimination, mapping of faults, fractures and joints, geomorphic analysis of land forms, hazard assessment, and resource exploration. Compositional information, however, another vital piece of the puzzle, cannot be obtained directly from aerial photography, and hence the interest in reflectance and emission spectroscopy as an additional tool.

In addition to the expanded regional perspective, the Landsat MSS provided the new spectral dimension previously lacking. The identification of iron oxides associated with hydrothermal alteration zones, by means of color ratio composites of digitally processed data, was the first major application of multispectral data to geologic mapping.[1,2] This new application stimulated interest in the acquisition of images in spectral regions beyond the sensitivity range of the eye, and required intensive ground and laboratory studies to verify the Landsat results and to develop requirements for new imaging systems. During the 1970s field spectrometers were developed that in turn led to a better understanding of the complexities of the interaction between electromagnetic energy and the surface, particularly averaged over the 79×79 m ground instantaneous field of view (GIFOV) of the Landsat MSS.[3]

The field and laboratory spectroradiometry measurements showed clearly that the Landsat MSS instrument with four spectral bands could not be used to identify the composition of surface materials because it severely undersampled the spectrum and did not cover wavelengths beyond 1.0 μm. In particular, the region 1.6–2.5 μm containing diagnostic vibrational overtone features important to geologic mapping, was not covered. A study of the Cuprite, Nevada mining district using airborne multispectral scanner data with bands in the 1.6 and 2.2 μm regions[4] made clear the importance of these regions for the identification of hydroxyl- and carbonate-bearing minerals. A follow-on instrument to the MSS called the thematic mapper (TM) incorporated channels at 1.55–1.75 μm and 2.08–2.36 μm. These bands made it possible to identify clay- and carbonate-bearing units independent of iron content. The first thematic mapper was launched on Landsat-4 in 1982.

By the late 1970s, it became increasingly apparent that higher spectral resolution data in the 0.4–2.5 μm region and multispectral data from the thermal infrared, 8–14 μm region would vastly improve remote mineral identification. The first airborne high-resolution instrument was a profiling spectroradiometer[5] operating in the visible and near-IR regions. This instrument provided new insight into geobotanical mapping based on anomalous spectral reflectance of trees stressed by metals in the soil.[6]

An experiment in profiling spectroradiometry of the earth was carried out from orbit with the shuttle multispectral infrared radiometer (SMIRR) in 1981.[7] Positive results from this experiment built confidence in the new program at NASA/JPL that had as its goal the development of airborne and spaceborne imaging spectrometers. Details of this program and the sensors will appear later in the chapter.

5.3. SPECTRAL PROPERTIES OF EARTH SURFACE MATERIALS

To understand the capabilities and shortcomings of high-spectral-resolution remote sensing and imaging spectrometery, it is necessary to understand the interaction between the electromagnetic energy and earth surface materials.

5.3.1. Minerals

The most commonly used term in remote sensing is bi-directional reflectance, $\rho(\lambda)$, which is

$$\rho(\lambda) = R_s/R_d \tag{5.1}$$

where R_s is the reflectance of the sample, and R_d is the reflectance of the standard. A common standard material is halon.[8] In the visible (0.4–0.7 μm), near-IR (0.7–1.0 μm), and short-wavelength IR (1.0–2.5 μm), $\rho(\lambda)$ is of primary interest. In the thermal IR region (8–14 μm), however, where over 99% of the energy received for the sensor is emitted from the earth's surface, the emittance $\varepsilon(\lambda)$ is the parameter of interest. Under conditions of thermodynamic equilibrium, Kirchhoff's law states that

$$\varepsilon(\lambda) = 1 - \rho(\lambda) \tag{5.2}$$

Light is scattered by the surface, approximately the upper 50 μm, and the diagnostic absorption features are created by absorptions within the particles. Light is transmitted through the particles on its way back out of the surface as shown in Fig. 5.1. Effects such as refraction, diffraction, and scattering can be described by classical theory (i.e., Maxwell's equations). The interaction of light with molecules and crystal

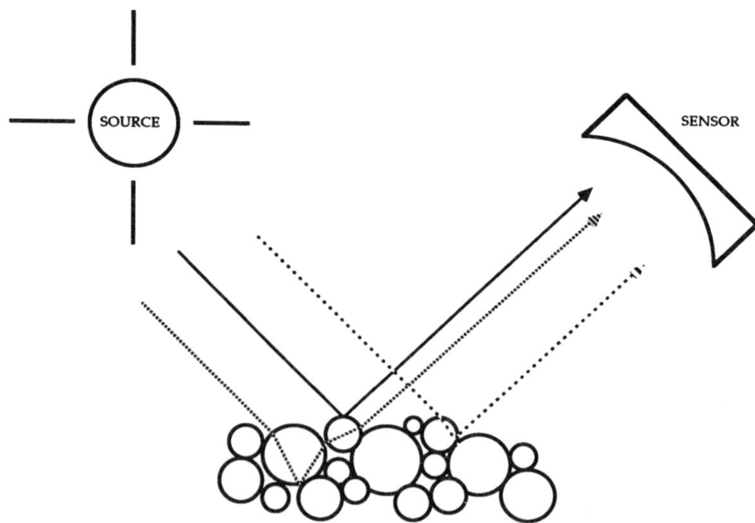

FIGURE 5.1. Paths taken by photons emitted from the source and collected by the sensor after being scattered by a solid surface.

lattices, however, requires quantum mechanics and the particulate view of electromagnetic energy.[9] The spectral absorption features seen in reflectance spectra of minerals depend not only on quantum-mechanical interactions but on the particle size of the surface and the single scatter albedo.[10–12]

5.3.1.1. Electronic Processes

There are two major types of energy transitions in crystal lattices, electronic and vibrational, which cause diagnostic absorption features in reflectance and emittance spectra throughout the optical region 0.4–14 μm discussed here. Electronic transitions generally fall short of 1.0 μm, while vibrational features usually occur at longer wavelengths.

Charge transfer is one of the electronic processes. In minerals, the most common charge transfer is electron sharing between iron and oxygen, which gives rise to a strong absorption in the UV, and absorption wings extend into the visible portion of the spectrum. The absorption spectra of rocks and soils almost uniformly exhibit a drop off in reflectance toward the UV caused by the ubiquitous presence of iron in nature.

The transition elements Fe, Cr, Ni, Ti, Co, Mn, Wo, and Sc all have unfilled 3-D shells. When these elements are imbedded in a crystal lattice and are under the influence of the crystal field, the symmetry of the field determines the energy levels and the transitions. The excited states of these electrons have energies corresponding to the visible wavelengths and are responsible for a wide range of colors. An example is the transition element chromium in corundum (Al_2O_3). In ruby (corundum) the substitution of chromium for a few percent of the aluminum ions creates the red color. The chromium ion has three unpaired electrons creating a complicated spectrum of excited states. The excited states form bands modified by the presence of the crystal matrix.[13] In corundum, each aluminum ion is surrounded by six oxygen ions in a distorted octahedron. The crystal field of corundum is brought about by the fact that the valence electron pairs are more closely coupled with oxygen ions than they are with aluminum and this gives rise to an electric field called the crystal field or ligand field. The characteristic absorption feature associated with the chromium ion is diagnostic of the crystal field in which it resides rather than of the ion itself. For instance, when the chromium ion is immersed in beryllium aluminum silicate (beryl), which also has six oxygen ions in octahedral coordination, the position of the absorption bands are different because the lattice has different chemical bonds and the magnitude of the electric field surrounding the chromium ion is reduced. Beryl containing chromium is called emerald, and has a violet absorption as well as an absorption in the yellow and red portion of the spectrum, allowing blue and green light to be transmitted and creating the characteristic emerald color.

Silicon, aluminum, and oxygen, which are the major constituents of crustal rocks, do not have electronic energy levels that show features in the visible and near-IR portions of the spectrum. The characteristic absorptions that allow these constituents to be identified are created by vibrational transitions.

5.3.1.2. Vibrational Processes

The vibrations of molecules and lattices result in small displacements of the atoms about their resting positions. These displacements are subject to quantum-mechanical rules, and the energy levels in the linear harmonic oscillator are given by

$$E_v = (v_i + \tfrac{1}{2})h\nu_i \tag{5.3}$$

where v_i is an integer, h is Planck's constant and ν_i is the oscillator frequency. Absorption features in reflectance spectra arise as a result of

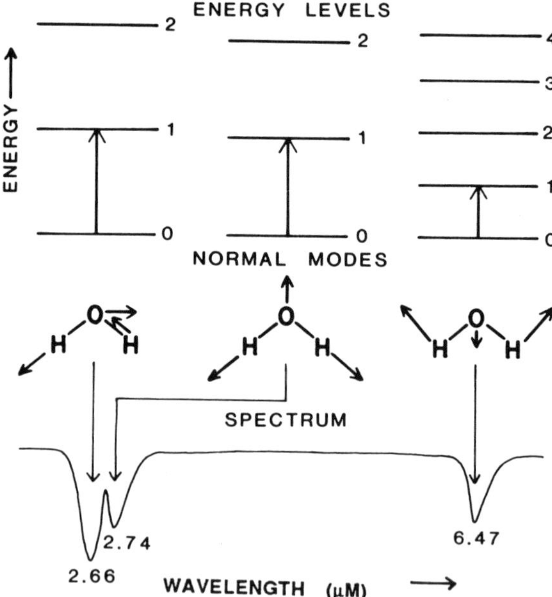

FIGURE 5.2. Vibrational energy level diagram for water vapor showing three fundamental vibrations along with the IR absorption spectrum corresponding to the ν_3, ν_2, ν_1 vibrations (from Ref. 9).

transitions from one state to another. Fundamentals are frequencies arising from a transition from the ground state $v_i = 0$ to $v_i = 1$. Overtones occur when there is a transition from the ground state to $v_1 = 2$ or more, and combinations occur when there is a transition from the ground state to the sum of two or more fundamental or overtone states. The case of water vapor is seen in Fig. 5.2. Overtones and combinations of these modes make up the well-known atmospheric water vapor absorption bands at 1.875 μm ($\nu_2 + \nu_3$), 1.454 μm ($2\nu_2 + \nu_3$), 1.38 μm ($\nu_1 + \nu_3$), 1.135 μm ($\nu_1 + \nu_2 + \nu_3$), and 0.942 μm ($2\nu_1 + \nu_3$). In general, the fundamental vibrational frequencies of molecules, including silicates, takes place at wavelengths longer than 2.5 μm. Overtones

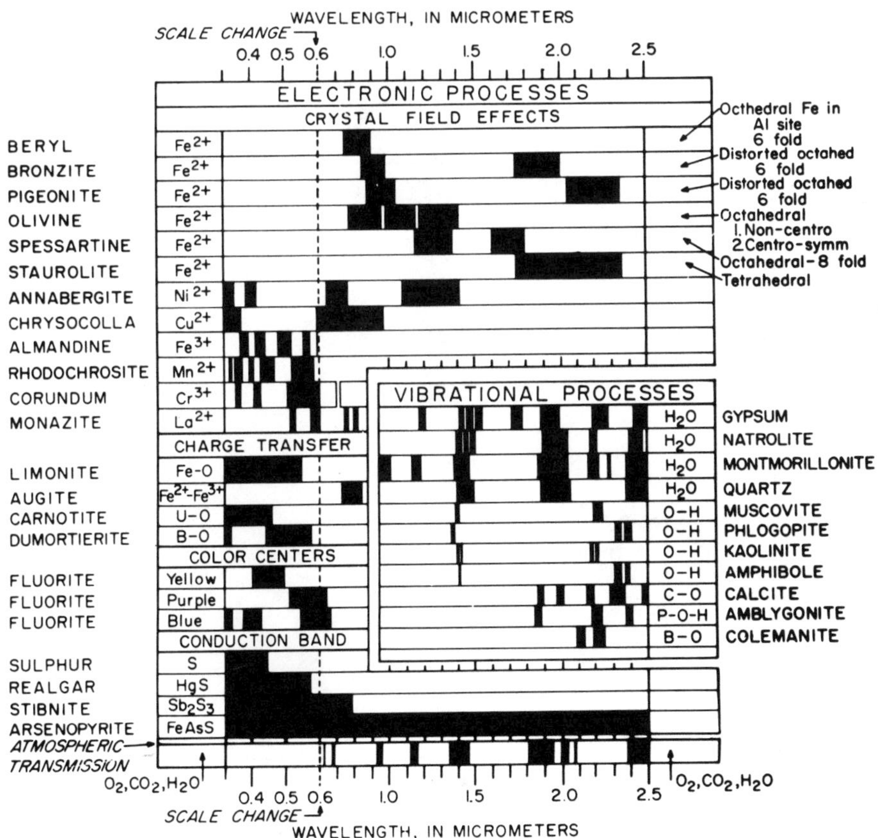

FIGURE 5.3. Compilation of the electronic and vibrational features seen in minerals arranged by wavelength (from Ref. 14).

and combinations reach into the SWIR region and overlap with the electronic transitions found mainly short of 1 μm. Figure 5.3 shows the location and types of transitions for various minerals in the region 0.4–2.5 μm.[(14)]

The OH ion has one fundamental stretching vibration at or near 2.77 μm depending on the site in the lattice. Several fundamental absorption features are possible at several different sites in the lattice. Features near 2.2 μm are created by Al–OH bonding coupled with fundamental OH stretching modes (Fig. 5.4). The 2–2.5 μm atmospheric window contains many diagnostic features for minerals (Fig. 5.5) and is the most important region for remote sensing of the earth's weathered surface layer.

Silicates, which make up the bulk of the crustal rocks, exhibit fundamental vibrational stretching modes in the 10 μm region (Fig. 5.6). The fundamental vibrational features of silicates are located in the

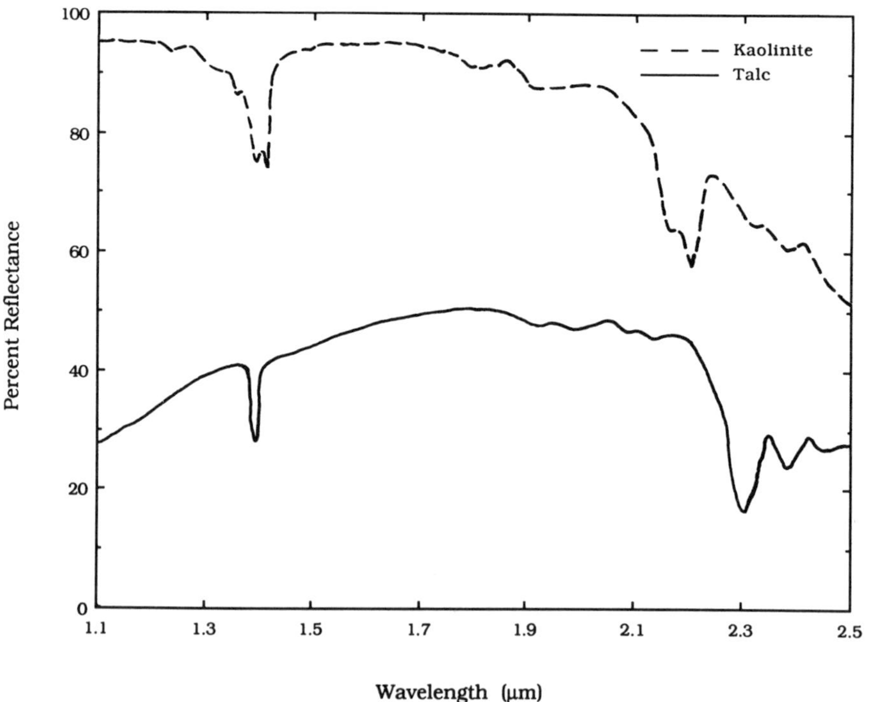

FIGURE 5.4. OH-bearing minerals showing an overtone at 1.4 μm and combination overtones in the 2.2 μm region that are diagnostic for remote measurements.

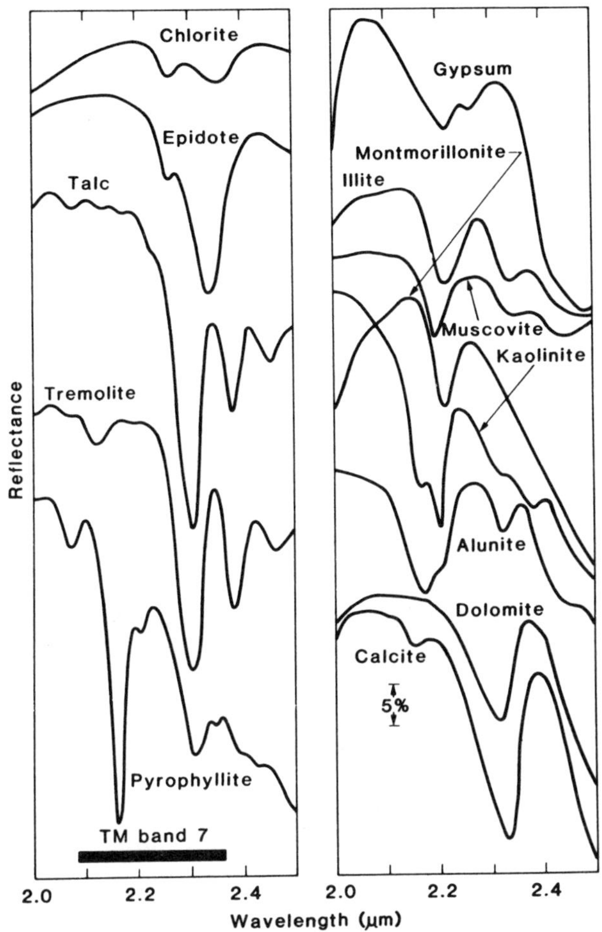

FIGURE 5.5. Selected laboratory spectra of minerals containing overtone vibrational absorption features for Al–OH (2.16–2.22 μm), Mg–OH (2.3–2.35 μm), and CO_3 (2.3–2.35 μm). The band 7 bandwidth of the Landsat TM is also shown (from Ref. 20).

8–14 μm atmospheric transmission window and, therefore, these are useful for remote sensing. Reflectance and emittance measurements in this region are complicated by particle size effects because the particles are in the range of the size of the observational wavelengths.

5.3.2. Vegetation

The spectral reflectance of vegetation in the 0.4–2.5 μm region is not as rich in diagnostic features as is the spectral reflectance of minerals.

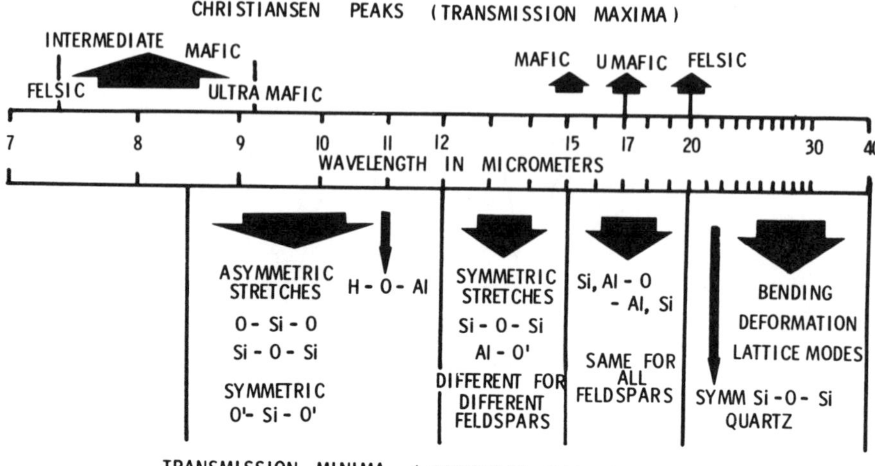

FIGURE 5.6. Location of features and the types of vibrations that produce the spectral signatures of silicates in the mid-IR region (from Ref. 9).

This is because the constituents of vegetation, the pigments and the biochemicals that make up the leaves, are nearly the same for all types of vegetation. Figure 5.7 is a typical spectral reflectance curve for an actively photosynthetic plant. The main absorption features in the visible portion of the spectrum at 0.48 and 0.68 μm are the result of chlorophyll absorption in the leaf. Absorption at 0.48 μm is the result of electronic transitions in carotenoid pigments which function as accessory pigments in the photosynthetic process in addition to chlorophylls.(15) The reflectance maximum at 0.55 μm lies between the major blue and red absorption features, and results in the green appearance of plants to the human eye. The rise in the reflectance at wavelengths longer than the 0.68 μm minimum is known as the red edge of the chlorophyll absorption, and the slope of the curve in this region is related to chlorophyll concentrations in the leaf. Collins et al.(6) note a blue shift consisting of a 0.007 to 0.010 μm shift of the chlorophyll shoulder or red edge to shorter wavelengths in plants influenced by geochemical stress. Such shifts can only be resolved using high-spectral-resolution measurements.

The near-IR plateau, that is, the region between 0.8 and 1.3 μm, has high reflectance associated with scattering in the mesophyll layer of the leaf. Subtle absorption features in the infrared plateau may be related to both cellular arrangement within the leaf and hydration state.(16,17) Longward of 1.3 μm the spectral reflectance of vegetation is dominated by the leaf water content and features associated with various leaf biochemicals that can only be detected readily when the leaves are desiccated.(18)

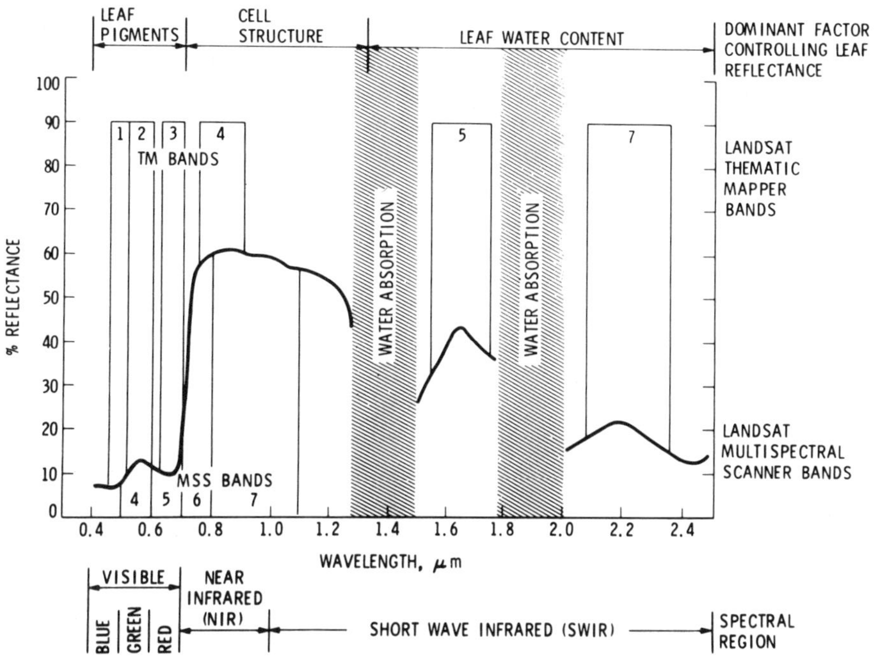

FIGURE 5.7. Typical vegetation reflectance curve acquired in the field. Landsat MSS and TM bands are indicated. Gaps in the spectral curve at 1.4 and 1.9 μm are due to atmospheric water absorption (from Ref. 15).

5.4. IMAGING SPECTROMETRY

Imaging spectrometry is the simultaneous acquisition of images in contiguous spectral bands. Because of the large number of spectral bands, the term "hyperspectral" has been used to describe the resulting data set.[20] Therefore, a complete reflectance or emittance spectrum can be derived for each picture element. At present, for observing the earth, it is only practical to consider reflectance spectroscopy in the region 0.4–2.5 μm. The reason for this is a combination of detector availability and the location of diagnostic spectral absorption features in earth surface materials that lie in atmospheric windows. The spectral resolution required depends on the natural width of the diagnostic absorption features. Unlike atmospheric spectroscopy, in which very high resolution is required to resolve the rotational line features, spectroscopy of solid materials is not so demanding of spectral resolution because rotational degrees of freedom are not present in solids. Almost all absorption features of interest in minerals have a full width at half-maximum of greater than 20 nm in the region short of 2.5 μm. Therefore, an imaging spectrometer system with a resolution of 20 nm and a sampling interval of 10 nm or less will be able to recreate those features in solids. Narrower features associated with OH overtones and the combinations can be seen in the 1.4 μm region.[19] Because the region coincides with a band of total atmospheric water vapor absorption, however, it cannot be used for remote sensing purposes.

The development of practical imaging spectroscopy of the earth began at the NASA Jet Propulsion Laboratory in 1980. The ultimate goal of that program was to create a spaceborne imaging spectrometer that would acquire images at a spatial resolution equivalent to the Landsat TM but in 200 or more spectral bands rahter than 7 as in the case of the TM. This capability would represent a major breakthrough in the technique of remote sensing because it would allow the direct identification of surface materials rather than thematic mapping based on statistical variations in the 7-channel data.[20] The program has since created two airborne systems, the airborne imaging spectrometer (AIS) and the airborne visible and infrared imaging spectrometer (AVIRIS). The spaceborne sensor, the high-resolution imaging spectrometer (HIRIS), has been selected as a facility instrument on the NASA Earth Observing System payload to be carried aloft in the mid-1990s on the Polar Orbiting Platform.

5.4.1. Sensors

The development of imaging spectrometry has depended very heavily on detector technology. Figure 5.8 shows the four approaches to sensors for multispectral imaging. The Landsat MSS and TM are optomechanical systems in which one or more discrete detector elements are scanned across the surface of the earth perpendicular to the flight path, and then the detectors convert the incoming photons from the scene to an electronic signal [Fig. 5.8(a)]. Band-pass filters with widths of 80–300 μm are placed in front of the detector elements. The MSS has four such sets of filters and detectors, and the TM has seven. Because the individual detectors are scanned across the scene, each ground instantaneous field of view is observed by the detector for a very short period of time, on the order of microseconds for the TM. The passbands must be wide enough to achieve an adequate signal-to-noise ratio without sacrificing spatial resolution.

One approach to increasing the residence time of the detector in each instantaneous field of view (IFOV) is to use line arrays of detector elements as shown in Fig. 5.8(b). In this case, there is a dedicated detector element for each cross-track pixel, and the residence time of the

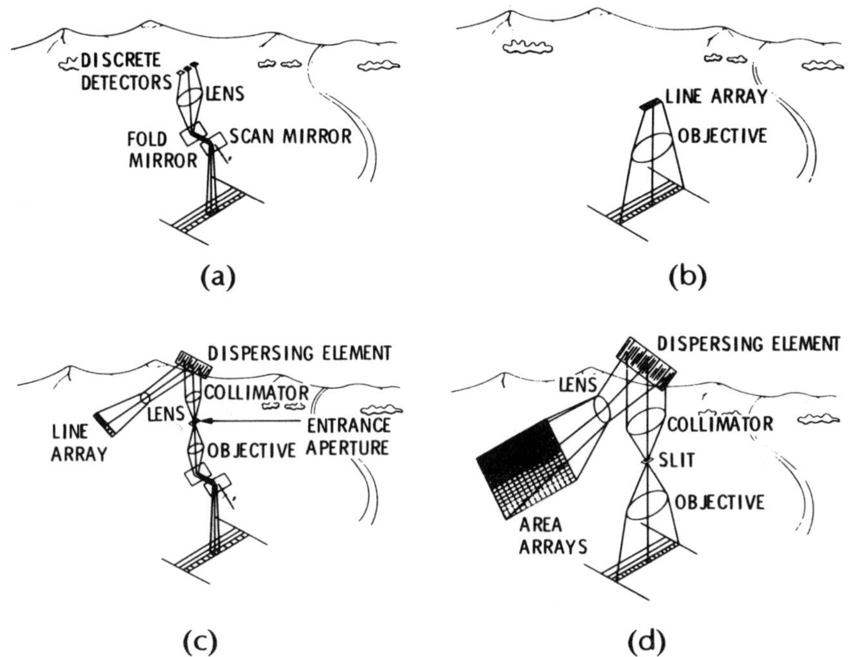

FIGURE 5.8. Four approaches to sensors for multispectral imaging: (a) multispectral imaging with discrete detectors; (b) multispectral imaging with line arrays; (c) imaging spectrometry with line arrays; and (d) imaging spectrometry with area arrays (from Ref. 20).

detector on a ground instantaneous field of view (GIFOV) is increased to the time interval required to move one GIFOV along track. Different spectral bands are implemented by using a beam-splitter band-pass filter system. A French satellite sensor called SPOT (Systeme Probatoire d'Observation de la Terre) uses line array detectors to achieve a 20-m GIFOV in three spectral bands and a 10-m GIFOV in a single broad band.[21]

There are serious limitations associated with the use of multiple line arrays, each having its own band-pass filter. The arrays can be placed side-by-side in the focal plane of the telescope, but then the same ground locations are not imaged simultaneously in each spectral band. If beam splitters are used to enable simultaneous data acquisition, the signal is reduced by 50% or more for each additional spectral band, and the instrument complexity increases substantially if more than a few spectral bands are desired.

In Fig. 5.8(c) and (d), two approaches to imaging spectrometry are shown. In the "whisk broom" approach shown in Fig. 5.8(c), a line array acts as a series of exit slits in a spectrometer. Scanning fore-optics are used as in the case shown in Fig. 5.8(a). Thus, each pixel is simultaneously sensed in as many spectral bands as there are detector elements in the line array. This approach is only suited for a high-flying aircraft platform flying at subsonic speeds. In this case, the readout time of the detector array is a small fraction of the integration time.

Imaging spectrometers designed for earth orbit require a different approach. Because of the high spacecraft velocities, imaging spectrometers require the use of two-dimensional area arrays of detectors at the focal plane of the spectrometer to obtain a sufficient signal-to-noise ratio. This implementation obviates the need for the optical scanning mechanism. Here, there is a dedicated column of spectral detector elements for each cross-track pixel in the scene.

To cover the 0.4–2.5 μm wavelength range, two types of detectors are necessary. For the region 0.4–1.0 μm, silicon CCD area arrays can be used. Beyond 1.0 μm, it is necessary to move to hybrid arrays using a HgCdTe array sandwiched with a silicon CCD multiplexer. Some 64×64 element devices, which can be abutted on opposite sides for

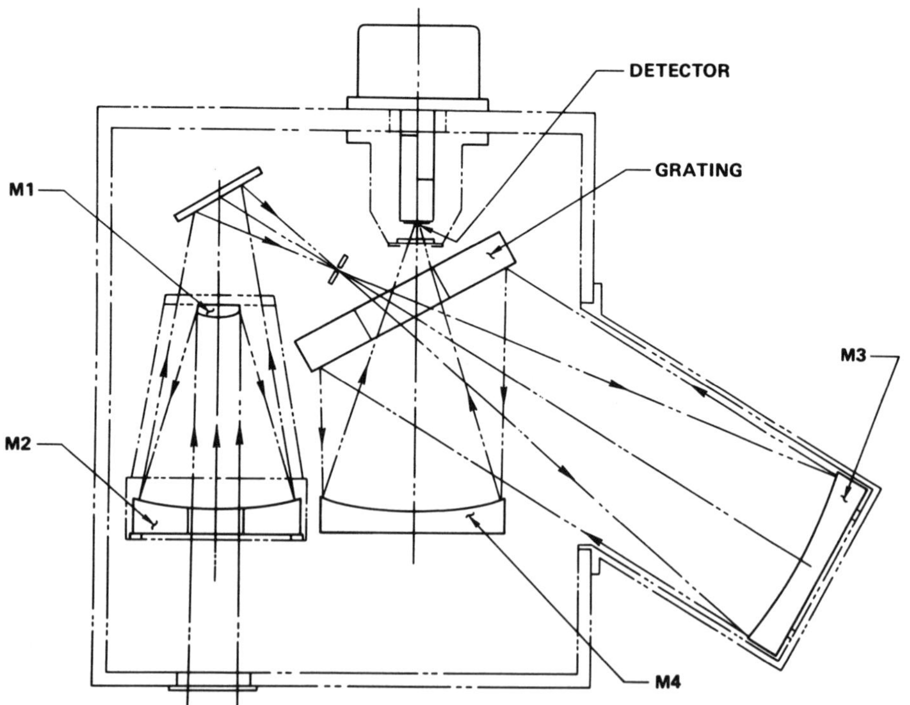

FIGURE 5.9. Schematic of the airborne imaging spectrometer (AIS).

purposes of mosaicking, have already been flown in imaging spectrometers,[22] and 128-element square arrays have been constructed. It appears that 256 × 256 arrays are feasible and would simplify the mosaicking of large arrays for increased swath width.

5.4.1.1. Airborne Imaging Spectrometer

At JPL, the airborne imaging spectrometer (AIS) (Fig. 5.9) was built to test the imaging spectrometer concept with IR area arrays.[23] AIS operates in the mode shown in Fig. 5.8(d). The spectral coverage of the instrument was 1.2–2.4 μm in contiguous bands 9.3 nm in width. The instrument was flown in the NASA C-130 aircraft at an altitude of approximately 4,000 m above the terrain. The IFOV of AIS is 1.9 mr, which is equivalent to a GIFOV of approximately 8 × 8 m from a typical operating altitude of 4,200 m. Continuous strip images 32 pixels wide and in 128 spectral bands were acquired. In spite of the narrow swath width, results from AIS showed that it was possible to directly identify minerals based on spectra in the 2.0–2.4 μm region (Fig. 5.10)[20] and, in fact, to

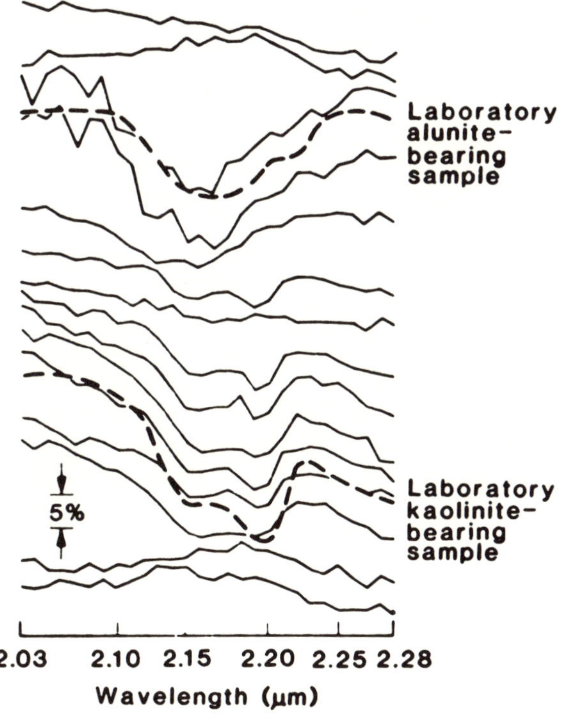

FIGURE 5.10. AIS 3 × 3 pixel spectra acquired over the Cuprite Mining District, Nevada, with laboratory spectra of samples acquired in the same area superimposed (after Ref. 20).

make the discovery of a rare mineral, buddingtonite, an ammonium feldspar, which had not been detected previously in the Cuprite Mining District, Nevada.[24]

5.4.1.2. Airborne Visible and Infrared Imaging Spectrometer

The airborne visible and infrared imaging spectrometer (AVIRIS) is planned as the workhorse instrument for the next decade. AVIRIS collects images in an 11-km swath with 224 spectral bands simultaneously covering the region 0.4–2.5 μm. The instrument is designed to fly in the NASA U-2 and ER-2 at an altitude of approximately 20 km. The instrument IFOV is 1 mr and, therefore, the GIFOV is 20 m. The implementation of the instrument is unique (Fig. 5.11) in that a "whisk broom" approach [Fig. 5.8(c)] is used, and the optical scanning head is separated physically from the spectrometers and connected to them with optical fibers.[25] The fibers are of silicon and fluoride glass and have a very high numerical aperture of 0.5. The optical scanning head is unique in that an oscillating mirror approach was used to gain the high scan efficiency (70%) necessary to obtain sufficient signal-to-noise ratio.[26] Four spectrometers are used to cover the approximate three-octave range in wavelength (Fig. 5.12). Again, very fast optics (f/1.2) are used to obtain sufficient signal-to-noise ratio.

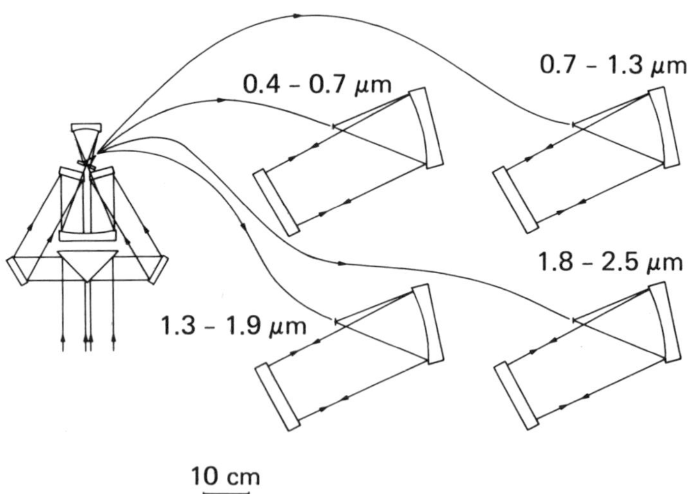

FIGURE 5.11. AVIRIS optics schematic (from Ref. 25).

A 32-element silicon CCD detector array is used in the visible portion of the spectrum and three 64-element indium–antimonide arrays, cooled to liquid nitrogen temperature, cover the rest of the spectrum.

Data are acquired at the rate of 17 Mbits/s and recorded on board. Approximately 45 min worth of data can be acquired during a flight. One 11 × 11 km scene (Fig. 5.13), which takes approximately 50 s to acquire, fills one standard 6,250-bpi computer-compatible tape. Therefore, each mission is capable of producing 50 CCDs. Therein lies the nightmare.

5.4.1.3. The High-Resolution Imaging Spectrometer

The next step beyond AVIRIS is into orbit. NASA is planning a major new initiative as part of the Mission to Planet Earth called the Earth Observing System.[27] This mission will consist of a series of instruments attached to a polar-orbiting space platform with a planned launch in 1995. The plans call for five facility instruments, two of which are imaging spectrometers. One is the moderate resolution imaging spectrometer (MODIS),[28] which will acquire images over a 1,500 km swath in 40 spectral bands, allowing it to cover the globe every two days

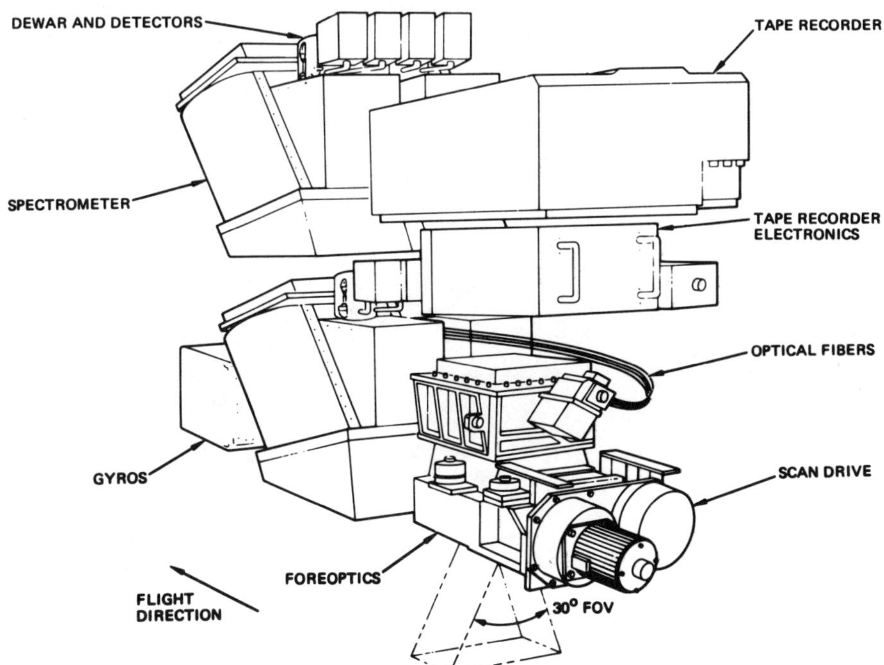

FIGURE 5.12. AVIRIS instrument (from Ref. 26).

FIGURE 5.13. AVIRIS image of Moffett Field, California, and south San Francisco Bay in one of the 224 spectral bands. The image cube depicts the relative spectral reflectance of each of the edge pixels encoded such that low values are dark and high values light. The wavelengths extend from 0.4 μm at the edge of the image to 2.45 μm at the back of the cube. The dark bands correspond to the 1.4 and 1.9 μm atmospheric water vapor absorptions.

with a GIFOV of 1 and 0.5 km depending on spectral channel. The other instrument is the high-resolution imaging spectrometer (HIRIS).[29] HIRIS will acquire images over a 30 km swath with a GIFOV of 30 m in 192 spectral bands covering 0.45–2.45 μm. The fore-optics of HIRIS are steerable +30°, −60° downtrack and ±24° across track. The crosstrack pointing allows any point on the earth other than the poles to be acquired every four days. The orbital altitude is 824 km. The nightmare becomes more real considering that HIRIS has a raw data rate of 512 Mbits/s. This corresponds to a complete CCT every 2 s. Not all the data can be returned at that rate, and provisions are made for on-board editing to reduce the transmitted data rate to 150 Mbits/s. In spite of the editing, enormous data volumes will be generated.

5.5. IMAGING SPECTROMETER DATA ANALYSIS

5.5.1. The Nightmare

Imaging spectrometry has provided a breakthrough in remote sensing of the earth. For the first time, it is possible to directly identify materials on a pixel-by-pixel basis in images taken from aircraft platforms and eventually from space. The increased information content, however, comes at the expense of an enormous increase in data rate. The nightmare is not confined to the data rate problem alone but rather it is magnified by the present-day lack of suitable techniques to extract the desired information from the data at reasonable rates without needing a supercomputer.

As discussed earlier, data are being acquired at sufficient spectral resolution to resolve all the characteristic absorption features in solids. This means that it is now possible to apply deterministic processing techniques to these data instead of the statistical, inferential techniques that are applied to undersampled multispectral images such as those acquired by the Landsat TM. The statistical techniques rely on determining a variance–covariance matrix for the image. In the case of AVIRIS, this would mean approximately 10^{13} floating-point operations, which, if carried out on a MicroVAX-II computer, for instance, would require a year of computation time.[20] Clearly, there are more efficient ways to reduce the dimensionality of the data.

5.5.2. Analysis Status

The main steps in hyperspectral image analysis are outlined in Fig. 5.14. Imaging spectrometers are designed to acquire data in contiguous

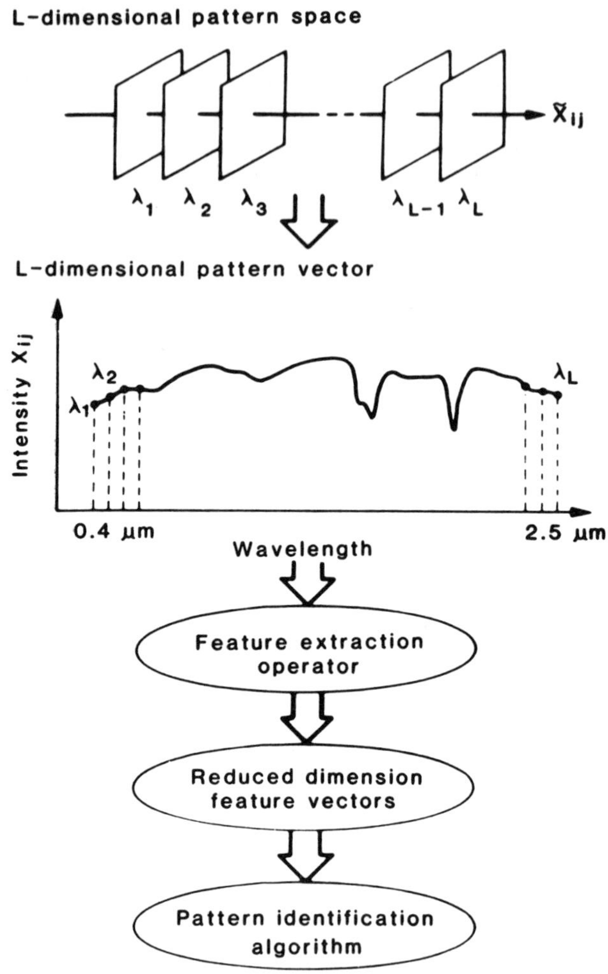

FIGURE 5.14. Generalized approach to the multispectral classification problem (from Ref. 20).

bands over the wavelength range of interest. The information contained in the spectrum usually can be derived from a small segment or several segments of the spectral region, and the entire spectrum is not required. Each application, however, may require a different portion of the spectrum. Therefore, complete spectral coverage of the region is required in the data acquisition phase. Efficient analysis of the spectral data requires identifying the portions of the spectral coverage to analyze, further reducing the dimensionality of the data, and, finally, applying pattern recognition algorithms in conjunction with spectral libraries.

The problem is made more complex because not all the signal returned comes from the surface, nor are the spectral absorption features all associated with the surface materials. Atmospheric scattering creates unwanted path radiance, although no sharp spectral features are associated with this additive contribution to the signal. In the region 0.4–2.5 μm, atmospheric absorption features are seen primarily for oxygen, water, and carbon dioxide. Both O and CO_2 are uniformly mixed in the atmosphere, but H_2O vapor varies greatly and creates absorption features that interfere with the interpretation of vegetation spectra which contain H_2O and also OH and H_2O bound in minerals.

Another major problem is associated with observing a mixture of materials on the earth's surface because of the finite size of the GIFOV. In the case of AVIRIS, the GIFOV area is 400 m^2, and in the case of HIRIS, 900 m^2. Therefore, within a single pixel, one would expect to see a mixture of spectral features associated with various material compositions. The question of mixtures has been addressed for multispectral data sets containing less than 10 spectral bands.[30,31] Two kinds of mixing must be considered. First, linear mixing will occur when there are groups of homogeneous cover types within a pixel. Second, intimate mixing, which is particles of different materials juxtaposed within the surface layer, is nonlinear in nature and consequently more difficult to unravel.[10] The problems of nonlinear mixing and grain size effects have limited the use of reflectance spectroscopy as a laboratory tool as well.[19] The problem of unmixing a reflectance spectrum into its component end members for imaging spectrometer data remains a major hurdle and challenge to imaging spectrometer data analysis.

Even if the unmixing problem can be addressed adequately, there remains the fact that each HIRIS image will contain 1 million spectra. Each line of the image contains data acquired from 192,000 individual detector elements. Each detector has its own gain and offset characteristics which must be accounted for before data analysis begins. A recurring nightmare revolves around the proper calibration, and the stability of the calibration, for this number of detectors. The problem is somewhat simplified in the case of AVIRIS because the instrument uses only 224 individual detectors.

5.5.2.1. Present Approach

The first problem that has been addressed is the visual interaction with hyperspectral images. Current graphics and image display technology allow implementation of a variety of tools for visual analysis. Among these, are the ability to display a time sequence projection of the images in the spectral direction, cursor-designated spectral plots of single pixels or averages over a spatial window, and rapid location of pixels having similar spectral signatures. In the case of AIS and AVIRIS, it is now possible to interactively locate and display spectra from a spectral library for visual comparison, and to interactively create spectral data sets from specified regions in the scene.[20,32]

To extract small spectral absorption features from the mass of highly correlated data, some type of normalization is required. One method is equal-energy spectral normalization given by

$$\sum_{l=1}^{L} A_{ij} X_{ij}(l) \quad \text{is constant} \tag{5.4}$$

where l is the wavelength index, (i, j) is the spatial location index, A_{ij} is the scalar multiplier for the intensity X_{ij} of pixel (i, j), and L is the total number of spectral measurements per pixel. This technique allows direct comparison of the shapes of spectral curves, and is particularly useful in identifying minerals with well-defined absorption features.

Fast pattern recognition techniques are required to be able to classify images into various surface compositions and to map them on the image. One successful technique used for identifying the dominant component in vegetation-free rock and soil surfaces relies on binary encoding of the spectral data in order to achieve fast cross-correlation for signature matching. In Fig. 5.15,[20] using the same notation as in equation (5.4), we find that each pixel can be written as an L-component vector

$$\mathbf{X}_{ij} = [X_{ij}(\lambda_1), X_{ij}(\lambda_2), \ldots, X_{ij}(\lambda_L)]^T \tag{5.5}$$

where λ denotes wavelength and T denotes transpose. The scalar μ_{ij} is the mean of the (i, j) spectral vector:

$$\mu_{ij} = \frac{1}{L} \sum^{L} X_{ij}(l) \tag{5.6}$$

A binary-valued L-element vector $\mathbf{Y}_{ij}$ is constructed by

$$\mathbf{Y}_{ij} = H(\mathbf{X}_{ij} - \mu_{ij}) \tag{5.7}$$

where $H(u)$ is the unit step operator

$$H(u) = \begin{cases} 1, & u \geq 0 \\ 0, & u < 0 \end{cases} \tag{5.8}$$

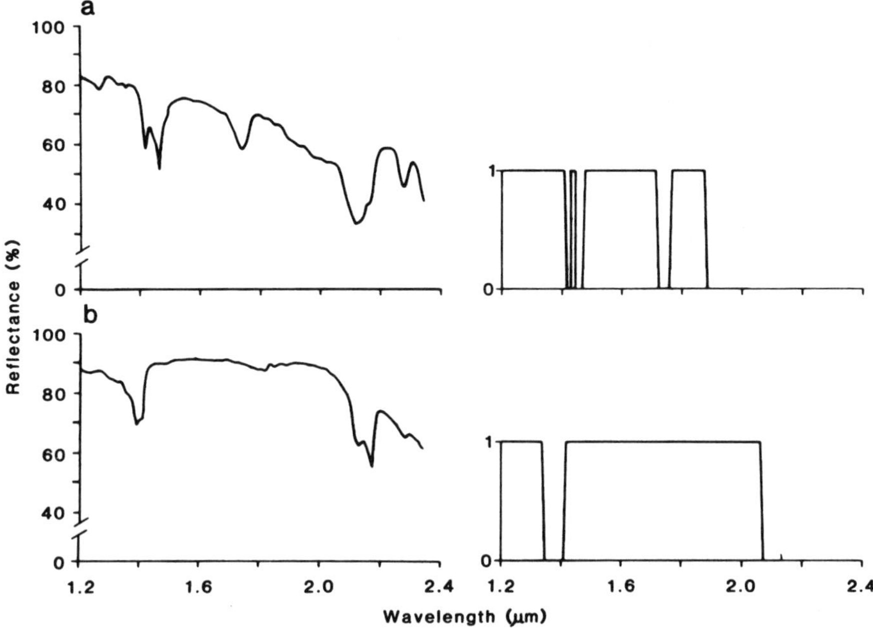

FIGURE 5.15. Binary encoding scheme for laboratory reflectance spectra of two minerals, (a) alunite and (b) kaolinite (from Ref. 20).

The results of this encoding are shown in Fig. 5.15. The data compression is greater than a factor of 100. The binary encoding technique is most accurate and efficient when applied to only that portion of the spectrum containing diagnostic spectral features.

To characterize the similarity between two binary spectral vectors, the Hamming distance[33] is used:

$$D_H(\mathbf{Y}_{ij}, \mathbf{Y}_{mn}) = L - \mathbf{Y}_{ij}\mathbf{Y}_{mn}^T \tag{5.9}$$

or

$$D_H(\mathbf{Y}_{ij}, \mathbf{Y}_{mn}) = \sum_{l=1}^{L} Y_{ij}(l) \oplus Y_{mn}(l) \tag{5.10}$$

where $\oplus$ denotes the logical exclusive OR operator. This distance measure is related to the more familiar Tanimato measure[33,34] used in taxonomic classification. There is a natural variability in spectra of similar minerals and the Hamming distance is chosen, $D_H = d$, such that

$$\mathbf{Y}_{ij} \doteq \mathbf{Y}_{mn} \quad \text{if and only if} \quad D_H \leq \bar{d} \tag{5.11}$$

Using a 200-entry library, a MicroVAX-II can classify an image at the rate of approximately 5,000 pixels per second.

Binary encoding is a nonlinear technique and by far not the only technique applicable to the problem. There is a large class of linear deterministic transformations having fast numerical implementations that are obvious candidates for dimensionality reduction. Examples include Fourier, Walsh–Hadamard, and Haar.[30] The Walsh–Hadamard transformation is particularly attractive from the standpoint of computational speed being even faster than the fast Fourier transform. A dimensionality reduction of eight or more is possible, and these linear transformations may be more applicable to spectral reflectance data such as from vegetation which does not have sharp spectral features and where the differentiation is made based on slopes and trends in the spectra.

Another approach proposed is a symbolic representation using space scale techniques.[35] A space scale image is a set of progressively smoothed versions of a spectral curve, leaving only the dominant spectral feature. The inflection points within the space scale image are plotted to produce a fingerprint. The fingerprint contains pairs of inflection points and a measure of importance related to the area contained within a spectral feature. This combination is designated a triplet and forms the symbolic representation of the hyperspectral data. The description is compact, quantitative, and hierarchical. The representation is also capable of describing subtle long-wavelength features or trends within the spectra.

The nightmare of data analysis will be converted into a dream, but only after the application of much ingenuity and hard work. The effort will be rewarding.

5.6. CONCLUSIONS

Imaging spectrometry is an exciting new technique for study of the earth from airborne and spaceborne platforms. Images are created in contiguous spectral bands, and the sampling is sufficient to reproduce all the information available in the solar reflected return. Plans are underway for creating massive data sets from the Polar Orbiting Platform, and a major challenge remains to overcome the nightmare of analyzing and interpreting high-dimensionality, hyperspectral images with modest computational means.

REFERENCES

1. L. C. Rowan, P. H. Wetlaufer, A. F. H. Goetz, F. C. Billingsley, and J. H. Stewart, "Discrimination of Rock Types and Detection of Hydrothermally Altered Areas in

South-Central Nevada by the Use of Computer-Enhanced ERTS images," USGS Professional Paper 883 (1974).

2. L. C. Rowan, A. F. H. Goetz, and R. Ashley, "Discrimination of Hydrothermally Altered and Unaltered Rocks in Visible and Near-Infrared Multispectral Images," *Geophys.* **42,** 522–535 (1977).
3. A. F. H. Goetz, F. C. Billingsley, D. Elston, I. Lucchitta, E. M. Shoemaker, M. J. Abrams, A. R. Gillespie, and R. L. Squires, "Applications of ERTS Images and Image Processing to Regional Geologic Problems and Geologic Mapping in Northern Arizona," JPL TR 32-1597 (1975).
4. M. J. Abrams, R. Ashley, L. C. Rowan, A. F. H. Goetz, and A. B. Kahle, "Mapping of Hydrothermal Alteration in the Cuprite Mining District, Nevada, Using Aircraft Scanner Imagery for the 0.46–2.36 μm spectral region," *Geology* **5,** 713–718 (1977).
5. H. Y. Chiu and W. E. Collins, "A Spectroradiometer for Airborne Remote Sensing," *Photogrammetric Engineering and Remote Sensing* **44,** 507–517 (1978).
6. W. Collins, S. H. Chang, G. Raines, F. Canney, and R. Ashley, "Airborne Biogeophysical Meaning of Hidden Mineral Deposits," *Econ. Geol.* **78,** 737–749 (1983).
7. A. F. H. Goetz, L. C. Rowan, and M. J. Kingston, "Mineral Identification from Orbit: Initial Results from the Shuttle Multispectral Infrared Radiometer," *Science* **218,** 1020–1024 (1982).
8. Weidner, V. R. and J. J. Hsia, "Reflection Properties of Pressed Polytetrafluoroethylene Powder," *J. Opt. Soc. Am.* **71,** 856–859 (1981).
9. G. R. Hunt, "Electromagnetic Radiation: The Communication Link in Remote Sensing," in: *Remote Sensing in Geology* (B. S. Siegal and A. R. Gillespie, eds.), Wiley, New York (1980).
10. B. Hapke, "Bidirectional Reflectance Spectroscopy:1. Theory," *J. Geophys. Res.* **86,** 3039–3054 (1981).
11. B. Hapke, "Bidirectional Reflectance Spectroscopy:3. Correction for Macroscopic Roughness," *Icarus* **59,** 41–59 (1984).
12. B. Hapke, "Bidirectional Reflectance Spectroscopy:4. The Extinction Coefficient and the Opposition Effect," *Icarus* **67,** 264–280 (1986).
13. K. Nassau, "The Causes of Color," *Sci. Am.,* October, 124–154 (1980).
14. G. R. Hunt, "Spectral Signatures of Particulate Minerals in the Visible and Near-Infrared," *Geophys.* **42,** 501–513 (1977).
15. A. F. H. Goetz, B. N. Rock, and L. C. Rowan, "Remote Sensing for Exploration: An Overview," *Econ. Geol.* **78,** 573–590 (1983).
16. H. W. Gausman, D. E. Escobar, and E. B. Knipling, "Relation of *Peperomia obtusifolias* Anomalous Leaf Reflectance to Its Leaf Anatomy," *Photogramm. Eng. Remote Sensing* **43,** 1183–1185 (1977).
17. H. W. Gausman, D. E. Escobar, J. H. Everitt, A. J. Richardson, and R. R. Rodriguez, "Distinguishing Succulent Plants from Crop and Woody Plants," *Photogramm. Eng. Remote Sensing* **44,** 487–491 (1978).
18. D. L. Peterson, J. D. Aber, P. A. Matson, D. H. Card, N. Swanberg, C. Wessman, and M. Spanner, "Remote Sensing of Forest Canopy and Leaf Biochemical Contents," *Remote Sensing of Environment* **24,** 85–108 (1988).
19. R. N. Clark, T. King, M. Klejwa, G. A. Swayze, and N. Vergo, "High Spectral Resolution Reflectance Spectroscopy of Minerals," *J. Geophys. Res.,* in press (1989).
20. A. F. H. Goetz, G. Vane, J. Solomon, and B. N. Rock, "Imaging Spectrometry for Earth Remote Sensing," *Science* **228,** 1147–1153 (1985a).
21. M. Courtois and G. Weill, "The Spot Satellite System," in: *Progress in Astronautics and Aeronautics Science,* Vol. 97, Monitoring Earth's Ocean, Land and Atmosphere

from Space—Sensors, Systems and Applications, pp. 493–523, American Institute of Aeronautics and Astronautics (1985).

22. J. B. Wellman, A. F. H. Goetz, M. Herring, and G. Vane, "An Imaging Spectrometer Experiment for the Shuttle," *Proceedings International Geoscience and Remote Sensing Symposium* (IGARSS'83), **11,** FA-5, 6.1-6.7 (1983).
23. G. Vane, A. F. H. Goetz, and J. Wellman, "Airborne Imaging Spectrometer: A New Tool for Remote Sensing," IEEE Transactions on *International Geoscience and Remote Sensing, GF-22,* 546–549 (1984).
24. A. F. H. Goetz and V. Srivastata, "Mineralogical Mapping in the Cuprite Mining District, Nevada," *Proceedings of the Airborne Imaging Spectrometer Data Analysis Workshop* (G. Vane and A. F. H. Goetz, eds.), JPL Publication No. 85–41 (1985).
25. S. A. Macenka and M. P. Chrisp, "Airborne Visible/Infrared Imaging Spectrometer (AVIRIS) Spectrometer Design and Performance," *Imaging Spectroscopy II* (Gregg Vane, ed.), Proceedings SPIE 834, pp. 32–43 (1988).
26. D. C. Miller, "AVIRIS Scan Drive Design and Performance," *Imaging Spectrometry II* (G. Vane, ed.), Proceedings SPIE 834, pp. 55–62 (1988).
27. D. M. Butler et al., "Earth Observing System," Technical Memorandum 86129, Vol. 1, NASA, Washington, D.C. (1984).
28. W. Esaias, *et al.,* "Moderate-Resolution Imaging Spectrometer," Instrument Panel Report, Earth Observing System, Vol. IIb, NASA, Washington, D.C. (1986).
29. A. F. H. Goetz et al., "High-Resolution Imaging Spectrometer: Science Opportunities for the 1990s," Earth Observing System, Vol. IIC, Instrument Panel Report, NASA, Washington, D.C. (1987).
30. J. B. Adams, M. O. Smith, and P. E. Johnson, "Spectral Mixture Modeling: A New Analysis of Rock and Soil Types at the Viking Lander 1 Site," *J. Geophys. Res.* **91,** 8098–8112 (1986).
31. M. O. Smith, P. E. Johnson, and J. B. Adams, "Quantitative Determination of Mineral Types and Abundances from Reflectance Spectra Using Principal Components Analysis," *J. Geophys. Res.* **90,** Supplement C797-C804 (1985).
32. A. S. Mazer, N. Martin, N. Lee, and J. F. Solomon, "Image Processing Software for Imaging Spectrometry," *Imaging Spectroscopy II* (G. Vane, ed.), Proceedings SPIE 834, pp. 136–139 (1988).
33. J. T. Tou and R. C. Gonzalez, *Pattern Recognition Principles,* Ch. 3, Addition-Wesley, Reading, MA (1974).
34. R. C. Gonzalez and P. Wintz, *Digital Image Processing,* Addison-Wesley, Reading, MA (1979).
35. M. A. Piech and K. R. Piech, "Symbolic Representation of Hyperspectral Data," *Appl. Opt.* **26,** 4018–4026 (1987).

Chapter 6

Computer-Enhanced Nuclear Magnetic Resonance Spectroscopy

George C. Levy

6.1. INTRODUCTION

Nuclear magnetic resonance (NMR) spectroscopy, discovered in the mid-1940s, has undergone explosive growth in technique development and application since the 1950s. Today, NMR studies contribute in six main areas: (1) identification of organic and inorganic molecules; (2) determination of three-dimensional solution structures of modest-sized proteins and nucleic acids as well as smaller molecules; (3) studies of changing chemical systems including chemical kinetics experiments and *in vivo* NMR; (4) quantitative analysis including impurity and mixture analysis; (5) physicochemical studies of stereochemical and electronic structures; and (6) evaluation of simple and complex molecular dynamics at atomic site resolution. Several of these application areas are included in metabolic and health-related studies and a few are directly involved in support of magnetic resonance imaging (MRI) modalities. The largest impact, however, of NMR spectroscopy on analytical chemistry (broadly

This chapter is adapted with permission from George C. Levy, *J. Chem. Inf. Comp. Sci.*, **28**, 167–174 (1988). Portions also appeared in *Computers in Physics*, Sept.–Oct. 1988.

George C. Levy • N.I.H. Resource and CASE Center, Bowne Hall, Syracuse University, Syracuse, New York 13244-1200.

defined) has been its unique power for elucidating complex molecular structures at atom-site resolution, coupled with its linear response for quantititative analysis (when experiments are designed for this).

Since the late 1960s, the computer has played a critical role in NMR developments. This resulted from two factors: (1) the precision and speed required in Fourier transform (FT) NMR experiments, and (2) the need for advanced data reduction technology to obviate spectral nonidealities arising from those experiments. Today, extraordinary advances are underway in computer technology, and NMR spectroscopy—along with only a few other laboratory techniques and computational chemistry—is positioned to rapidly utilize these advances. The chapter discusses the history *and* future of computing technology in NMR spectroscopy, with the emphasis on its use to enhance the results of those experiments qualitatively and quantitatively. The article will *not* cover computer-aided NMR spectral assignment, but that is also an area expected to undergo major enhancements over the next several years.

6.2. HISTORY

The first significant use of digital electronics in NMR was in the mid-1960s when hard-wired time-averaging computers allowed spectral signal-to-noise improvements by accumulating the results of repetitive scans—each scan typically requiring from 50 to 500 s. With the availability of initial commercial FT NMR instrumentation at the end of the 1960s, computers—then including programmable machines (e.g., PDP-8, Varian/Sperry 620i)—did a similar task, albeit with each scan of the FT experiment requiring only ca. 1 s. At that time, it was not unusual to do experiments averaging 50,000 or more scans (signal-to-noise ratio increase with the square root of accumulated scans, or by a factor of ca. 220 in this case).

During the early and mid-1970s, digital computers in NMR instruments acquired new responsibilities along with data acquisition: experiment control, intermediate storage of data, and some spectral data reduction, such as phasing, simple baseline corrections, plotting, etc. By the end of the 1970s it was clear that the computers within NMR instruments were woefully overworked and, further, that this hardware, which was optimized for NMR data acquisition, was not flexible or powerful enough for significant NMR data processing tasks. The first NMR laboratory computer network with separate data acquisition hardware and a general-purpose multiuser computer was thus born over the period 1978–1979.[1] Table 6.1[2,3] summarizes some major developments since that time.

TABLE 6.1. Recent History of Computer Technology in NMR Spectroscopy

Development	Date	Comments	Reference or Source
Commercial off-line data processing computer	Early 1980s	Generally identical to instrument computer	Jeol, Bruker, Nicolet, Varian
Workstation computer introduced for NMR	1984	Apollo DN-3000 system demonstrated	New Methods Research, Inc.
Use of Sun workstations for NMR data processing	1985	Sun-2 system demonstrated in early 1985	New Methods Research, Inc.
Use of Sun in NMR instrument console	1987	SCSI-coupled design	Varian
	1988	Bus-coupled design	GE NMR Instruments
Use of "single-user" supercomputer data station	1987–1988	Stellar GS/1000, Ardent Titan, others	New Methods Research, Inc.
Linking of 2-D NMR and 3-D molecular modeling	1987–1988	D-SPACE (1987) NMR2/QUANTA (1988)	Hare Research, Inc., New Methods Research, Inc., with Polygen Corp.
Hierarchical computer networking	1988		Ref. 2
Use of parallel computers	1988	Alliant FX/80	Ref. 3

Significant advances in general computer technology were largely unavailable in NMR instruments themselves until the early 1980s when the JEOL GX spectrometer included the DEC LS1-11 computer. This trend accelerated recently with incorporation of Sun Microsystems computers into Varian and GE NMR nuclear magnetic resonance instruments (Table 6.1).

Today's trends portend significant improvements for the scientist. The use of *advanced, industry-standard computer workstations* running a *standard,* general-purpose multiprogramming operating system (e.g., Sun OS or UNIX) as the central controller of an NMR instrument provides the user with unprecedented capabilities. These computers also provide the nearly universal networking protocol TCP/IP, facilitating off-loading of NMR data, and many other possibilities.

The current trend is for instrument manufacturers to provide fast, standard data processing on the instrument and in off-line data stations. Other vendors and independent laboratories are offering more advanced data-processing software and networking solutions.

6.2.1. Modern Computer Technology and NMR Data Acquisition

It is easy to dismiss the role of computer technology in NMR data acquisition, but this would be a major mistake. The newest NMR experiment designs always challenge our ability to control and acquire NMR data. Time events with nanosecond accuracy and simultaneity requirements, along with the need for wider spectral widths (>100 kHz) *and* better digitization (to 15 or 16 bits!), challenge designer skills. Furthermore, use of general-purpose hardware and operating systems such as UNIX makes tasks of real-time response far more difficult. Multi-cpu designs have resulted—with attendant interprocess communications difficulties—in the most complex designs.

The advances underway in design of instrument data acquisition computers are largely relegated to manufacturers. The remainder of this chapter will describe current and future trends in data-processing hardware and software and their impact on the practice of NMR spectroscopy.

6.3. CURRENT ENHANCEMENT OF DATA REDUCTION IN NMR SPECTROSCOPY

Modern NMR experiments require, at a minimum: (1) pre-Fourier transform operations such as apodization and FID-data baseline offset corrections; (2) the Fourier transform itself (which for the largest multidimensional data sets can be a challenge!); (3) postprocessing such as phasing, peak finding, and integration; and (4) output—plotting and peak table listings.

A variety of advanced procedures can also be invoked, such as baseline conditioning,[(4)] "deconvolution", and quantitation of overlapping features by curve fitting,[(5)] Fourier deconvolutions, and so on. Methodologies such as maximum entropy deconvolutions,[(6,7)] image processing techniques,[(8,9)] or multivariate analysis[(10)] are also being proposed as superior to conventional processing for specified applications and tasks. Several examples follow.

6.3.1. Curve Fitting

For NMR spectra having overlapping bands, it can be useful to perform curve fitting to Lorentzian or other lineshapes. For example, a complex band from the ^{13}C spectrum of a synthetic polymer is shown in Fig. 6.1, along with the results of a fully automatic curve-fitting procedure. (Compute time on a Sun workstation ~<1 min). Such "deconvolutions" into individual lines, arising from stereochemical configuration along the polymer chain, allow scientists to determine characteristics of the polymerization and resulting polymer.

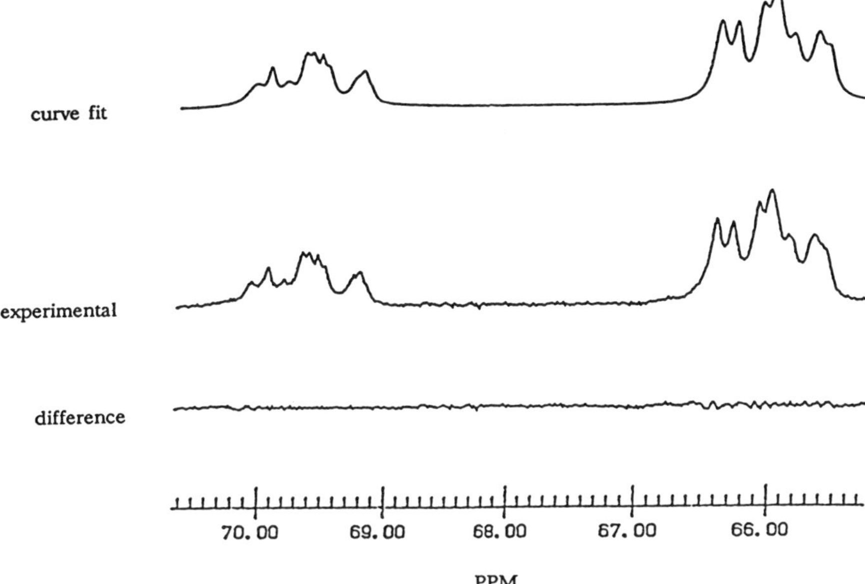

FIGURE 6.1. Example of automated curve fitting, as applied to a ^{13}C spectrum of polyvinyl alcohol.

Other applications of spectral curve fitting include dynamic NMR and quantitation of quadrupolar spectra such as in ^{23}Na NMR, where lineshapes may be bi-Lorentzian. It is possible to automatically curve fit 30 or more overlapping lines, but compute time becomes a significant factor with a large number of lines. Also, the method will of course lose accuracy when lineshapes are not well characterized, as with *in vivo* ^{31}P NMR, where tissue heterogeneity results in non-Lorentzian and even asymmetric resonance bands. One alternative approach for deconvolution of overlapping lines uses an iterative maximum entropy method (MEM).[6,7]

6.3.2. Maximum Entropy Processing

The maximum entropy method (MEM) can be very useful in applicable cases, but limitations on the spectral dynamic range prevent direct use of MEM on many NMR free induction decays (FIDs).[11] In this latter case, it is still possible to obtain the benefits of MEM deconvolution, starting from spectral segments after conventional Fourier transformation of the FID.[11] Thus, a low signal-to-noise section of a spectrum can be used, avoiding a very large solvent line. Alternatively,

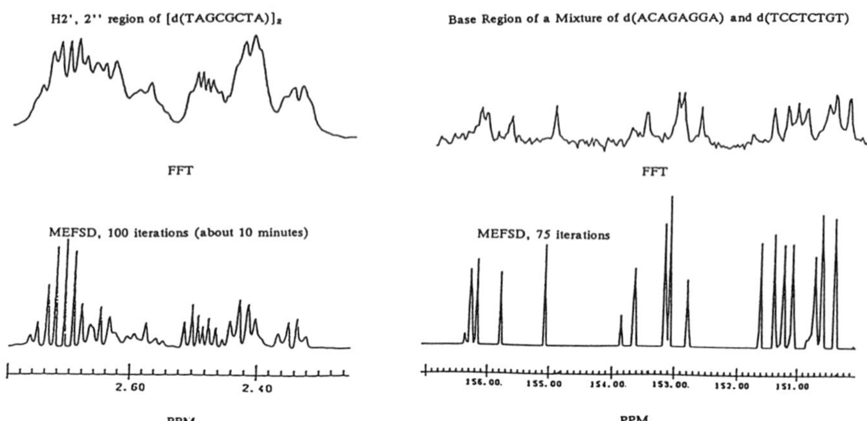

FIGURE 6.2. Maximum entropy reconstructions of ^{1}H and ^{13}C NMR spectral segments compared with results of conventional FFT processing: (Left) proton data on a self-complementary octamer; (right) base carbon region of an octamer. Note that the MEFSD spectra, at the bottom, show effective line deconvolution with no degradation of the signal-to-noise ratio (the apparent signal-to-noise ratio increases dramatically versus the low signal-to-noise ratio of the ^{13}C FFT spectrum).

compute time may be minimized for low signal-to-noise spectra if only small regions are of primary interest. Two examples are shown in Fig. 6.2, one an incompletely resolved 500 MHz ^{1}H spectrum of a DNA oligomer, and the second a low signal-to-noise spectral segment from a ^{13}C spectrum of a similar molecule.

Proponents of maximum entropy processing claim significant sensitivity increases for MEM spectra, compared with results from apodization and conventional FFT processing. They also point out that MEM is able to simultaneously "optimize" the spectral signal-to-noise ratio *while* narrowing resonance lines and eliminating peak overlaps. Detractors of this approach point out correctly that without additional input information, the MEM is not able to detect peaks that are not detectable by careful inspection after FFT processing. Nonetheless, it is clear that MEM dramatically increases the contrast (i.e., the apparent signal-to-noise ratio) between resonance signals and noise, while simultaneously affording effective spectral deconvolution. This would not be possible by conventional self-deconvolution, even for moderately high signal-to-noise spectra (e.g., order of one hundred).

It is also possible that the MEM can, in the future, be combined with ancillary data information, or that similar but distinct signal analysis methods (such as the maximum likelihood method) will demonstrate superior signal discrimination and spectral quantitation. In tests done thus far[11] on synthetic and real mixtures of compounds, the MEM used on spectral segments of low signal-to-noise has been shown to give quantitation ranging from equal to that from Lorentzian curve fitting to 2.5 times more precise. Quantitation of small peaks in the MEM approach is limited by coexisting large peaks, but a dynamic range exceeding 10 is achievable with ordinary iterative MEM calculations and, as stated above, much higher spectral dynamic ranges can be accommodated by inverse transformation of spectral segments. Figure 6.3 shows an example of a maximum entropy quantitation compared with quantitation via automated curve fitting. In this case the MEM treatment proved quite useful.

6.3.3. Linear Prediction Processing

Another technique that has been proposed for determination of noisy spectra, the linear prediction method (LPM),[12,13] can dramatically enhance the appearance of spectra, and also afford complete deconvolution of overlapping lines. An example is given in Fig. 6.4. This method searches for Lorentzian lines (exponential decays) and calculates a best fit to the input data as a set of theoretical Lorentzians. The method is quite good for contrast enhancement, but is not generally reliable as the spectral signal-to-noise ratio approaches 1.0. Further, calculation times for even moderately complex spectra can be quite excessive. LPM spectra are generally not as quantitative for poor signal-to-noise cases as are MEM spectra (at least from results obtained in this laboratory). LPM determinations, however, can be quite useful for free induction decays having signal dynamic ranges exceeding 1,000, where direct MEM treatment of the FID would be useless.

6.3.4. Baseline Conditioning

In NMR, as in other analytical spectroscopies, correction of baselines is critical to spectral quantitation. Several approaches have been utilized in NMR: (1) modification of the experiment to minimize baseline imperfections; (2) calculation of integral corrections via interactive user-input zero and first-order corrections; (3) automated calculation of baseline segments as polynomials, with subsequent subtraction; and (4) use of a finite automation to determine the location of the baseline and subtract it from the data.

1. Pulse sequence modifications can minimize initial data point corruption, with consequent straightening of a rolling baseline. Judicious use of ADC filter type and settings and postpulse acquisition delay time can also facilitate baseline control.

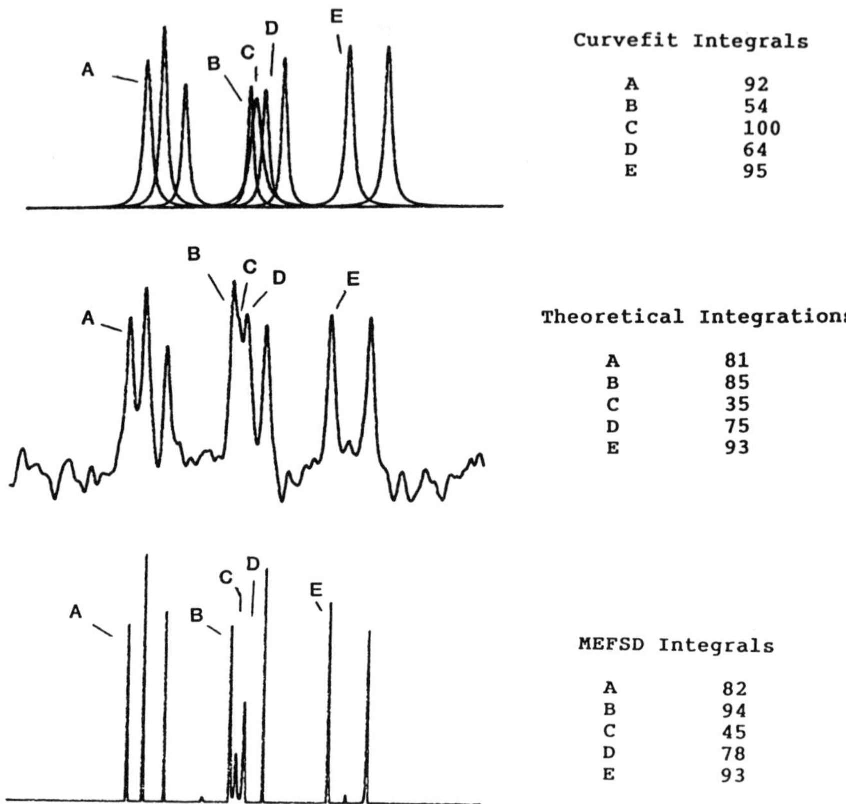

FIGURE 6.3. A synthetic test spectrum. An expanded spectral segment with known peak integral values is quantified by curve fitting and MEFSD, followed by a Simpson's rule integation. MEFSD is shown to give significantly improved quantitation. This dramatic improvement in quantitation is not always realized; however, MEFSD has been shown to give quantitative results for low signal-to-noise spectra that are at least as good as, and in many cases better than, FFT–curve-fitting methods.

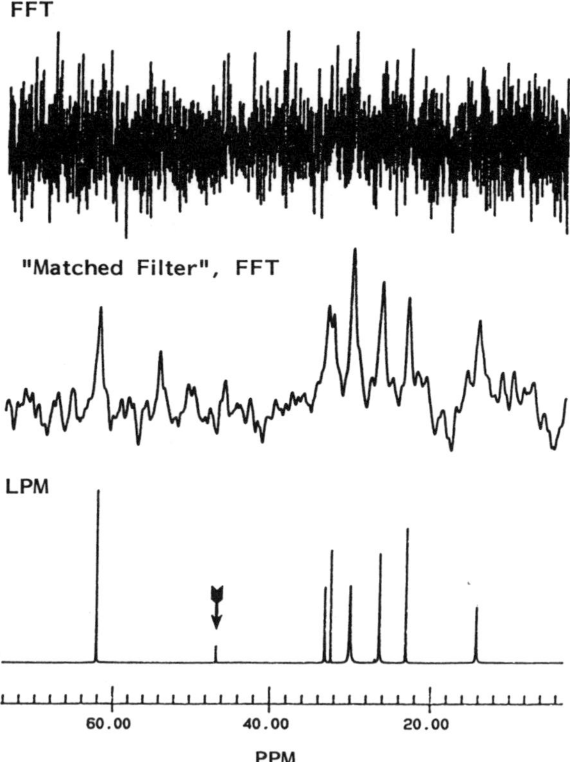

FIGURE 6.4. The ^{13}C spectrum of 1-octanol obtained at conditions of low signal-to-noise and poor resolution. At top is the spectrum using FFT alone. Application of a "matched filter" exponential apodization (middle) increases the signal-to-noise ratio but degrades resolution. Forward/backward linear prediction with 5-Hz linewidth reduction gives a reconstruction with enhanced resolution and virtually no noise. The arrow indicates a remaining "noise" peak which LP was unable to discriminate due to the low signal level. Also, the two resonances near 30 ppm were too close to be resolved by the experimental conditions, and thus are represented by one peak in both conventional and LP methods.

2. On most commercial FT NMR instruments, integration is performed by the user, while canceling effects of baseline problems via interactive adjustment of knobs. This method can be effective for high-resolution spectra, but is subject to user bias.
3. The third method, polynomial fitting to a deviant baseline,[4] can be quite effective except for spectra that have peaks overlapping across significant segments of the baseline. This method can be extended to make practical baseline conditioning of spectra missing baseline sections, but results are not always ideal (Fig. 6.5).

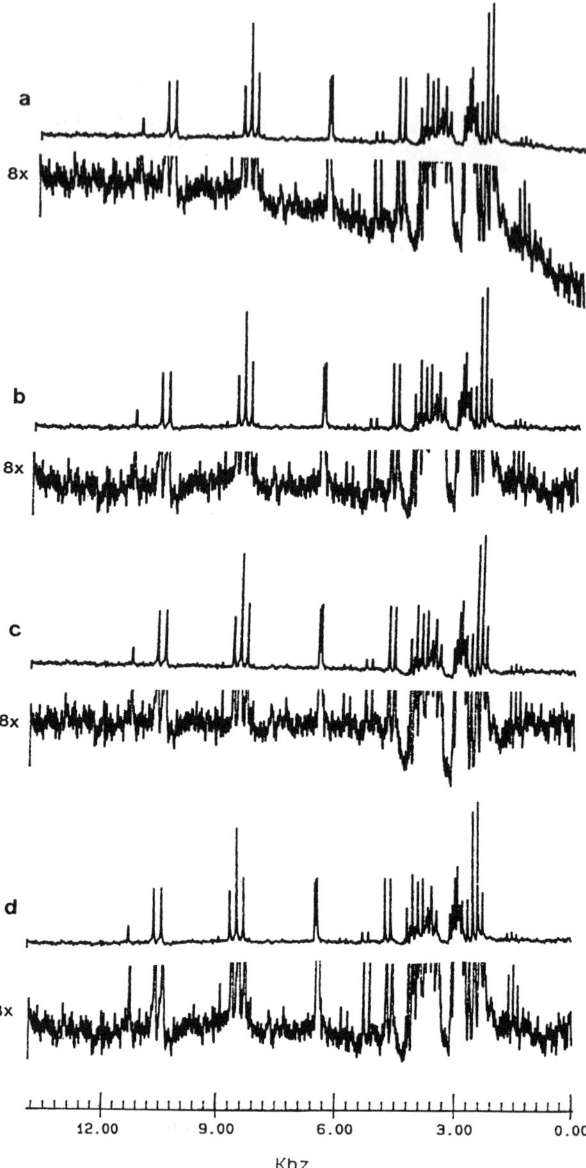

FIGURE 6.5. (a) A proton-coupled ^{13}C spectrum with a poor baseline. (b) Result from automatic baseline flattening (fourth-order polynomial fit). (c) Baseline flattening performed by fitting a fourth-order polynomial to each of four segments of the spectrum. This causes an overcorrection on the right side of the spectrum, where peaks overlap heavily. (d) Selection of four segments of baseline for a fourth-order fit, but with elimination of correction for the region of heavy peak overlaps. The baseline flattening performance is improved over the case shown in part (c).

4. There are two cases where conventional baseline conditioning can be quite inadequate: (a) NMR of very broad signals, as with some quadrupolar nuclei; and (b) *in vivo* NMR. Figure 6.6 shows an example of a human brain ^{31}P spectrum, obtained at 1.9 T. The large baseline rise seen in the full spectrum is typical of some types of *in vivo* ^{31}P spectra. Correction of the baseline is complicated by the fact that the actual ^{31}P lines overlap significantly. Further, the ^{31}P resonance lineshapes are markedly non-Lorentzian, as a result of distributions of chemical shift environments for each band (representing chemical species in different cellular—and chemical—environments). Figure 6.6 also shows the result of automated baseline conditioning of this spectrum. This algorithm[14] utilizes a finite automaton that determines baseline and resonance band segments and calculates a smooth function connecting the automatically determined baseline segments. The program then subtracts the calculated "complete baseline" from the spectrum.

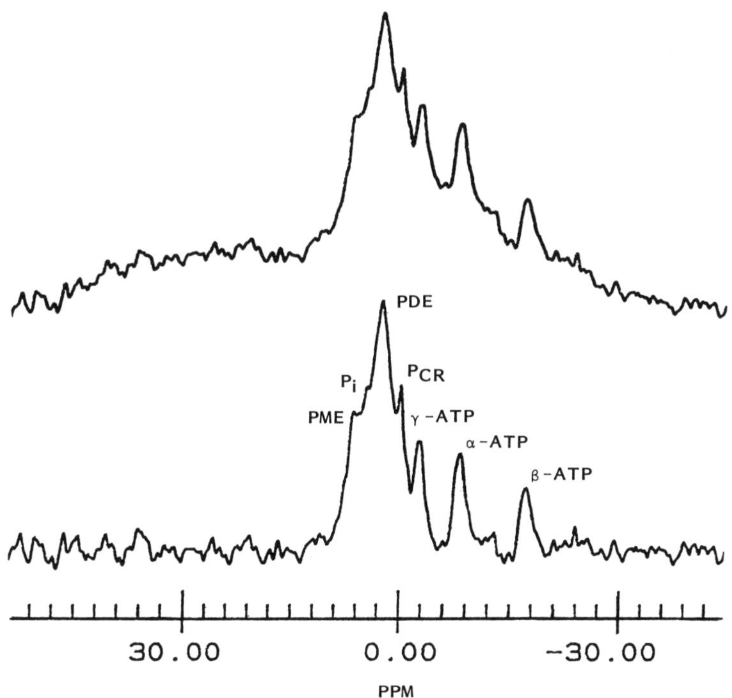

FIGURE 6.6. *In vivo* ^{31}P NMR at 1.9 T. Human brain spectrum: (top) after apodization, FFT, and phasing; (bottom) after automated baseline correction using a finite automaton to determine baseline segments.

The finite automaton method alone can give poor results when large sections of a poorly determined "baseline" are completely absent due to overlaps between groups of broad peaks. Improvement can be obtained when baseline segments are declared only for spectral data exceeding a minimum number of points (e.g., 5 or 10). In this way, a "sharp" valley between two overlapping peaks may be excluded as a potential baseline segment.

6.3.5. Other Methods for One-Dimensional NMR Spectra

Scientists need a variety of tools for NMR applications: for example, spin simulations, spectral assignment, chemical kinetics from time series NMR spectra, spin relaxation value determination (a special case of kinetics—first-order or exponential decay). One example of a "kinetics" experiment is shown in Fig. 6.7, which shows both an automated

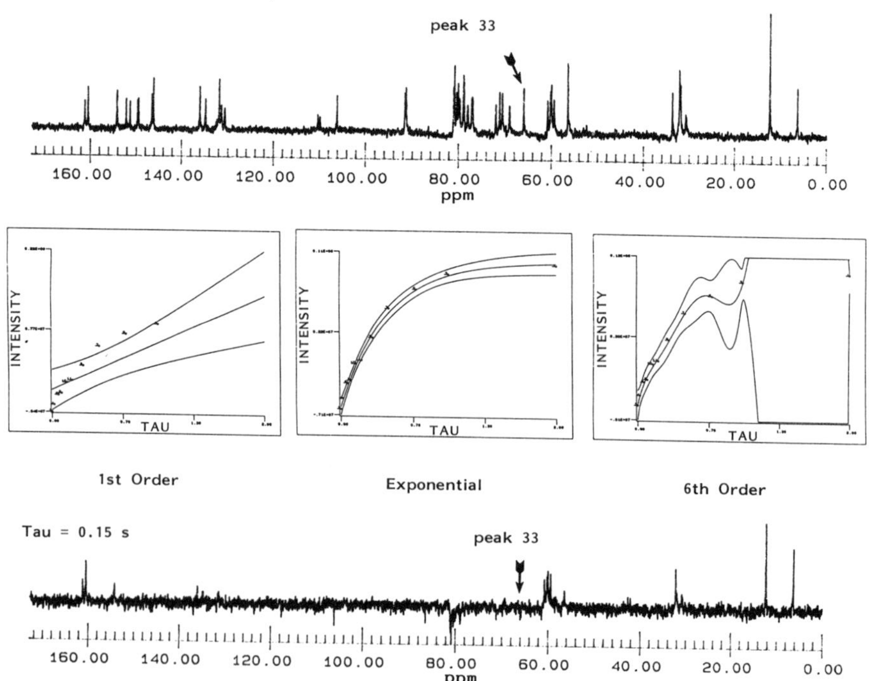

FIGURE 6.7. Regression analysis of an NMR relaxation (T_1) data set using a kinetics feature tracking utility. The thirty-third peak is selected for T_1 analysis. Of the three regression equations applied to the data, only the exponential function (i.e., the theoretically correct equation) gives a satisfactory fit, as evidenced by the 95% true interval confidence limits, shown on each fit.

determination of spin lattice relaxation times and a statistically checked kinetics-type determination for a single line that appears in the eleven spectra of this experiment. At the top of Fig. 6.7, the ^{13}C spectrum (spin relaxed) of a DNA oligomer is shown and at the bottom of Fig. 6.7, a single short "tau" value (nine other partially relaxed spectra from the inversion recovery pulse experiment are not shown). It is facile to calculate *fully automatically* the T_1 value for all 44 peaks in the full spectrum, and this can be done in a few seconds. For more demanding kinetics applications, however, it is useful to have statistical checking of fits to assumed kinetics equations. For example, in Fig. 6.7, three fits are given, but only the proper exponential fit is shown to be significant, because the 95% true interval confidence curves (see Fig. 6.7) clearly indicate overfitting to the sixth-order polynomial, despite a perfect fit to the data points.

6.3.6. 2-D Spectroscopy

The methods described above have analogies in multidimensional NMR spectroscopy and imaging. Use of automated 2-D surface fitting on crowded cross peak regions (Fig. 6.8) gives excellent quantification for spectra with reasonable signal-to-noise ratios. This type of scheme[15] is not simple to implement, however, because errors in determination of

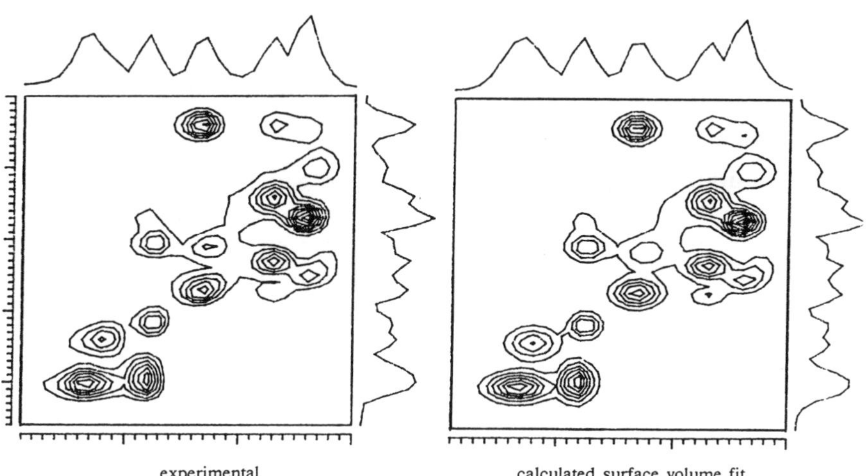

FIGURE 6.8. Example of 2-D surface fitting to quantitate 20 peaks in an overlapping region. The software makes initial guesses based on determination aided by cluster analysis. The left spectral segment is the experimental data; on the right is displayed the simulation. This calculation took 5 min on an Alliant FX/80 computer.

both spectral baseplane and resonance bandshape can, unless well accounted for, give rise to significant errors. In the example shown, the lineshapes are apodized appropriately and corrections are applied for deviations in the local baseplane.[15]

As an alternative to surface fitting, maximum entropy and other constrained restoration techniques can effectively deconvolute 2-D spectra, reducing spectral overlaps while simultaneously increasing (dramatically) the apparent signal-to-noise ratio. Figure 6.9 shows an example, a section of a 2-D NOESY spectrum obtained on a DNA duplex oligomer (minihelix) processed conventionally and also by maximum likelihood deconvolution.[16] The stacked plots shown in Fig. 6.9 give a good perspective of the utility of these techniques. The deconvolution succeeds in defining ^1H–^1H spin–spin coupling multiplets. Also, in test[16] of this 2-D deconvolution module, it was determined that along with line sharpening and noise suppression, spectral quantitation improved considerably, compared with conventional processing. The compute time for these calculations is approximately 15 min to 1 h on Sun-3 or Sun-4 workstations for a single 256 × 256 data point cross-peak region, but processing of an entire 2k × 2k spectrum was achieved using an Alliant FX/80 parallel/vector minisupercomputer in under 6 h. Considerable time savings can result from use of vector processors, because such deconvolutions require 250–400 2-D Fourier transforms. Other 2-D restorations can also be used that do not require FT calculations, and these are generally faster.

Recent proposals[17] to extend 2-D NMR spectroscopy to three and even four dimensions offer exciting possibilities to simplify spectral elucidation of very complex molecules. Just as 2-D FT NMR increases dispersion of signals by placing them in two dimensions, the higher-dimensional spectra will have significantly enhanced information content. The limits of current data processing, however, with restrictions on our abilities to utilize the new data, will limit these studies for some time. In addition, the experiments themselves require greatly extended periods, except in cases where only selected frequencies are sampled.[18] It is expected that data processing restrictions will be obviated by next-generation cpu's with integrated digital signal processing (DSP) chips capable of performing 512 × 512 FFT calculations in the order of 1 s or less. Also needed, however, will be very large RAM memories and/or greatly improved disk I/O (10 + MB/s *sustained* rate).

6.4. NEW TRENDS IN NMR SOFTWARE

NMR Data Processing in 1988 showed at least three major new developments: (1) use of image processing methods in 2-D

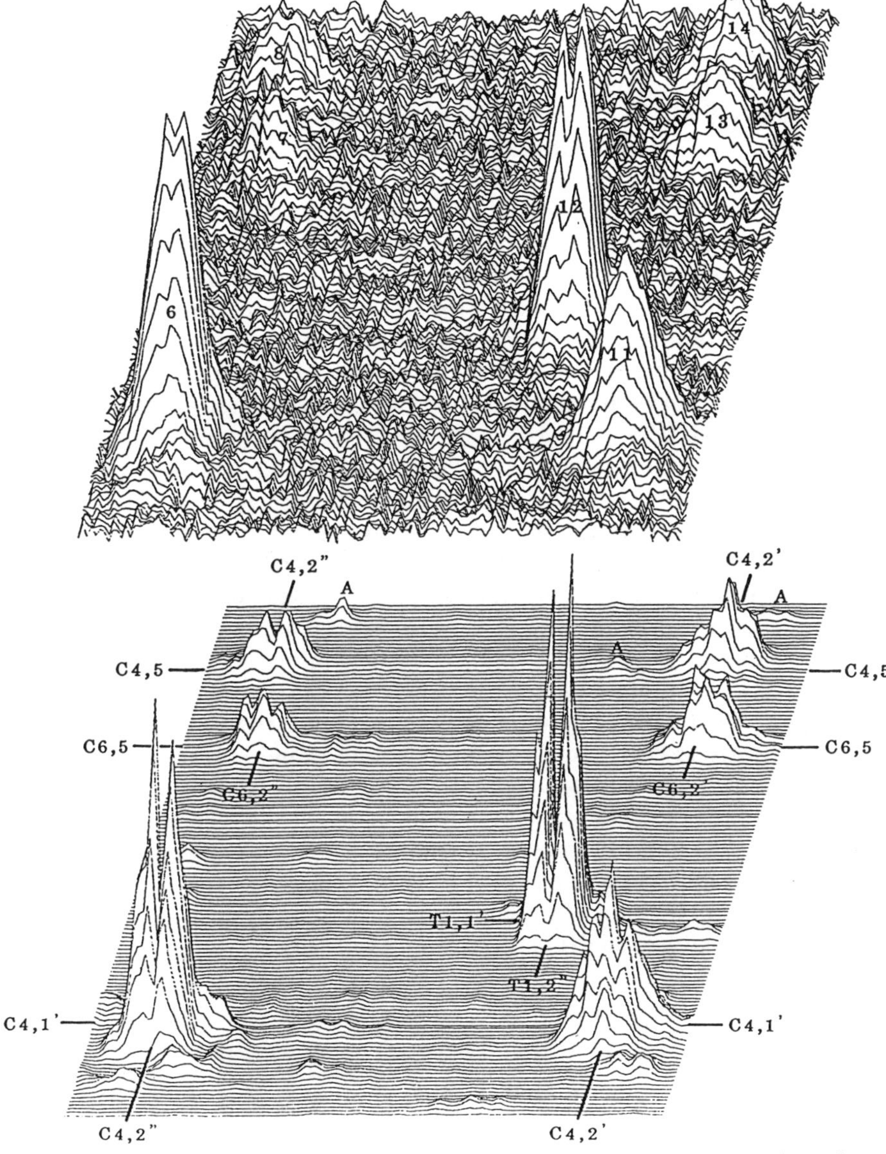

FIGURE 6.9. Two-dimensional spectral deconvolution using the criterion of maximum likelihood (ML). The top spectral segment is a portion of the 2′, 2″-1′ and base-5 proton NOESY cross peak region for the molecule $[d(\mathrm{TAGCGCTA})]_2$. Note the ^1H–^1H spin–spin coupling in the ML reconstruction. Over one order of magnitude of signal amplitudes, quantitation in the ML reconstruction was seen to be twice as precise as in the conventionally processed spectrum (comparing cross peak volumes above and below the diagonal).

spectroscopy;[8,9,19] (2) use of multivariate analysis methods for signal enhancement and pattern recognition;[10] (3) use of distance geometry[20] or restrained dynamics[21] calculations with NOESY 2-D NMR data to determine a 3-D solution structure of small proteins or nucleic acids. Not yet demonstrated, but expected, is the advent of true expert system software for totally automated NMR data reduction—even where spectra have significant irregularities.

Examples of image processing technology used for spectroscopy include use of morphological filters,[22,23] which are superior to traditional symmetrization for noise suppression in diagonal-symmetry 2-D spectra,[22] and use of segmentation methods[19] to automatically set low-level contours for noisy and complex 2-D spectra. The use of a morphological filter is demonstrated in Fig. 6.10, showing superior suppression of both random and nonrandom (simulated T1) noise in a COSY spectrum.[22]

Principal component analysis (PCA) and partial least squares (PLS) modeling are multivariate analysis techniques that can classify and identify data trends under conditions where traditional methods fail. PCA and PLS can be used for suppression (separation) of random *and* systematic spectral noise *or* specified spectral features.[10] These techniques can also be used as powerful data compression tools for complex 2-D spectra, and they may provide a practical way to abstract information from 3-D and 4-D spectra. Combination of multivariate data analysis with other techniques, such as logic programming, may give rise to a new type of expert software in NMR spectroscopy.[24] One example is the preliminary study[10(c)] showing automated separation and identification of the patterns resulting from spin topology in amino acids. This technique may be applied to develop automated assignments of 2-D spectra of small proteins and other molecules.

Another application of principal component analysis is for separation of systematic features in 2-D spectra; for example, T1 ridges.[10(b)]

6.5. COMING TRENDS IN COMPUTER HARDWARE

The impact of the workstation computer is just now being felt in NMR spectroscopy. Current development of two new classes of workstation hardware, however, will greatly extend this influence. In 1989, PC systems from IBM and Apple will qualify as true workstations, with 32-bit performance, good high-resolution graphics, and networking (and with UNIX operating systems). At the high end, a new class of workstation has arrived, *the graphics supercomputer.*

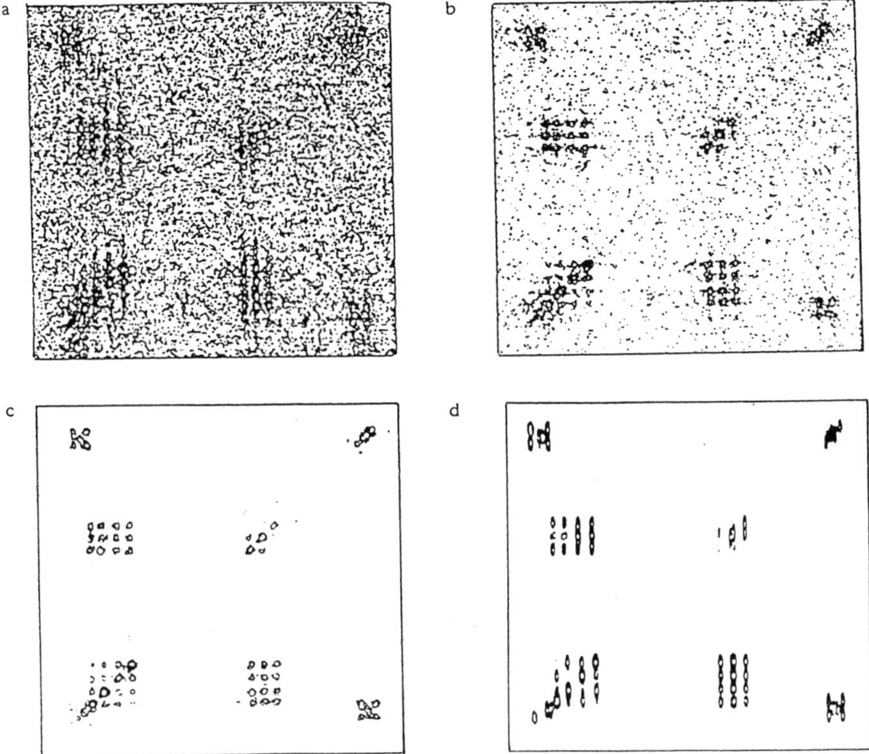

FIGURE 6.10. COSY spectra of fluoranthene with Gaussian noise added (○ = 10% data maximum): (a) Original data plus noise. (b) Traditional symmetrization, replacing point and image with minimum. (c) Morphological filter. (d) Data with no noise added. *All contour plots made with levels equally spaced starting at 10% of data maximum.*

Three graphics supercomputers were announced in early 1988: the Stellar GS1000, Ardent Titan, and Apollo DN 10000. This new class of computers combines near-Cray compute power with extraordinary graphics capability—and with an extremely high bandwidth between computation and the screen that was not previously possible. The graphics supercomputer will find applications across a broad spectrum of science and engineering, combining theoretical and empirical calculations with experimentation and providing high-speed interactive graphics of very complex molecules using 3-D solid atom representations. In chemistry, however, the most significant applications will be in multi-dimensional NMR data processing and integrated molecular modeling.

All of the graphics supercomputers derive some of their speed by exploiting parallelism in their central processing units. This is one of several important trends in computer hardware design, but full realization of the promise of parallelism requires extension of current software methodologies.[3]

6.6. LABORATORY COMPUTER NETWORKING

Today's NMR spectrometers are beginning to support high-speed computer networking, with those based on UNIX or DEC computers making available standard networking protocols, TCP/IP (common on UNIX-based computers) and DECnet (developed by Digital Equipment Corporation, but now offered by third parties for many UNIX-based computers). Over the past year, a large number of NMR laboratories have begun to integrate their local data links to departmental or remote computer networks. One network design, the hierarchical laboratory computer network at Syracuse University,[2] is shown in Fig. 6.11, as configured in late 1988. In Fig. 6.11, the top row of computers, not local to the laboratory, are equally available as a result of the campus FASTNet system. In the NMR laboratory network, where it is available, the NMR instruments are linked by Ethernet and, in other cases, by network-style RS-232 links.* Use of the X-graphics windows, which are standard throughout this network, allows users to have several processing sessions simultaneously on a local Sun or other workstation, with multiple windows on a single screen. The X-graphics protocol further allows users to open high-speed graphics windows, where actual processing takes place on remote, networked computers. Thus, in the hierarchical network shown in Fig. 6.11, computationally intensive calculations such as 2-D deconvolutions (or even very large 2-D FFT calculations) have often been performed on the network-linked University (NPAC) Alliant FX/80 minisupercomputer, to achieve up to a two orders of magnitude reduction in processing time[3] versus a microVAX. Almost all of the network nodes can support X-windows, including all of the laboratory workstations and the PC (PC supported as of 1989).

For the Syracuse University Natural Science Computing network (NSCnet), the arrival in 1988 of two graphics supercomputers† allowed

* NMR Instrument data transfers, X-windows implementation, and SpecStation software, from New Methods Research, Inc., Syracuse, New York. SpecStation™ is a trademark of New Methods Research, Inc.

† One to New Methods Research, that is networked by optic fiber.

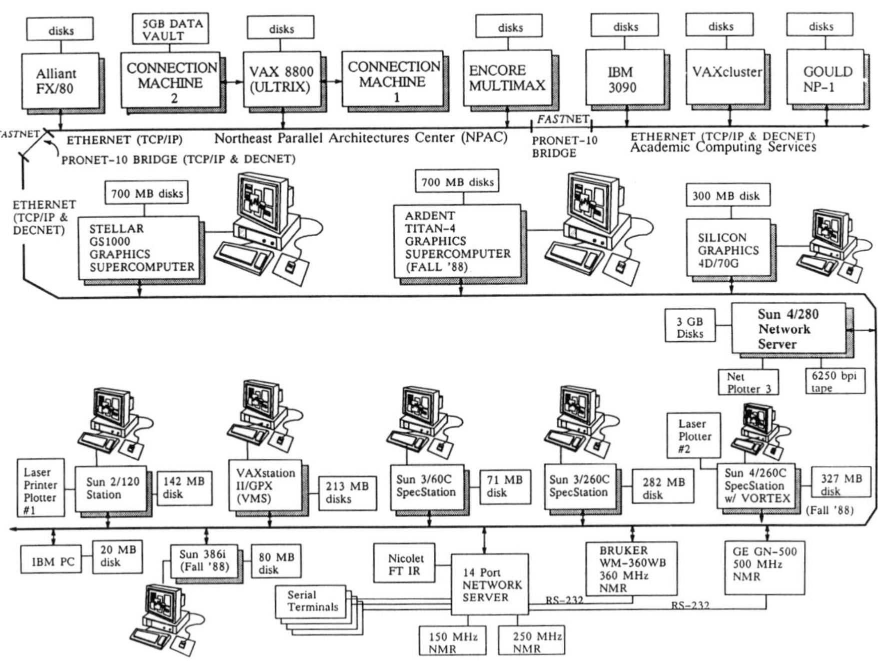

FIGURE 6.11. The hierarchical computing network at Syracuse University. The top level of computers are operated by central academic computing services and the Northeast Parallel Architectures Center. The Natural Science Computing Network, operated by the Syracuse University NMR and Data Processing Laboratory, receives support from the Division of Research Resources, N.I.H., and Syracuse University, as well as from corporate sponsors.

most compute-intensive calculations to migrate to the laboratory's Stellar and Ardent Titan systems, with middle-sized jobs running on the Sun-4 or Silicon Graphics nodes. Some algorithms, easily converted for concurrency, will be ported to the University's Connection Machine-2 (CM2) from Thinking Machines. Inc. (Cambridge, MA). CM2 is an example of a massively parallel computer with up to 65,536 1-bit processors and a 1,000-element floating-point accelerator; the system gives a peak performance of 3,500 MFlops. CM2 is controlled by a front-end computer, for our use, a VAX 8800 (running Ultrix). To carry out processing, data is first loaded onto each processor and, at the end, it is read back to the front-end. For small processes, such as a 32k FFT, the Connection Machine is slower than the front-end machine. But for very large processes, such as a 16-Mword 2-D FFT or 2-D deconvolution, this will not be the case. At present, the Connection Machine does not time share, and so it may be tied up for several hours before access is obtained. A time-sharing operation system is due to be launched in 1989 and this will allow the sort of interactive use vital for NMR data processing.

6.7. CONCLUSIONS

Over the past 20 years, the computer has had a profound impact on NMR spectroscopy. Current trends point to extraordinary developments in instrument control, and more importantly to giving us the ability to extract *all of the information* from NMR experiments. With accelerating technological changes in computing, revolutionary changes will develop for the role of the computer in NMR. These will make possible accurate and efficient structure and quantitation studies on systems far more complex than can be successfully examined today.

ACKNOWLEDGMENTS

Support from the Division of Research Resources, N.I.H. (Grant RR-0317), the National Science Foundation (partial purchase of the GN-500 spectrometer), Syracuse University, and several corporate sponsors (General Electric, Digital Equipment, Star Technologies, Apollo, Sun, Silicon Graphics, Stellar, Ardent, New Methods Research, Inc., and others) is gratefully acknowledged, as are technical contributions over the last several years from Syracuse University: Roy Hoffman, Anil Kumar, Mark Roggenbuck, William Curtiss, Marc Delsuc, Hans Grahn, Phil Borer, Anthony Mazzeo, Steve LaPlante, Karl Bishop, Mats Josefson, Feng Ni (Cornell University) and New Methods Research, Inc.: Alex Macur, Frank Delaglio, John Begemann, Elliot Dudley, Molly Crowther, Pascale Solé, John Stanley, David Sternlicht, and others.

REFERENCES

1. G. C. Levey and D. Terpstra, eds., "WARPATH, A Computer Network for a Multi-User Laboratory Environment," in: *Computer Networks in the Chemical Laboratory,* Wiley-Interscience, New York (1981).
2. G. C. Levy, "Hierarchical Computer Networks for the Spectroscopy Laboratory," *Spectrosc.* **3,** 30–36 (1988).
3. R. E. Hoffman and G. C. Levy, "Optimization of NMR Data Processing with Coarse-Grain Parallel Computers," *Comput. Chem.* **13,** 179–184 (1989).
4. G. A. Pearson, "A General Baseline-Recognition and Baseline-Flattening Algorithm," *J. Magn. Reson.* **27,** 265–272 (1977).
5. A. Kumar, C. H. Sotak, C. L. Dumoulin, and G. C. Levy, "Spectral Peaks and

Quantitative Analysis by ^{13}C Fourier Transform NMR Spectroscopy," *Comp. Enhanced Spectrosc.* **1,** 107–115 (1983).
6. S. Sibisi, "Two-Dimensional Reconstructions from One-Dimensional Data by Maximum Entropy," *Nature* **301,** 134 (1983).
7. F. Ni, G. C. Levy, and H. A. Scheraga, "Simultaneous Resolution Enhancement and Noise Suppression in NMR Signal Processing by Combined Use of Maximum Entropy and Fourier Self-Deconvolution Method," *J. Magn. Reson.* **66,** 385–390 (1986).
8. P. H. Bolton, "Image Processing of Two-Dimensional NMR Data," *J. Magn. Reson.* **64,** 352–355 (1985).
9. P. Sole, F. Delaglio, A. Macur, and G. C. Levy, "Interactive Image Processing for Nuclear Magnetic Resonance Imaging and Spectroscopy," *Am. Lab.* 48–54 (Aug. 1988).
10. (a) Abstracts, H. Grahn, F. Delaglio, M. W. Roggenbuck, and G. C. Levy, 29th ENC, Rochester, NY, 1988; (b) H. Grahn, F. Delaglio, M. Roggenbuck and G. C. Levy, "Multivariate Techniques for Enhancement of Two-Dimensional NMR Spectra," unpublished; (c) H. Grahn, F. Delaglio, M. A. Delsuc, and G. C. Levy, "Multivariate Data Analysis for Pattern Recognition in Two-Dimensional NMR," *J.Magn. Reson.* **77,** 294–207 (1988).
11. A. R. Mazzeo, M. A. Delsuc, A. Kumar, and G. C. Levy, "Generalized Maximum Entropy Deconvolution of Spectral Segments: Resolution Enhancement for Low Signal-to-Noise Features in Spectra Having Dynamic Range," *J. Magn. Reson.* **81,** 512–519 (1989).
12. H. Barkhuijsen, R. De Beer, W. M. J. Bovee, and D. Van Ormondt, "Retrieval of Frequencies, Amplitudes, Damping Factors, and Phases from Time-Domain Signals Using a Linear Least-Squares Procedure," *J. Magn. Reson.* **61,** 465–481 (1985).
13. M. A. Delsuc, F. Ni, and G. C. Levy, "Improvement of Linear Prediction Processing of NMR Spectra Having Very Low Signal-to-Noise," *J. Magn. Reson.* **73,** 548–552 (1987).
14. (a) M. Josefson, unpublished; (b) NMR1 User Manual (New Methods Research, Inc.).
15. F. Delaglio, H. Grahn, P. Sole, A. Kumar, R. E. Hoffman, J. Begemann, A. Macur, and G. C. Levy, unpublished.
16. K. D. Bishop, S. LaPlante, P. Borer, and G. C. Levy, unpublished.
17. (a) C. Greisinger, O. W. Sorenson, and R. R. Ernst, "A Practical Approach to Three-Dimensional NMR Spectroscopy," *J. Magn. Reson.* **73,** 574–579 (1987); (b) G. W. Vuister and R. Boelens, "Three-Dimensional *J*-Resolved NMR Spectroscopy," *J. Magn. Reson.* **73,** 328–333 (1987); (c) R. E. Hoffman and D. B. Davies, "Three Dimensional *J*-Resolved and Correlated NMR Spectroscopy," *J. Magn. Reson.* **80,** 337–339 (1988).
18. Abstracts, Ray Freeman, European Experimental NMR Conference, Bad Ausee, Austria, 1988.
19. P. Sole, F. Delaglio, and G. C. Levy, "A Segmentation Technique for Automated Contour Selection in 2D NMR Spectroscopy," *J. Magn. Reson.,* **80,** 517–519 (1988).
20. D. J. Patel and L. Shapiro, summarized in *Ann. Rev. Biophys. Chem.* **16,** 423–454 (1987).
21. G. M. Clore, A. T. Brunger, M. Karplus, A. M. Gronenborn, "Application of Molecular Dynamics with Interproton Distance Restraints to Three-Dimensional Protein Structure Determination: A Model Study of Crambin," *J. Mol. Biol.* **191,** 523–551 (1986).
22. Abstracts, P. Sole, G. C. Levy and F. Delaglio, 29th ENC, Rochester, NY, 1988.
23. R. M. Haralick, S. R. Sternberg, and X. Huang, "Image Analysis Using Mathematical Morphology," *IEEE Trans. Pattern Anal. Mach. Intel.* **4,** 532 (1987).
24. G. C. Levy, "Current Trends in Computing Hardware, Software, and Nuclear Magnetic Resonance Research," *J. Mol. Graphics* **4,** 170–177 (1986).

Part II

Supervised Methods: Expert Systems, Modeling, and Quantitation

Chapter 7

Computer Identification of Mass Spectra

Fred W. McLafferty, Stanton Y. Loh, and Douglas B. Stauffer

7.1. INTRODUCTION

The efficient identification of specific organic compounds in complex mixtures is a key problem in many important areas such as pollution, insect chemical communication, forensic evidence, drug overdoses, drug metabolites, body fluid diagnoses, and chemical taxonomy. Despite recent advances in a variety of analytical instrumental techniques, mass spectrometry (MS) is by far the most widely used method[1] for such identifications, especially when coupled to an efficient separation device such as the gas or liquid chromatograph (GC or LC). As summarized in Fig. 7.1, electron-ionization (EI) mass spectra, measurable in less than 1 s on subnanogram samples, have an unusually high information content, with from scores to hundreds of peaks determined to unit mass accuracy. On the negative side, peak abundances of available reference spectra are poorly reproducible, and most common methods of sample preparation (e.g., GC separation) cannot guarantee the sample purity demanded by the three orders of magnitude abundance range commonly used in recording mass spectra. Understanding the source of these problems has provided the primary basis for recent improvements to computer

Fred W. McLafferty, Stanton Y. Loh, and Douglas B. Stauffer • Department of Chemistry, Baker Laboratory, Cornell University, Ithaca, New York 14853-1301.

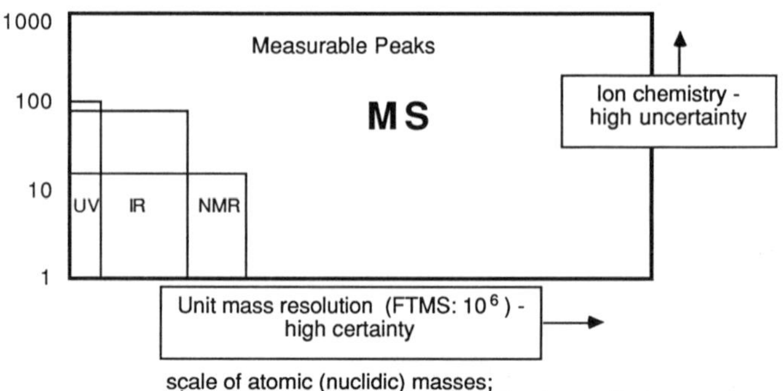

FIGURE 7.1. Data content of EI mass spectra.

methods for identifying unknown mass spectra, and these improvements have reduced the proportion of wrong answers by several fold.

7.1.1. Computer Identification Algorithms

With the current data base containing mass spectra of 118,000 different compounds, it is not surprising that computer techniques are increasingly valuable aids in structure determination.[2–13] The design of such computer algorithms, however, must take into account problems not important in other common spectroscopic techniques. Exact identification of structure from mass spectra[4–6] is made difficult by the higher rearrangement proclivity of ions versus neutrals (Fig. 7.2). Further, a specific functional group does not necessarily produce the same peaks, even in closely related compounds (Fig. 7.3).[14] Although the presence of characteristic peaks can indicate the presence of a substructure, their

Compound type	Spectral similarity
Different (e.g., MW)	Very different[a]
Similar (added $-NH_2$)	Can be different[a,b]
Some isomers (alkenes, m-/p-)	Poorly distinguishable[c]

(a) High data content

(b) Ion chemistry: competitive and synergistic effects; absence of a characteristic peak does not show absence of that substructure

(c) Ion chemistry: higher rearrangement tendency of cations vs. neutrals

THUS:
1) MS best for 'global' unknown to identify compound type, not exact structure
2) Spectral prediction difficult; need comprehensive spectral file
3) Correct "Class IV match" possible for ~10X the number of reference compounds

FIGURE 7.2. The effect of structure on mass spectra.

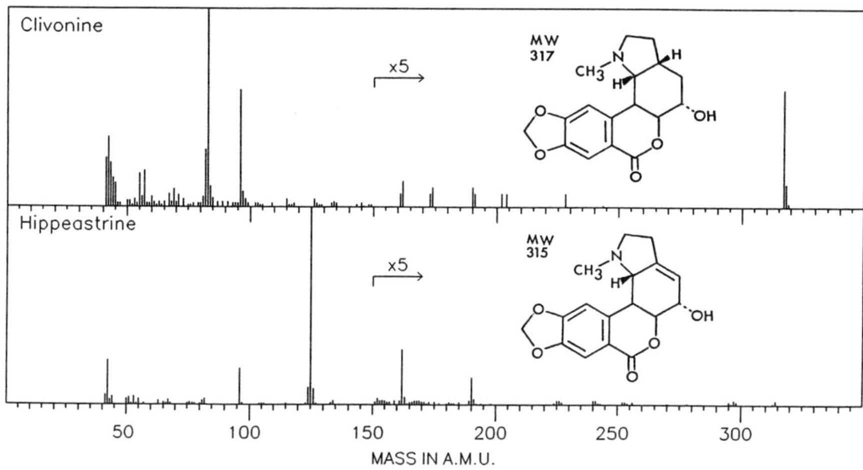

FIGURE 7.3. The EI mass spectra of clivonine and hippeastrine.

absence is *not* a reliable indication of the absence of the substructure. As illustrated in Fig. 7.4, matching of an unknown and reference spectrum of the same compound can be complicated by impurity and other artifact peaks. Further, spectra in large data bases have been run under a variety of experimental conditions, so that peak abundances can vary by more than an order of magnitude for spectra of the same compound.

This chapter will discuss mainly the Cornell algorithms known as probability-based matching (PBM)[(7)] and the self-training interpretive and retrieval system (STIRS),[(5)] which among such systems appear to exhibit the best performance as well as the widest use.[(11)] PBM and, to a much lesser extent, STIRS are now being used routinely on literally thousands of GC/MS systems.

Random errors: [a,b] mass calibration, amplifier saturation, incomplete mass or abundance range, data transcription

Mass dependent: [a,c] instrumental mass discrimination, sample pressure change during scan (over GC peak), temperature (extent of fragmentation), impurities [a,b,d]

Aids (a) Compare other spectra of the same compound
Reference: others in data base
Unknown: others of same GC peak or remeasure
(b) Reference peak flagging: remove and rematch
(c) Unknown spectrum scaling (quadratic fitting)
(d) Impurities. Reference: no peaks >MW, illogical neutral losses
Unknown: reverse search, spectrum subtraction

FIGURE 7.4. The effect of experimental conditions on mass spectra.

7.1.2. Matching versus Interpretative Systems

Of identification algorithms, PBM and STIRS represent the two main categories, "matching" (or retrieval) and "interpretive". The former is generally used first to match the unknown spectrum against the reference file. If the unknown compound is not in the data base, hopefully an acceptable match will not be retrieved; then the interpretive program can be used to provide structural information on the unknown compound. PBM and STIRS have been designed for the identification of total unknowns; in many cases there is insufficient sample to use other analytical methods, and in many problems the sample source provides little or no structural information. Because of the basic limitations of mass spectrometry, such as distinguishing particular types of isomers, the chemist should recognize that the main utility of these algorithms is to reduce the possible number of compounds from such a universe of original possibilities to a smaller number (Fig. 7.2). This means that such usefully close (class IV)[(7)] matches are possible for an order-of-magnitude more compounds than the 118,000 represented in the reference file. Many research organic chemists make little use of mass spectrometry for structural characterization because their knowledge of the chemical conditions producing the unknown has already provided this high reduction in the possible number of answers. As a further caveat, the user should remember that most reference spectra have not been measured under conditions identical to that of the unknown; thus, for more rigorous identification the reference spectrum of the indicated compound should be measured under identical MS conditions, and complementary analytical techniques should be applied where possible. Substantial portions of the material in this chapter are incorporated in a revised book (see ref. 25) on spectral interpretation.

7.2. EXPERIMENTAL

Testing and specification determination for the implementation and improvements to both PBM and STIRS over the years have been based on extensive statistical measurement of the algorithms' performance.[(12,13)] This measurement used large numbers (approximately 400 for PBM, 900 for STIRS) of "unknown" spectra selected randomly from the data base,[(14,15)] but excluded from it in the test. These data were also used to calculate the actual matching reliability of the predicted compound or substructure, based on the performance values exhibited by the unknown compound.[(12,13)]

PBM and STIRS have been implemented on a variety of main frame, mini-, and microcomputer systems; an updated list is available from the data base supplier.[16] These include centralized capabilities such as the Chemical Information System available over data networks, commercial GC/MS systems such as those of Hewlett–Packard, Finnigan, and Nicolet, and the Wiley data base on CD-ROM with PBM and STIRS for use with a personal computer.[16]

7.3. DISCUSSION

7.3.1. The Reference File

The performance of a retrieval program is dependent directly on the suitability of the reference file. The collection of reference spectra used has been the Wiley/NBS Registry of Mass Spectral Data, whose 1989 edition contains 140,000 different mass spectra of 118,000 different compounds.[14,15] This is the outgrowth of a project initiated in the mid-1960s by the senior author with Professors Stenhagen and Abrahamsson, now deceased, and the publisher.[17,18] The actual measurement of these 128,000 different spectra (140,000 with the NBS collection) resulted from the cooperation of literally thousands of mass spectrometrists worldwide. The data base incorporates valuable individual collections such as those of Dr. D. Henneberg of the Max Planck Institute, Dr. C. Djerassi of Stanford University, and Dr. B. H. Kennett of the CSIRO Division of Food Research. Because these spectra have thus been measured with a wide variety of instruments and experimental conditions, abundance values are also widely variable. To offset this, other spectra of the same compound are included in the reference file if these spectra do not match one another well by PBM.[19] Reference inaccuracies are a further problem. The fact that peaks of abundances as low as 0.25% are useful for identification also means that reference compound impurities at this level can produce spurious peaks in reference mass spectra, in contrast to most types of spectra. We have corrected some 40,000 errors in the reference file, indicative of the inherent problems of measuring and transposing such a huge amount of data.

7.3.2. Probability-Based Matching (PBM)

7.3.2.1. The Statistical Basis of PBM

Key requisites of any library search program are the incorporation of quantitative matching probabilities.[20] This has been done in PBM through data weighting, performance evaluation, and reliability ranking.

7.3.2.1a. Data Weighting. For the two main types of mass spectral data, abundance and mass, their statistical weighting values are determined directly from their probability of occurrence in the data base. Abundance value probabilities closely follow a log normal distribution, while the weights of mass values show a regular increase at higher mass because of their less frequent occurrence.[7]

7.3.2.1b. Performance Evaluation. Statistical tests of performance are also necessary in designing improvements to a library search algorithm. Such tests should be two-dimensional,[20] evaluating the reliability of matching as a function of recall, and the proportion of possible correct answers retrieved.[21] Using a "condensed" version of the reference spectra gives improved matching performance; the optimum number of peaks (as determined statistically) varies from 15 to 26 per spectrum, increasing with the molecular weight of the compound. This has the further advantage of greatly reducing the reference data requirements for computer storage.

7.3.2.1c. Reliability Ranking. Quantitation of matching probabilities using a readily interpretable value scale has been another goal of PBM. For most retrieval systems the matching value has little quantitative relationship to the probability that the answer is correct. Often the inexperienced user may conclude that the best matching spectrum represents the correct compound, oblivious of the fact this represents a poor match and/or that the correct one might not actually be in the reference file. For PBM each of the important matching parameters, including those discussed below, have been evaluated statistically to predict the actual reliability of the match, that is, the probability that the retrieved compound is the same as the unknown compound.[21,22] Values are calculated for two separate classes; class I reflects the probability that the retrieved spectrum is of the same compound or a stereoisomer, while a class IV match can also be that of a compound differing structurally in ways that should cause little (relative to that caused by the experimental error) effect on the mass spectrum.[7] The difference in these values then serves to remind the inexperienced mass spectrometrist that matching against this general reference file cannot give as reliable answers as using reference spectra measured under the same experimental conditions, and/or using a complementary technique for further structural confirmation.

7.3.2.2. The Impurity/Artifact Problem

The high number of peaks and high dynamic range of abundances in mass spectra create a substantial probability of false or misleading data

(Fig. 7.3). For reference spectra this problem is substantially alleviated by adding other spectra of the same compound to the data base. For unknown spectra, impure samples present additional matching problems, and unequivocal independent evidence of sample purity is often difficult to obtain. PBM contains several other aids for these problems.

7.3.2.2a. Peak Flagging. If the reference spectrum indicates incorrectly that the compound produces a peak at a specific mass—such as from an impurity, pyrolysis, or an error in mass measurement—the absence of this peak in the unknown spectrum will indicate incorrectly that this is not a match. In PBM the effect of such artifact peaks is minimized in a peak-flagging operation, in which a better match is sought after removing peaks of low abundance in several stages.[(22)] The reliability value is lowered to reflect the uniqueness and abundance of the peaks that had to be removed to improve the match.

7.3.2.2b. Reverse Searching. A basic method (Fig. 7.5) of PBM to minimize interferences from sample impurities, and even to identify other components in mixtures, is "reverse searching."[(7)] For this, PBM requires the peaks of the reference spectrum to be in the unknown spectrum, but not vice versa. Non-overlapping peaks of other components are thus ignored, improving the ability of the algorithm to match two or more components in the mass spectrum of a mixture. Even with GC introduction it is hard to guarantee sample purity at the 99 + % level required to eliminate impurity interferences from mass spectra.

7.3.2.2c. Subtraction of Retrieval References. Forward searching can, however, provide additional matching information. If it can be determined that the unknown spectrum represents a pure compound, then the matching can include the further criterion that all peaks of the unknown spectrum also occur in the reference, a "forward search." To incorporate[(12)] such forward-searching capabilities into PBM, the "percent contamination" value (i.e., the proportion unmatched of the abundance of the ten most important peaks) is used to modify the observed reliability. A low value for a correct match obviously indicates high unknown purity. Statistical data were gathered on the 392 unknown spectra showing how much more (less) reliable than predicted are matches with low (high) percent contamination values. Using these new reliability values substantially improved the overall performance of PBM for the unknown spectra of *pure* compounds, as expected for such a forward search approach.

To apply this approach to the unknown mass spectrum of a mixture, its component spectra must be "purified."[(12)] It is assumed that two or more of the best-matching compounds retrieved in the normal PBM

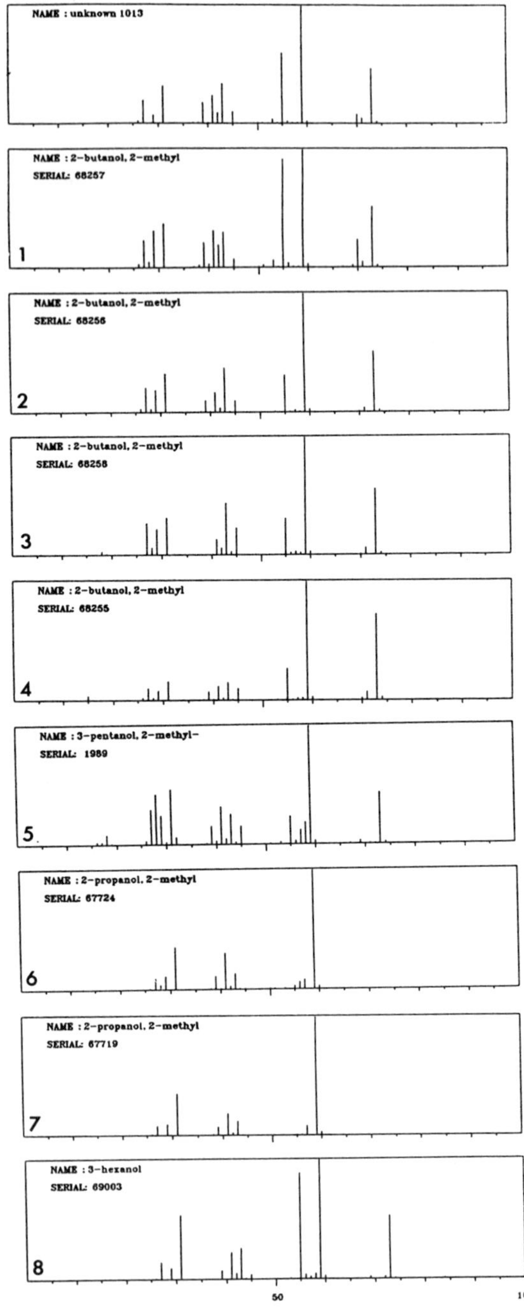

FIGURE 7.5. "Unknown" mass spectrum of 2-methyl-2-butanol (top) and spectra matching it retrieved by PBM.

search will represent components of the unknown mixture. The full spectrum of the best match is subtracted from the unknown spectrum, and the residual spectrum is rematched against the other best-matching spectra. If the subtracted spectrum represents one of the mixture components, the residual spectrum should then represent a higher concentration of the other component(s). Thus, this residual spectrum should produce a better match with any reference spectrum actually representing another component of the mixture; further, this should reduce the "percent contamination" value, which in turn will increase the predicted reliability value under the forward search modification. This subtraction/rematch process is repeated for each of the best-matching spectra, testing again for improved matching reliability.

The resulting overall improvement in PBM performance is impressive; for 60% mixture components, two-thirds of the wrong answers were eliminated. Although this forward search system was designed to aid the matching of mixture spectra, even for pure compounds the wrong answers were reduced by almost half.[(12)] This approach alleviates impurity peak problems and is also especially effective in reducing the predicted reliability of incorrect reference spectra that match well a part of the unknown spectrum, such as the spectrum of 3-octene matching the lower masses of the spectrum of octyl phthalate.

7.3.2.3. The Problem of Abundance Value Variation

In reference spectra[(14)] of the same compound, abundance values of specific mass peaks can vary over a large range. This can be caused by instrumental conditions such as mass discrimination or ion source temperature, or by recording the spectrum from a gas chromatographic effluent (GC/MS) in which the sample concentration in the MS is either increasing or decreasing rapidly (along the sides of the GC peak) during the spectral scan. Of course, including multiple spectra of the same compound in the data base reduces the probability that an unknown compound will not be matched for this reason. Some of these variations are mass dependent (Fig. 7.4); we have found these can be minimized by a least squares quadratic fitting of the unknown compound and reference abundances as a function of mass.[(22)] The significant performance improvement resulting from the incorporation of quadratic scaling was due in part surprisingly to cases in which the unknown and reference abundances differed by more than an order of magnitude.

7.3.2.4. Examples

Figures 7.5 and 7.6 show the unknowns (top) and best-matching spectra found by PBM for 2-methyl-2-butanol and morphine, respec-

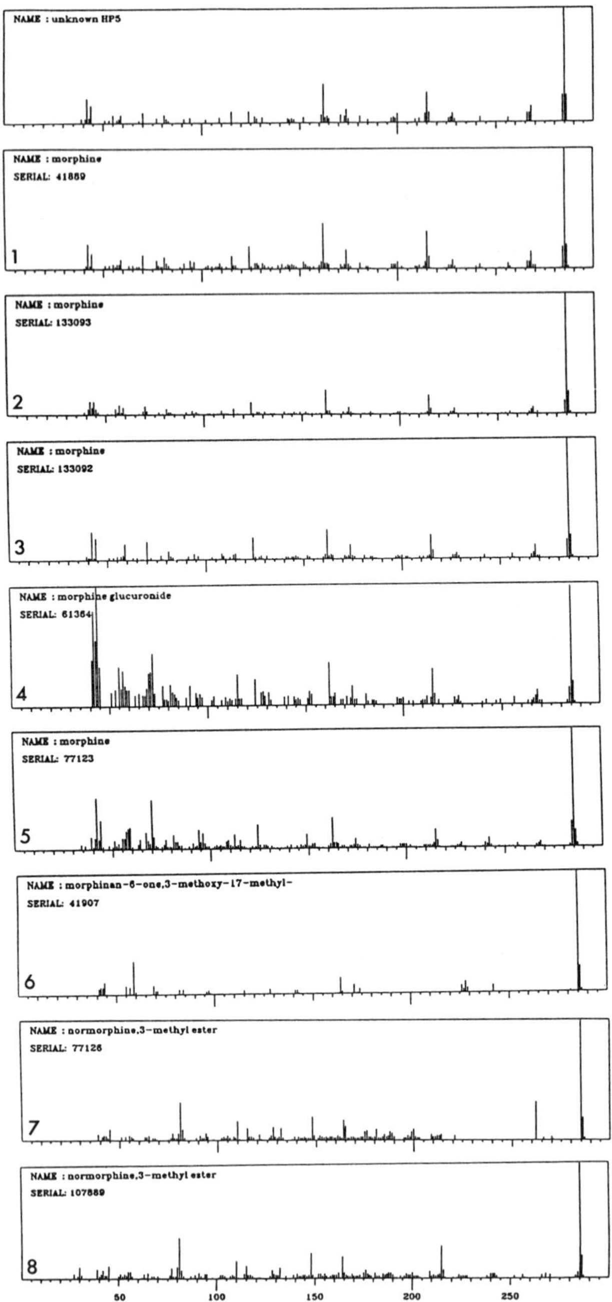

FIGURE 7.6. "Unknown" mass spectrum of morphine (top) and spectra matching it retrieved by PBM.

tively. The latter unknown spectrum was from an actual overdose case in which the matching algorithm used failed to retrieve morphine. Figures 7.7 and 7.8 show the effect on the matching reliability (class IV) of the stepwise inclusion in the PBM algorithm of flagging, reliability ranking, quadratic scaling, and forward searching. The low-molecular-weight 2-methyl-2-butanol produces a relatively few number of peaks and no molecular ion (m/z 88), making the distinction of homologs difficult. The presence of a matching molecular peak would have improved the final reliability of the best match to 95%.

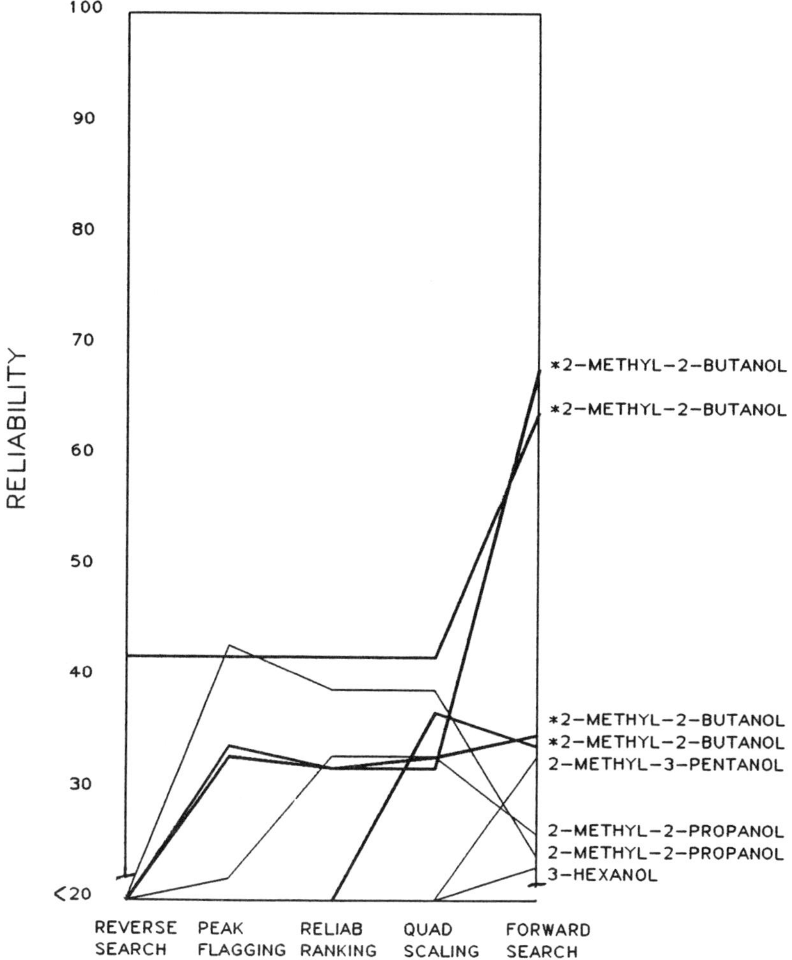

FIGURE 7.7. Effect on PBM improvements on the class IV reliability values for the 2-methyl-2-butanol unknown.

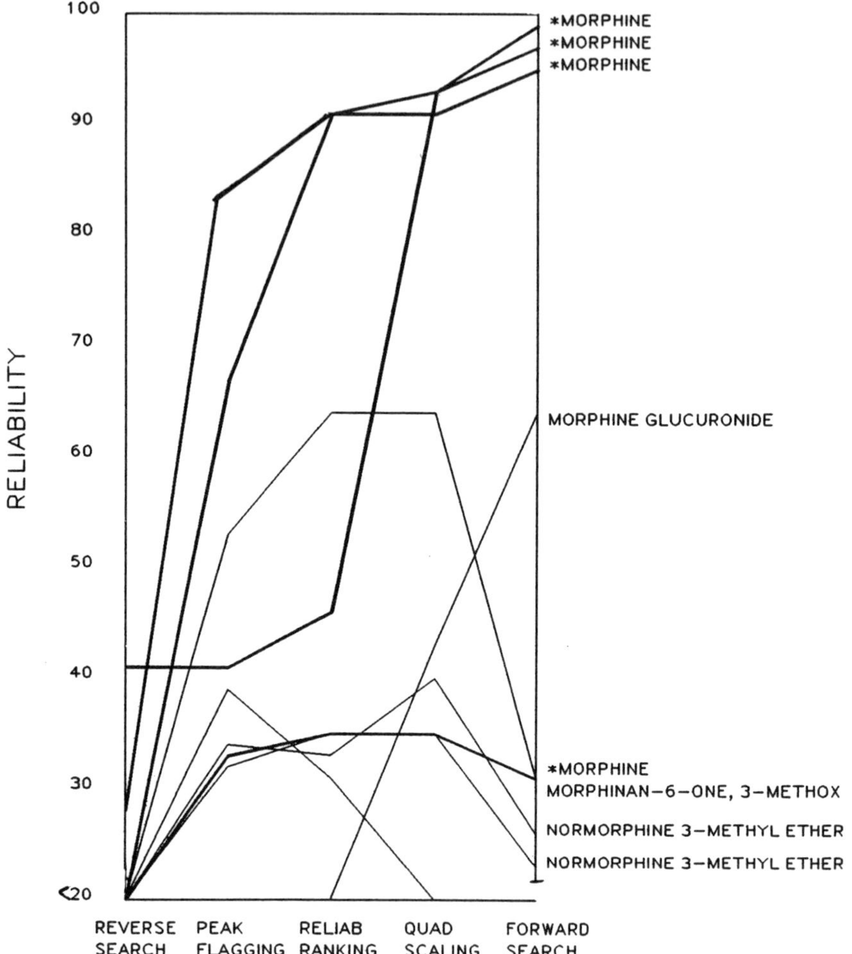

FIGURE 7.8. Effect of PBM improvements on the class IV reliability values for the morphine unknown.

The effect of peak flagging is shown dramatically by the results of both unknown spectra. For the first, the number of correct answers retrieved with reliability values >20% increased from one to three, and for the second unknown this increased from two to four. The first reference spectrum of 2-methyl-2-butanol has peaks at *m/z* 29, 42, 69, and 70 that are relatively much higher than those of the unknowns; *m/z* 70 probably arises from pyrolysis (H_2O loss), so that this would represent an impurity in both spectra. The flagging improvement for reference 3 results from removal of *m/z* 29, 43, and 57; note that the situation for *m/z* 70 is now reversed, and its presence only in the unknown is treated as an impurity (which it is) by PBM reverse search. For morphine, references 2 and 3 show small peaks (although at statistically important masses) not in the unknown spectrum, such as *m/z* 202, 203, 254, and 270, for which flagging improves the matches.

On the other hand, reliability ranking has a significant effect only on the morphine unknown; on average this method improves PBM's capabilities only for matches of reliability values >70%. (Of course, for individual matches this calculation can either increase or decrease the value predicted.) Thus, for the first and third morphine matches the predicted probability of a *wrong* answer is reduced from 17% to 9% and from 33% to 9%, respectively.

Quadratic scaling leads to the retrieval of the fourth correct match for 2-methyl-2-butanol; note that the low-mass (<*m/z* 59) abundances of this spectrum are less than half of those of the unknown spectrum. Similarly, scaling improves the second-best morphine match from 46% to 93% reliability; its low-mass peaks are similarly less abundant than those of the unknown spectrum. Note that the wrong answer for the morphine unknown found with 38% reliability using flagging was not retrieved with the further application of reliability ranking and quadratic scaling.

Forward searching had a generally beneficial, even dramatic, effect. The two top matches of the first unknown had their class IV reliability values increased from 32% to 68% and 42% to 64%, respectively, while the probability of *incorrect* identifications for the first three morphine matches went down from 7%, 7%, and 9% to 1%, 3%, and 5%, respectively. In all cases this was due largely to the fact that such a high proportion of the important unknown peaks (low percent contamination) had been used in matching. Similarly, the decreases in reliability (39% to 24%, 33% to 26%) of the incorrect 2-methyl-2-propanol answers for the 2-methyl-2-butanol unknown, and the decreases (64% to 31%, 40% to 26%, 35% to 23%) of incorrect answers 6, 7, and 8 for the morphine unknown resulted from the opposite effect. However, for this unknown the other incorrect retrieval, morphine glucuronide, number 4, has had its predicted reliability increased dramatically by quadratic scaling and forward searching. Apparently, the spectral peaks due to the glucuronide portion of the molecule appear to PBM with these modifications as resulting from another component of the sample because the morphine portion of the molecule produces spectral features so closely resembling those of morphine itself. If it were known that these unknowns represented pure compounds, this knowledge would have increased the predicted reliability values 1–2% for values >85%, and by ~5% for values of 55–75%.[12]

7.3.2.5. Exact-Mass PBM (EPBM)

A variety of GC/MS instruments are now available in which masses can be measured with sufficient accuracy to define the elemental composition. For example, although both acetyl ($C_2H_3O^+$) and propyl ($C_3H_7^+$) ions have the same nominal mass (m/z 43), their actual masses are 43.1084 and 43.0547, respectively. Because almost all EI reference spectra have only nominal mass data, we have developed[(23)] an EPBM approach for exact-mass unknowns to use our nominal-mass reference file by restricting the elemental compositions of each reference peak to those possible from the elemental formula of the molecule. This increases the weighting values, which have been assigned from statistical correlations; fortunately, the weighting values show quite predictable trends for homologous series of ions. Although a test as extensive as those with normal PBM has not been made on the performance of EPBM because of the restricted number of exact-mass "unknown" spectra available, the number of wrong answers using EPBM is approximately half that obtained using normal PBM.

7.3.2.6. PBM Speed

The time required for matching has been improved to the point that PBM can match several unknown spectra per minute in a commercial GC/MS/computer system,[(10,24)] using background (low-priority) calculations of its dedicated computer. By choosing the spectra corresponding to the top of each GC peak, PBM can thus display matching results from most of the GC-separated components by the end of the GC/MS run. The matching speed has been increased by nearly two orders of magnitude through weighted ordering of the PBM reference file according to the mass of the spectrum's most important peak. This makes it possible to search only the most relevant reference spectra with a minimal loss (<3%) in retrieval recall.[(24)] The matching speed has been enhanced further for low-molecular-weight compounds by a secondary ordering of the mass value of the highest mass peak of significant abundance.[(10)] A recently-introduced commercial PBM system (see Ref. 29) requires 10 s to search the 140,000 reference spectra using a personal computer.

7.3.3. Self-Training Interpretive and Retrieval System (STIRS)

"Intelligence-based" interpretive systems attempt to teach the computer the necessary rules of spectral correlations to elucidate structural information.[(4)] On the other hand, "pattern recognition" techniques use the computer itself to derive and apply such correlations based on which spectra do, and do not, have specific structural features.[6] STIRS can be viewed as a combination of these techniques; as such, it was one of the first intelligence-based systems to be applied to the routine solution of real problems.[5]

7.3.3.1. The STIRS Approach

General information concerning spectral rules and correlations[25] has been used to define 26 data classes that are combinations of masses or mass differences characteristic of different types of structural features. For the unknown data in each of these classes, the computer finds the most similar reference spectra. If the proportion of best-matching references containing a specific substructure is high relative to the proportion in the data base, that substructure should also have a high probability of being in the unknown compounds. The most basic way to utilize these data is for the user to examine the structures of the retrieved compounds to find such common substructures. Computer techniques have been developed to aid this laborious task.

7.3.3.2. Automated Substructure Prediction

For 589 substructures found to be best identified by STIRS, a "reliability value" (i.e., the statistical probability of the prediction) can be assigned using techniques similar to those described above for PBM.[13] The substructure occurrence in the reference file is known, and a statistical study has shown the relationship of this to the actual reliability of the predicted substructures. An alternative approach, which should be applicable to a much wider range of functionalities, utilizes the maximal common substructure (MCS) algorithm.[26] This employs graph theory to compare the connection table descriptions of the molecular structures of two of the best-matching compounds to find the largest overlapping segments, ignoring stereochemistry. In tests of complex systems, such as steroids, the MCS algorithm often does better, but never worse, than experienced chemists. This has not been used routinely for STIRS, however, because of the large computational requirements to compare each best-matching compound against every other best match. Utilization of specialized new computer hardware, however, looks promising for this problem.

7.3.3.3. Prediction of Molecular Weight

Approximately 15% of EI mass spectra in the reference file do not exhibit an appreciable abundance (>0.2%) of the molecular ion. Without supplementary information, such as from chemical ionization mass spectra, this seriously compromises the determination of molecular structure. A special STIRS algorithm for predicting molecular weight[27] reflects the use of "neutral loss" information in interpreting mass spectra.[25] If no molecular ion is present, this is often due to the facile loss of a small neutral fragment, such as the loss of water (mass 18) from alcohols. This program[27] predicts the molecular weight of the unknown compound by adding the actual neutral losses of the best-matching compounds found by STIRS to the masses of the highest unknown peaks. For randomly selected unknown spectra the first selected value of molecular weight was found to be correct in 91% of the cases, and that of the first or second choices to be correct in 94% of the cases. STIRS also identifies combinations of chlorine and bromine atoms present in the peaks of the unknown spectrum by a PBM-type matching of their isotopic abundances.[28] A future improvement of STIRS will incorporate exact mass data, as described above for PBM. For example, this should help in distinguishing assignments such as a large m/z 105 peak as resulting from the substructures $C_6H_5CH(CH_3)$— versus C_6H_5CO—.

7.4. CONCLUSIONS

7.4.1. Performance

No matching or interpretive programs for unknown mass spectra have been shown by thorough tests[21] to be comparable in performance to the present versions of PBM and STIRS. This would appear to be due to PBM/STIRS program features of statistical evaluation (e.g., data weighting and reliability ranking) and of compensation for the serious problems of mass spectra, such as an unusual sensitivity to impurities and wide variations in recorded peak abundances. The basic reason that such adjustments followed by rematching increase the number of correct answers more than incorrect answers is the unusually high information content of mass spectra, allowing some data of the unknown to be ignored or even changed incorrectly without making it a better match for an incorrect compound.

Further, the comprehensive file "Registry of Mass Spectral Data,"[14,16] which contains 140,000 reference spectra, appears to be by far the best for use with both PBM and STIRS. The recent 65% increase in the size of this file decreased the average number of PBM wrong answers to less than half the extent expected statistically. This file is stored easily on a single CD-ROM disk, and PBM searching[24] requires only 2 s on a DEC MicroVAX-2 and 10 s on a personal computer.[29] The most common 10,500 compounds of the file are represented by nearly

28,000 different reference spectra. Using these extra spectra in matching actually reduces the number of PBM wrong answers by 35%, on average.

The growing availability of both PBM/STIRS and the Registry file on commercial MS/computer systems should make possible more extensive evaluations under a broad range of real conditions. Key past improvements resulted from user criticisms. The authors welcome further suggestions.

7.4.2. Future

It would appear that the application of programs such as PBM and STIRS for the identification of unknown mass spectra is entering a new era. The number of scientists using these programs on a daily basis has grown tremendously with the ready availability of smaller "user friendly" GC/MS and LC/MS systems. Further, these users have their main expertise in a much wider variety of scientific areas, and thus have much less formal training and experience in the interpretation of mass spectra. It follows that an important emphasis for future improvements to PBM and STIRS should be in making their results more readily understandable and directly applicable to specific problem areas. With the cooperation of Chemical Abstracts Service,(30) structural images can now be displayed for 83% of the compounds in the current reference file.(14) For STIRS, images of predicted substructures and user-controlled methods for assembling these into probable molecules would also be helpful. Specialized reference files for particular applications, such as pollutants or drugs, would increase search speed. Also, the greatly increased number of scientists using the reference data in a specialized area should make it much more possible to engender collective efforts to improve both the quality and the quantity of such reference data.(14) Obviously, these attributes of the reference file are a primary limitation to the performance of programs such as PBM and STIRS.

7.4.3. Significance

The recent exciting resurgence of analytical chemistry is in substantial measure due to the unusual contribution of such automated techniques to the solution of important societal problems. Here the United States has been in the forefront through its support of basic research in such instrumental methods, its pioneering and innovative instrument manufacturers, and its leadership in the computer field. Scientific progress is always the greatest in those areas that lie between fields of current major interest; by making the results of such modern analytical methods as mass spectrometry readily available to scientists in a wide variety of areas, analytical chemists are now much more widely recognized as catalysts to important progress in the leading areas of technology.

Acknowledgments

Many people contributed to the development of these PBM and STIRS programs and the Registry file of reference mass spectra, including R. Venkataraghavan, K.-S. Kwok, G. M. Pesyna, H. E. Dayringer, I. K. Mun, K. S. Haraki, B. L. Atwater (Fell), J. L. Serum, C. Wesdemiotis, D. R. Bartholomew, W. Staedeli, R. D. Ellis, D. W. Peterson, M. Sharaf, R. B. Spencer, and A. B. Twiss-Brooks. Generous financial support was provided by the National Science Foundation under grants CHE-7910400, CHE-8303340, and CHE-8620293.

REFERENCES

1. K. Levsen, *Org. Mass Spectrom.* **23,** 406–415 (1988).
2. F. W. McLafferty and R. S. Gohlke, *Anal. Chem.* **31,** 1160 (1959).
3. H. S. Hertz, R. A. Hites, and K. Biemann, *Anal. Chem.* **43,** 681 (1971).
4. D. H. Smith, B. G. Buchanan, R. S. Engelmore, A. M. Duffield, A. Yeo, E. A. Feigenbaum, J. Lederberg, and C. Djerassi, *J. Am. Chem. Soc.* **94,** 5962 (1972).
5. K.-S. Kwok, R. Venkataraghavan, and F. W. McLafferty, *J. Am. Chem. Soc.* **95,** 4185 (1973).
6. J. B. Justice and T. L. Isenhour, *Anal. Chem.* **46,** 223 (1974).
7. G. M. Pesyna, R. Venkataraghavan, H. G. Dayringer, and F. W. McLafferty, *Anal. Chem.* **48,** 1362 (1976).
8. L. Dokomos, D. Henneberg, and B. Wiemann, *Anal. Chim. Acta* **150,** 37 (1983).
9. P. Cleij, H. A. van't Klooster, and J. C. van Houwelingen, *Anal. Chim. Acta* **50,** 23 (1983).
10. F. W. McLafferty and D. B. Stauffer, *J. Chem. Inf. Comput. Sci.* **25,** 251 (1985).
11. W. M. Shackelford, D. M. Kline, L. Faas, and G. Kurth, *Anal. Chim. Acta* **146,** 15 (1983).
12. D. B. Stauffer, F. W. McLafferty, R. D. Ellis, and D. W. Peterson, *Anal. Chem.* **57,** 1056–1060 (1985).
13. K. S. Haraki, R. Venkataraghavan, and F. W. McLafferty, *Anal. Chem.* **53,** 386–392 (1981).
14. F. W. McLafferty and D. B. Stauffer, *Wiley/NBS Registry of Mass Spectral Data,* Wiley-Interscience, New York (1989).
15. F. W. McLafferty, D. B. Stauffer, A. B. Twiss-Brooks, and S. Y. Loh, *J. Am. Soc. Mass Spectrom.,* submitted.
16. Electronic Data Division, John Wiley, 605 Third Avenue, New York, NY 10158.
17. S. Abrahamsson, E. Stenhagen, and F. W. McLafferty, *Atlas of Mass Spectral Data,* Wiley-Interscience, New York (1969).
18. E. Stenhagen, S. Abrahamsson, and F. W. McLafferty, *Registry of Mass Spectral Data,* Wiley-Interscience, New York (1974).

19. F. W. McLafferty and D. B. Stauffer, *Int. J. Mass Spectom. Ion Processes* **58,** 139–149 (1984).
20. G. Salton, *Dynamic Information and Library Processing,* Prentice Hall, Englewood Cliffs, NJ (1975).
21. F. W. McLafferty, *Anal. Chem.* **49,** 1441–1443 (1977).
22. B. L. Atwater (Fell), D. B. Stauffer, F. W. McLafferty, and D. W. Peterson, *Anal. Chem.* **57,** 899–903 (1985).
23. R. B. Spencer, S. Y. Loh, D. B. Stauffer, and F. W. McLafferty, *Adv. Mass Spectrom.* **10,** 1213–1214 (1986).
24. I. K. Mun, D. R. Bartholomew, D. B. Stauffer, and F. W. McLafferty, *Anal. Chem.* **53,** 1938–1939 (1981).
25. F. W. McLafferty and F. Turecek, *Interpretation of Mass Spectra,* 4th ed., University Science Books, Mill Valley, CA (1990).
26. M. M. Cone, R. Venkataraghavan, and F. W. McLafferty, *J. Am. Chem. Soc.* **99,** 7668–7671 (1977).
27. I. K. Mun, R. Venkataraghavan, and F. W. McLafferty, *Anal. Chem.* **53,** 179–182 (1981).
28. I. K. Mun, R. Venkataraghavan, and F. W. McLafferty, *Anal. Chem.* **49,** 1723–1726 (1977).
29. Palisade Corporation, 2189 Elmira Road, Newfield, NY 14867.
30. Chemical Abstracts Service, American Chemical Society, P.O. Box 3012, Columbus, OH 43210.

Chapter 8

A Distributed Expert System for Interpretation of GC/IR/MS Data

Bo Curry

8.1. INTRODUCTION

8.1.1. Background

Over the last 25 years, rapid advances in computer technology have led to large increases in data rates of analytical instruments. It is now common for a GC/MS system, for example, to record spectra of several distinct compounds per minute on a continuous basis. Efforts to use computers to selectively retrieve and interpret these vast quantities of data have continued in parallel with their use in data acquisition. Early efforts concentrated on storage and fast retrieval of spectra and chemical structures.[1] Statistical methods of feature selection, cluster analysis, and linear classification were applied to index and classify spectra by compound type.[2] Attempts were also made to automate more traditional interpretation methods as practiced by spectroscopists. This approach, pioneered by the Dendral project,[3] uses spectra–structure correlations embodied in rules of spectral interpretation.

Some reported spectral interpretation systems have concentrated exclusively on the interpretation of a single type of spectrum, although

---

Bo Curry • Hewlett-Packard Laboratories, Palo Alto, California 94304-1209.

most such systems allow the user to specify constraints external to the input spectrum. These systems have employed a wide variety of techniques. Some examples are PAIRS, which uses decision trees to interpret IR spectra,[4] STIRS, which interprets mass spectra using a library correlation method,[5] rule-based systems for ^{13}C NMR interpretation,[6,7] and a rule-based system for interpretation of MS/MS data.[8] Bremser has reported library correlation methods for ^{13}C NMR data.[9] Unsupervised learning machine techniques have also been used to interpret spectra, though with somewhat less success.[10,11] Such systems rarely attempt to determine complete structures, because there is seldom enough information in a single spectrum to make this feasible.

Other groups have attempted to propose complete structures of unknowns. No single source of structural information is sufficient in general to completely determine structure, so these systems combine data from different sources. One type of spectra, however, usually ^{13}C NMR, is treated as the primary information source, and other input sources are used as constraints or filters. A structure generator is used to combine the substructures and constraints derived from the data to form candidate molecules. Prominent examples are DENDRAL,[12] CHEMICS,[13] and CASE.[14]

We have approached the problem of structure elucidation from a somewhat different standpoint. No single technique is sufficiently powerful to justify ignoring complementary methods. Yet the effort involved in developing a competent interpreter for a single type of spectrum is so large that it is unlikely that any one group will be able to develop more than one or two. We have therefore concentrated on the design of a flexible system architecture, which allows multiple spectral data interpreters, of differing origin and design, to cooperate in the analysis of a single unknown compound.

8.1.2. Design Philosophy

The power of computers arises from their speed, tirelessness, and attention to detail, not from their intelligence. It is unlikely that any computer program will in the near future be able to interpret a spectrum as well as a human expert with experience of the general class of compounds being analyzed. Automatic spectral interpretation systems can hope to address automatic screening of "trivial" problems, interpretation of spectra for nonexperts, relief from the more tedious aspects of interpretation (freeing the expert for more interesting tasks), and automatic confirmation of the expert's conclusions. The purpose of our project, which we call Plato, is to develop a software tool which can easily and reliably carry out these functions.

To be successful as a practical tool, a structure eludication system must either automatically solve or significantly expedite the solution of real analytical problems. Rarely is an analyst required to determine the structure of a total unknown from spectral data alone. More often, the task is to confirm a proposed structure. Even when no candidate structure is assumed, the source of the unknown and its prior treatment usually provide important clues to its structure. Also, time and the amount of sample available often preclude extensive use of the more powerful structural methods, such as ^{13}C and 2-D NMR. To be generally useful, an automatic system should be able to make full use of whatever data, spectroscopic or chemical, is available.

Real unknowns are usually approached by teams of experts, each specializing in one of the analytical techniques employed (IR, NMR, MS, etc). The team is led by the chemist/chromatographer, who knows what result he or she expects to find. The chemist generally controls the interpretation and bears final responsibility for the conclusion, knowing what data to collect, what questions to ask, and what constitutes a satisfactory solution.

Plato is an expert system that tries to identify unknown organic compounds from spectra and other available information. Our approach to structure elucidation attempts to mimic, so far as possible, the procedure followed by human experts. The philosophy that has guided the design and development of the Plato system rests on the following assumptions:

1. The system design should impose no limitations on the types of data that can be used to elucidate a structure. No particular type of data should be required or preferred over another in general. The degree to which each data type contributes to the system's conclusions should depend only on its information value in the context of a particular analysis.
2. Different algorithms may be most appropriate for different types of input data.
3. The system should be designed to take maximum advantage of spectral interpretation programs developed by others. For example, it should be easy to modify an existing NMR interpreter so that it functions as an integral component of Plato.
4. The system should permit, but not require, user intervention and control of all stages of an analysis, including spectral assignments, functional group constraints, and substructure assembly.
5. The system should allow maximum flexibility in its presentation of results. It must be easily configurable to solve different types of analytical problems.
6. The interactions among the components of the system and the user should be guided by the way teams of human experts interact.

8.1.3. System Architecture

The above considerations have led us to the system architecture shown in Fig. 8.1. The system consists of four major components: a controller-reasoner module, a librarian, a molecular structure editor, and a number of data experts. The data experts acquire and interpret individual data types, analogous to the team of spectroscopists. The controller plays the role of the chemist. The librarian can access spectral libraries by spectrum, structure, or partial spectrum. The structure editor knows how to assemble, decompose, and display structures.

The system architecture lends itself naturally to a distributed implementation. Plato was originally implemented as a single Lisp program, separated into packages corresponding to the functional modules.[15] More recently, we have divided the system into a set of independent programs. Each data expert module can have its own user interface, and thus can function independently as an interactive spectral interpretation tool. The library search and structure editor modules are also complete, stand-alone programs. These programs run as separate processes on the same or different processors, and communicate using a standard command protocol.

This implementation has several advantages. It allows for concurrency, which could greatly speed up interpretations. It is flexible, so that it is easy to add or replace the data experts. It is easy to adapt most

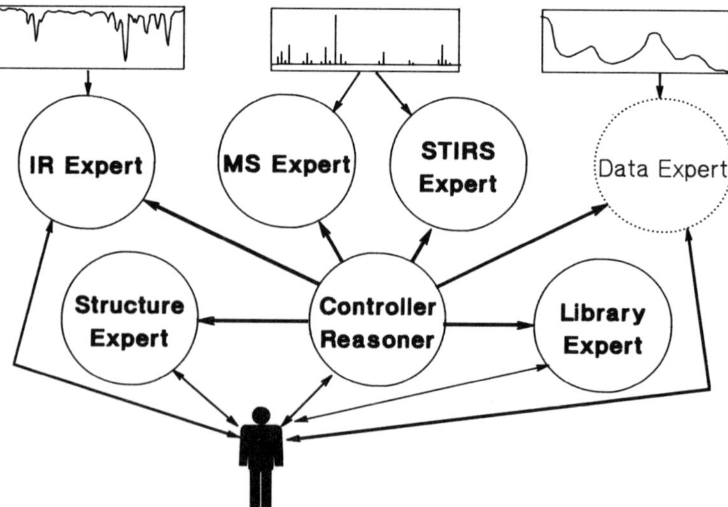

FIGURE 8.1. Schematic of the Plato system architecture. Each circle represents a program module. Queries can be initiated in the direction of the arrows.

spectral interpretation programs to function as Plato data experts, so the best available technique can be used to interpret each spectrum. Since each module is a self-contained process, it is relatively easy to debug and test the components separately. Finally, the architecture allows the use of standardized user interface and communications libraries.

The components of the Plato system can be used to perform several related tasks. The individual data experts can be used interactively to confirm proposed structures, interpret spectra, or as a teaching tool. Data from one source can be used to prune and/or confirm library search results of data from a different source.[(16)] The data experts can also be used as servers by a more specialized expert system. For example, an expert process control system may periodically collect IR spectra and direct specific queries to an IR data expert. The system can be used to classify an unknown by functional group, or to screen batches of samples for compounds containing a particular functionality. When the amount of data permits, complete candidate structures can be proposed for an unknown.

8.2. SPECTRAL DATA EXPERTS

8.2.1. Role of the Data Expert

A "data expert" is a stand-alone program that interprets data bearing on the structure of an unknown organic compound. Each data expert specializes in a single type of data. It incorporates all the knowledge and logic required to interpret its data. A typical data expert can be run independently of the rest of the Plato system in an interactive mode. In this mode the data expert assists the user with an interpretation of the spectrum. When used in conjunction with Plato or another expert system, the data expert functions as a server.

Data experts in the current Plato system include an IR interpreter, a rule-based mass spectrum interpreter, the STIRS mass spectrum interpreter, and an "element expert" that counts the number of each substructure allowed by the molecular formula. In the future, other data experts could interpret ^{13}C NMR, ^{1}H NMR, more exotic NMR spectra, a GC retention indices, chemical stability, UV/vis absorbance, etc. There is no proscription against having multiple data experts interpreting the same spectrum using different methods.

8.2.2. Requirements as Servers

A server is a program that runs continually in the background, either on the chemists's workstation or on a machine accessible over a network.

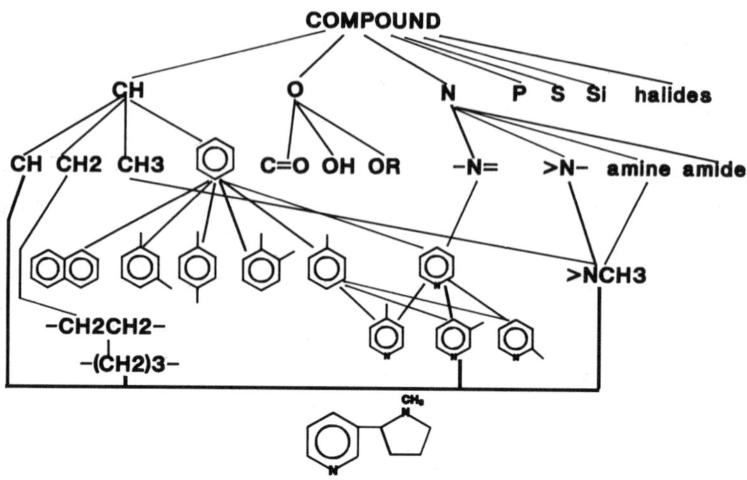

FIGURE 8.2. A subset of the lattice of chemical classes.

When another program has a need for the service provided by a server, it connects to the server, issues requests, and then disconnects. Multiple simultaneous connections to the same server may be allowed. We think of a server as a program that is always available as a consulting expert.

Many interpretive programs described in the literature perform the functions of data experts in interactive mode. There are three requirements for an interpretation program to be able to serve as a consulting expert: (1) it must speak a common chemical language with the other experts and the controller; (2) it must support a standard interface protocol for opening connections and exchanging information; and (3) it must recognize and respond to a standard set of commands and queries. A data expert is most effective as a server if it is able to adapt its interpretation to an externally supplied chemical context.

The common chemical language used by the Plato system is represented by a hierarchy of over 1,000 molecular substructures (Fig. 8.2). All queries, conclusions, and interpretations made by the data experts or the controller are expressed in terms of these predefined substructures. The database of substructures is discussed in more detail in Section 8.3.1.

Each of the Plato modules incorporates a command interpreter, which is a library of functions supporting network interconnections and command-line parsing. These functions work in a Unix* environment under X Windows. The command interpreter supports the definition of

* Unix is a trademark of AT&T.

TABLE 8.1. Data Expert Commands

Command	Arguments	Action
SETUP	Sample ID	Read new sample data. Respond TRUE if data exist.
SUGGESTIONS	None	Return list of likely substructures.
CONTEXT	Description	Update active context for future queries.
TEST	Substructure	Return the confidence that the substructure is absent (0 = can't deny it, −100 = certainly absent).
CONFIRM	Structure	Evaluate the likelihood that the given structure produced the observed data.
SPECIALIZE	Substructure	Return a list of more specific substructures suggested by the data.
RECONSIDER	Substructure	Reconsider the presence or absence of a substructure after contradiction by another data expert.
EXPOUND	Substructure	Return the reasons supporting the presence or absence of the substructure.

application-specific commands, which can then be executed either remotely (by a client process) or locally (via interactive invocation).

To function as a Plato data expert, an interpretive program must include a command processor, and respond appropriately to a standard set of commands from the controller. The minimum set of commands currently required for all data experts is listed in Table 8.1.

In normal operation, a data expert is first asked to set up a new case. The raw data is acquired by the data expert either directly from an instrument, from a library or archive, or from user input. An initial context is then established. The context includes the molecular weight (MW), the molecular formula (possibly expressed as a range of possibilities) and a probability assigned to each of the predefined substructures. The initial default context for all data experts assumes that the MW and the formula are unknown, and all functional groups have their prior probabilities (the frequency of occurrence in a large library). The controller then sends a series of queries to the data expert, interspersed with updates to the context.

A query of the form "TEST pyridine" is interpreted by the data expert to mean "given the current context, give me any evidence you have that the unknown does *not* contain pyridine." The data expert will issue a reply of the form "pyridine 80", which means "based on my data, it is 80% likely that pyridine is absent." The query "SPECIALIZE pyridine" means "given that pyridine is present (probability 1.0 in the current context), rate the likelihood that each subclass of pyridine is present." A typical response might be "2-substituted-pyridine 10, 3-substituted-pyridine 40, disubstituted-pyridine 10." The queries listed in Table 8.1 have proved to be sufficient to express the results obtained from interpreting the IR and mass spectra.

8.2.3. The IR Expert

Of the existing Plato data experts, the infrared expert has received the most attention to date. The IR expert is a rule-based interpreter of gas-phase IR data. Its input data is a set of IR peaks and intensities, a global peak width parameter, and a global noise parameter. Intensities are normalized to percent of maximum absorbance (1–100%). Once these parameters have been determined, the raw IR absorbance spectrum is not used further except for display purposes.

8.2.3.1. Design

After evaluating the strengths and limitations of existing IR interpreters, we decided to develop our own. Since the IR expert was designed to adapt its interpretations to a dynamic, externally supplied context, we decided on a rule-based expert system approach. This has the further advantage of expressing its results and justifications in terms readily accessible to practicing spectroscopists.

The major problem with rule-based interpretation schemes is the difficulty associated with creating, verifying, and maintaining the required knowledge bases. Essentially the same statistical requirements go into identifying and verifying valid IR rules as would be used by a cluster analysis program. We were able to take advantage, however, of the considerable effort that historically has gone into determining and verifying IR correlations.[17,18] Development and testing of new rules is still a laborious process, but the requirement for expert human intervention prevents the kind of spurious correlations that plague automatic statistical methods.[19]

8.2.3.2. IR Rules

The IR rules used by the Plato IR expert are simply descriptions of the frequency and intensity ranges associated with particular vibrational modes of specific functional groups. There are over 1,050 IR correlations, each of which has been verified against 3,009 compounds from the EPA vapor-phase library. With each rule are associated two measures of its information value. The "nprob" is the support given to the proposition that the functional group is absent from the compound, given that there is no band in the rule range. It is computed from the library as

$$\text{nprob}(r) = 1.0 - p(\bar{r} \mid c)/p(\bar{r} \mid \bar{c}) \tag{8.1}$$

where $p(\bar{r} \mid c)$ is the probability that the rule fails for compounds containing functional group c, and $p(\bar{r} \mid \bar{c})$ is the probability that the rule fails for compounds that do not contain the associated functional group c. A high value for nprob(r) indicates a large degree of discrimination when the rule fails. The justification for this equation is discussed in Section 8.4.3.3. The "pprob" is the probability that a compound containing the functional group shows the band, normalized by the width of the rule range. It is computed as

$$\text{pprob}(r) = p(r \mid c). \tag{8.2}$$

Unlike the nprob, the pprob does not directly measure the discriminating power of the rule. Rather, it indicates how much weight should be given to this rule as a possible assignment for the band when it appears.

The wave number ranges of the rules are sufficiently wide that the rules should rarely incorrectly fail. Sometimes bands appear at unusual frequency or intensity because of specific intramolecular perturbations. The C=O stretching vibration, for example, normally appears between 1650 cm^{-1} (conjugated ketones) and 1880 cm^{-1} (carbonate esters). It can appear, however, at a higher frequency in beta-lactones, or at a lower frequency in internally hydrogen-bonded ketones. These "abnormal" cases are noted as exceptions to the normal rule.

When equation (8.2) is used to compare the relative probabilities of competing assignments for a band, the implicit assumption is made that the vibrational bands are equally likely to appear anywhere within the rule range. Because the rule ranges are so broad, this assumption is often not satisfied. Therefore, the rule ranges are divided into up to twelve equal subranges, between 5 and 30 cm^{-1} in width, each of which is assigned a separate pprob. Some authors have assumed a Gaussian distribution of peaks within the range of each rule.[(20)] We have found, however, that the actual peak distributions in our library are multimodal, and adequately described by a step function (see Trulson and Munk).[(21)]

Upon receiving a spectrum, the IR expert assigns all significant peaks to vibrational modes. All possible assignments consistent with the external constraints are noted. At this stage, a peak may have a large number of possible assignments. These are ranked by probability, considering the peak intensity, the pprob for the rule, the probability that the functional group is present in the compound, and the presence or absence of other bands expected for the functional group. The probability that a band b is due to vibration r of class c is given by

$$p(r \mid b) = \text{pprob}(r) * \text{hit}(r, b) * p(c), \tag{8.3}$$

where hit(r, b) is a function (ranging from 0 to 1) describing how well the band agrees with the expected intensity, and $p(c)$ is the prior probability that class c is present in the unknown. The relative probabilities of alternative assignments are used to determine how much support the observation of the band provides to each of the alternative classes (see Section 8.4.2.3). The IR expert refers to these recorded alternative assignments when it responds to queries from either the user or the Plato controller.

8.2.3.3. Results

The IR expert module has undergone extensive testing at three levels. First, the individual rules have been compared against 3,009 compounds from the library. Rules that failed to provide significant information were eliminated. Some rules are highly informative when they fail, but not when they succeed, or vice versa. Examples are methyl-deformations (1,400–1,450 cm^{-1}) and C=C stretches of olefins, respectively. These rules were retained. The results of these tests are reflected in the nprob and pprob values stored with the rules.

Complete interpretations of the IR spectra of the 3,009 library compounds were carried out, and the success rates were recorded for each functional group. Rules for functional groups whose success rates were indistinguishable from (or worse than) random guessing were modified or dropped. Because of correlations among the rules, this can occur even for rules that, when tested individually, appear to provide significant information. Some representative results for functional groups showing high, moderate, and low IR selectivity are shown in Fig. 8.3. Note that the relatively low reliability of SO_2 assertions (30%) is nonetheless far better than would be expected by chance (1.6%), indicating a high information content in the rules for SO_2. On the other hand, the relative high reliability of the methyl group (87%) is only moderately better than would be obtained by random guessing (69%).

The data in Fig. 8.3 were taken at an arbitrarily chosen confidence level of 45%. Figure 8.4 illustrates the dependency of average recall and reliability on the reported confidence level. Plotted results are an average for 134 "primary" IR classes, which were chosen because they each had at least one IR rule with an nprob $\geq$ 90%.

The results of interpreting the IR spectra of over 100 individual compounds were examined in detail. For each compound, the ranking of

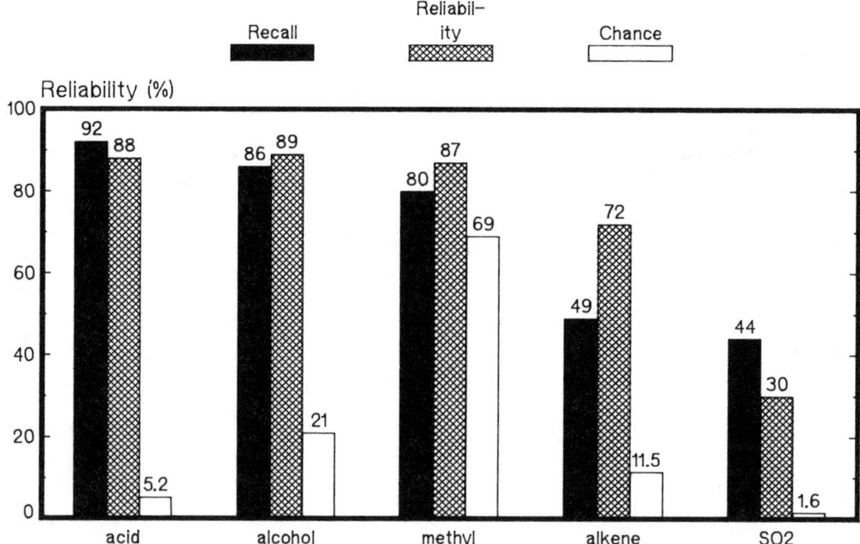

FIGURE 8.3. IR discrimination of five selected substructures among 3,009 library compounds. Substructures assigned a confidence level >45% were scored as reported. Recall (%) = 100 * number correctly reported/total number present. Reliability (%) = 100 * number correctly reported/total number reported. Chance (%) = 100 * total number present/total compounds. Chance is the reliability that would be obtained if the functional group were always reported to be present.

suggested assignments for each of the major peaks was compared with the "correct" assignments (as determined by the author). The correct assignment was ranked first by the program for over 80% of the peaks. In all cases, the correct assignment was among the top five possibilities.

We have compared our results for selected compounds with results from the PAIRS program using its vapor-phase rules.[22] While PAIRS results seem to show somewhat higher recall and lower reliability, the two programs are comparable. We interpret this to mean that we are asymptotically approaching the limits imposed by the inherent distinguishability of functional groups in vapor phase IR spectra. Major improvements in functional group discrimination will therefore require the additional information available from other types of data.

8.2.4. The MS Expert

Our current version of Plato uses STIRS for MS interpretation. Although STIRS does rather well as a stand-alone MS interpreter,[23] it

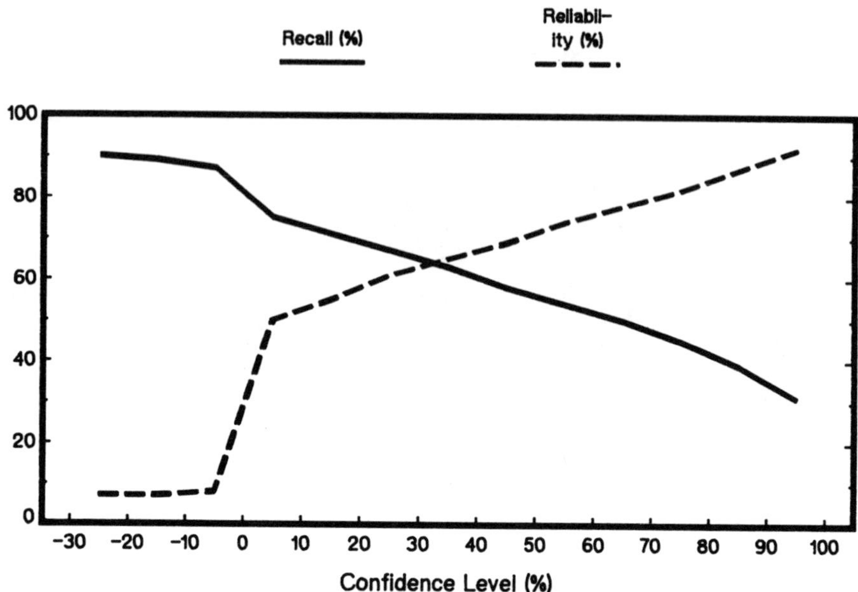

FIGURE 8.4. Average IR recall and reliability for 134 selected substructures having IR rules, as a function of reported confidence level. The 3,009 library compounds were interpreted using IR spectra only. The "average" compound contained 9.6 of the 134 substructures.

has several features that make it unsuitable as a server. It is not sensitive to context, and therefore cannot make use of information provided by other data experts. This is in distinct contrast to the way in which teams of human spectroscopists interpret combined IR and MS data, since the mass spectrum can be reliably interpreted in much greater detail if the major functional groups have been identified. Many substructures are never reported by STIRS because they cannot be reliably detected *ab initio,* although they *can* be reliably identified if interfering groups are known to be absent. Also, STIRS cannot relate its conclusions to particular features of the spectrum, which is important for interactive use. We are therefore developing a rule-based mass spectrum interpreter to supplement STIRS.

The interpretation scheme used by the MS expert follows roughly that proposed by McLafferty.[24] First, an attempt is made to determine the molecular weight. Our current version uses the molecular weight suggested by STIRS. If the molecular ion cannot be identified, and no reasonable guess at the MW can be made from user input or other data, the MS expert cannot proceed. Once a molecular weight is assumed, possible ion formulae are determined for significant ions in the spectrum, using isotopic peak ratios, primary neutral losses, and tables of probable ion occurrence.[25] Assignments are then made for low-mass ion series, characteristic ions, and primary neutral losses.

This work is still at an early stage. Results so far appear promising, and we are in the process of compiling a full-scale knowledge base that will permit comparison with STIRS, both with and without assistance from the IR expert.

8.2.5. Structure Confirmation

A major analytical task Plato is designed to address is the confirmation of proposed structures. A structure is proposed for the unknown either from chemical data, library search results, or the chemist's expectations. The spectra are then examined to see if they could reasonably be accounted for by the proposed structure. It is sufficient to show that the spectra are consistent with the proposed structure. This function is supported by the data experts via the command "CONFIRM structure," where "structure" is a connection table representing a complete molecular structure. Some of the proposed data experts, such as a GC retention index expert, can function only to confirm complete structures.

8.2.6. The Librarian

The Plato librarian acts as a server for the other data experts, as well as for the controller. It is able to retrieve matches for structures or substructures, complete spectra, or particular features of spectra. A possible command to the librarian might take the form "RETRIEVE (and aromatic amide [*IR-peaks 1640:1700])", which is interpreted as "Retrieve all compounds that are both aromatic and amides, and have IR bands between 1640 and 1700 cm^{-1}."

At present, the librarian is not directly invoked by any data experts during the interpretation of an unknown spectrum. It is used by the IR expert to suggest and confirm the validity of new IR rules. Our current vapor-phase IR library is too small (3,009 spectra) to lend statistical validity to the interpretive library correlation techniques used by STIRS for mass spectra and by Passlack and Bremser et al.[26] for condensed-phase IR spectra.

8.3. THE CONTROLLER

The controller module is the heart of the Plato system. In the team of interacting experts, it plays the role of the chemist looking for an answer. The controller sets the goal of the analysis, assembles an appropriate team of experts, and asks them a series of directed queries leading to the goal. As responses to queries are received, they are combined together to form an increasingly detailed solution. This evolving solution forms the current context of the analysis, which is established by the controller and communicated to the data experts. The controller is responsible for maintaining consistency in the solution. It determines when the goal has been reached, or cannot be reached without more data, and terminates the analysis.

8.3.1. Chemical Knowledge

The centralized information repository for the Plato system is a hierarchical lattice of predefined chemical structures and functional groups (Fig. 8.2). Each item in this database defines a "chemical class." There are currently over 1,000 chemical classes defined, including most important organic functional groups composed of the elements C, H, O, N, S, P, Si, F, Cl, Br, and I. The highest-level class, at the root of the lattice, is the class CHEM-CLASS, which includes all organic compounds. Traversing the lattice downward from the root leads to ever more specialized classes with larger and more precise structures. This knowledge is distributed among all of the Plato modules, but is maintained and updated by the controller.

The lattice of chemical classes serves three distinct functions. First, it contains the chemical knowledge of the system. Second, it defines a vocabulary in which statements about structural units can be expressed. Finally, it serves as a central repository of the current state of knowledge about the unknown. The current probability associated with each class forms the chemical context in which the data experts interpret their data. The reasoner also associates with each class a list of the reasons for believing the unknown belongs or does not belong to the class.

The chemical knowledge embodied in the class definition can be interpreted in one of two ways. A chemical class is a structural unit of a molecule, defined by its constituent atoms and their connecting bonds. The CARBONYL class, for example, is defined as the structural unit —C(=O)—. The hierarchy defines the structural relationships among the classes, and records chemical stability and frequency of occurrence (prior probability). When thought of in this sense, the classes are usually referred to as "substructures" or "functional groups." A chemical class may also be thought of as the set of all compounds that contain the defined structural unit. In this sense, the problem of determining the structure of an unknown compound is a classification problem.

The appropriate lanaguage in which to express the results of interpreting a spectrum depends on the type of data being interpreted. Structure elucidation systems described in the literature that define a substructure language have been strongly biased towards a particular data set. For example, ^{13}C NMR data is naturally expressed in terms of local carbon environments,[27] mass spectral data in terms of elemental compositions,[28] and IR data in terms of functional groups. The Plato scheme can simultaneously accommodate structural units expressed in any or all of these forms. There is no restriction on the definition of a chemical class other than the requirement that it have a clearly defined structure. Structural interrelationships are expressed directly by links to other substructures in the lattice. Our current system includes all substructures for which significant IR correlations were found, as well as all the primary STIRS substructures. The knowledge base can be readily extended to include any additional structures required by a different data expert, such as a ^{13}C NMR expert. The advantage of this scheme is that it allows each data expert to express its conclusions in a natural way, and at the same time allows the controller to easily coordinate results from different experts.

It is also necessary to be able to express that the unknown compound is one of a set of possibilities. Features of spectra often cannot be attributed unambiguously to a particular functional group. For example, a strong IR band at 3040 cm^{-1} may be due to the C—H stretch of either an aromatic or an alkene group. The IR expert expresses this by supporting the disjunction "AROMATIC OR ALKENE." The set of compounds represented by such a disjunction is the union of the sets represented by its members. Some disjunctions are based on purely structural relationships among the classes, and are predefined in the knowledge base. For example, if a compound is an alcohol it must be either primary, secondary, tertiary, phenol, or enol.

8.3.2. Interface with Data Experts

When the controller determines that the services of a particular data expert are required to meet its goals, it attempts to establish a communication link with that data expert. Subsequent communications with the data expert take the form of commands issued by the controller (see Table 8.1). Communication with the data experts is always initiated by the controller.

The results obtained by one data expert can influence the interpretations of the others indirectly by modifying the context of the analysis. The context consists of the molecular weight, the molecular formula, and, for each of the chemical classes, a probability that the substructure is present and a range of how many examples of the substructure are present. When answers to queries are returned by data experts, they are checked for consistency with previous conclusions, and then used to update the global context. Changes to the context are broadcast to the data experts, which then presume the new context when responding to subsequent queries.

8.3.3. Search Strategy

The problem of structure elucidation can be viewed as a search through the set of all possible organic compounds for the one that is most consistent with the input data. In practice, the task is usually considerably more limited. The set of possibilities is limited by the origin of the unknown and the methods used to separate it. Also, it is often sufficient to classify the unknown as a member of a small set of compounds.

The Plato controller uses a simple search strategy to carry out this task (Fig. 8.5). The major steps are establishment of an initial context, determination of functional groups, formulation of a more specific hypothesis, iteration until the current hypothesis is specific enough to define a small number of candidate structures, and testing and ranking of candidates. If the data are not sufficient to restrict the possible structures to a manageable number, the analysis terminates and reports the final functional group classification as its answer.

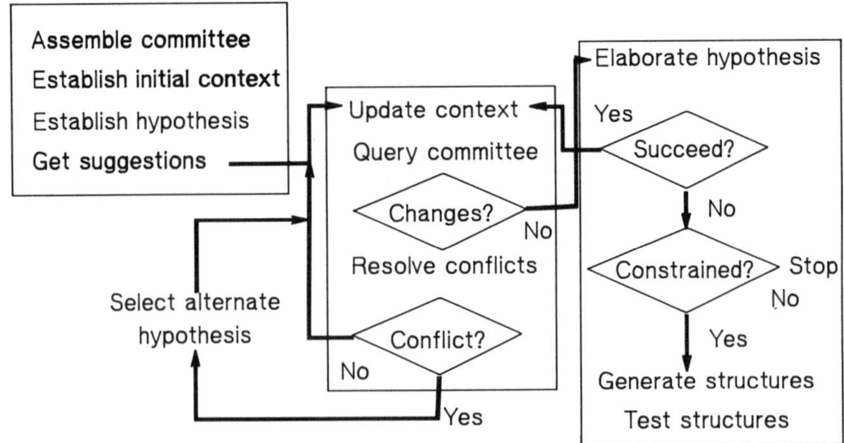

FIGURE 8.5. Schematic of the search strategy used by the controller.

8.3.3.1. Initial Context

The first step in an analysis is to assemble an appropriate group of data experts (the "committee") to address the given problem. The initial context is established from user-supplier information. This should include any information not available to one of the data experts. Initial context information must be supplied by the user in the form used by the data experts: a molecular weight, molecular formula, or functional groups assumed to be present or absent. The user may also specify a default initial context to be used for a series of interpretations. This allows a chemist to express any general prejudices about the unknowns expected to be encountered.

8.3.3.2. Functional Group Analysis

The goal of the functional group analysis is to determine, as precisely as possible, the classes to which the unknown compound belongs; or, alternatively, to determine the most specific substructures present in the compound. The hierarchical organization of the lattice of chemical classes facilitates this goal. In the course of an analysis, the controller follows the lattice downward, collecting evidence at each node until the deepest, most specialized level that can be supported by the data is reached. When no supporting evidence can be collected for any of the children of the most specialized nodes considered so far, the functional group analysis is complete.

As the first step in the functional group analysis, the controller asks each expert on the committee for suggestions. The suggestions returned consist of assertions, with confidence levels, of the presence or absence of chemical classes and disjunctions of chemical classes. After receiving the suggestions from the data experts, the controller passes them on to the reasoner, which makes deductions and updates the global context to reflect the new information. Each data expert is requested to TEST all suggestions received from any other data expert. If there is conflict in the evidence, it is detected by the reasoner and resolved if possible.

At the end of the functional group analysis, each functional group has assigned to it a confidence level, indicating the total support given by all the data experts to its presence in the unknown. The confidence levels range from −100% (substructure definitely absent) to +100% (definitely present). If the data provides enough information about the unknown, some functional groups that were *a priori* unlikely will have strong support, and some that were *a priori* more likely will have strong negative support. Further progress can then be made by *assuming* those classes with strong support to be true and those with strong negative support to be false. These assumptions form the basis for a new, more specific hypothesis about the structure of the unknown. If none of the confidence levels change significantly as a result of the functional group analysis, it indicates that the information content of the data has been exhausted, and more data must be acquired before further progress can be made.

8.3.3.3. Hypotheses

A hypothesis is simply a context presumed to be true. The initial hypothesis is established from the initial context. After completing a functional group analysis, a new hypothesis is formulated by assuming, tentatively, that the most probable functional groups are in fact present. The functional group analysis is then repeated to explore the consequences of this assumption.

For example, assume a low-mass ion series has an ambiguous interpretation in the absence of other information. The initial functional group analysis by the MS expert fails to allot significant support to any functional group. At the same time, however, the IR expert finds strong evidence (70%) supporting the presence of an aliphatic alcohol group. A new hypothesis is generated, which is a refinement of the previous hypothesis with the further stipulation that the compound is an aliphatic alcohol. The functional group analysis is then repeated, with the probability of aliphatic alcohol set to 1.0 in the new context. The mass spectrum expert may now be able to definitely assign the low-mass ion series, and consequently provide support for other functional groups.

When a new hypothesis is generated, it is assigned a probability equal to the product of the probabilities of the parent hypothesis and of its additional assumptions (in the context of the parent hypothesis). At the same time, the contradictory hypothesis is also formulated and assigned a probability. Thus the number of competing hypotheses proliferates as the analysis progresses. The leaves of the hypothesis tree are mutually exclusive and exhaustive of all possibilities.

At any time, the currently most probable hypothesis determines the context. If unresolvable contradictions arise while considering this hypothesis, its probability is decreased (and the probabilities of its rivals are increased). If, as a result of accumulated contradiction, the probability of the current hypothesis falls below that of an alternative, that alternative becomes the current hypothesis, and the context is revised accordingly.

The process of refinement of an hypothesis continues until one of the following conditions occurs:

1. The functional group analysis fails to add significant additional support to any class. This implies there is insufficient data to

further elaborate the hypothesis. If one of the alternative hypotheses seems viable (i.e., has a "sufficiently high" probability), it will be elaborated in turn. Otherwise, the analysis terminates.

2. The sum of accumulated contradictions is severe enough to lower the probability of the hypothesis to less than that of some alternative hypothesis. In this case, an attempt is made to further elaborate the competing hypothesis.
3. There are too many hypotheses, none of which stands out as distinctly most probable. In this case, the analysis terminates.
4. A hypothesis is specified in sufficient detail to make it feasible to generate complete candidate structures. A completely specified molecular structure is the ultimate elaboration of a hypothesis, and terminates the analysis with a solution.

8.3.3.4. Structure Generation

If the current hypothesis being considered by the controller is restrictive enough, it may be feasible to generate a complete list of the molecular structures that are consistent with the hypothesis. This requires that the molecular formula be precisely specified, and that the total number of substructures be sufficiently small. If these conditions are met, the controller passes the molecular formula, along with lists of required and forbidden substructures, to the molecular structure editor. The molecular structure editor uses algorithms modeled on those developed by the DENDRAL group[(12)] to generate complete molecular structures.

The controller can request each data expert to evaluate a candidate structure by issuing the "CONFIRM structure" command. The final list of structures is then ranked, and, if it is short enough, presented to the user for evaluation. If no satisfactory structures can be generated under the constraints of the working hypothesis, it is devalued and the next-ranked hypothesis is considered.

8.3.3.5. Search Results

The final results of the search will be a list of mutually exclusive hypotheses about the structural components of the unknown, ranked by their probabilities of being correct. Each alternative hypothesis has been elaborated as much as the data permit, possibly to the extent of proposing a complete molecular structure. This search scheme extracts the total information contained in the input data, limited only by the power and accuracy of the data experts at interpreting their spectra.

8.4. THE REASONER

8.4.1. Requirements for the Reasoner

The function of the reasoner is to combine information from different sources (data experts) into a single consistent set of beliefs about the structure of an unknown compound. A belief is a statement about the likelihood that a proposition is true. The propositions that Plato can believe are statements about the structure of the unknown compound, expressed in the language of chemical classes. Examples of propositions understood by the Plato reasoner are:

- The compound contains an AROMATIC group.
- The compound is not a member of the ESTER class.
- The compound contains no more than three METHYL groups.
- The compound is either AROMATIC or an ALKENE.

When the controller receives a reply to a query from a data expert, it translates the response into evidence supporting a proposition. This evidence is then passed on to the reasoner, which uses it to create beliefs. Beliefs are created not only for propositions directly supported by evidence but also for propositions that can logically be deduced from the evidence. The reasoner draws conclusions from the evidence and computes the probabilities and degree of support for each of the propositions that is believed.

The reasoner maintains two distinct measures of the likelihood that the unknown compound belongs to a particular chemical class: the *support* for the proposition, derived from evidence, and the *probability* that the proposition is true. These measures each serve a different function in controlling the analysis and evaluating the results.

The concept of support for a proposition derives from the Dempster–Shafer theory of evidence.[29] Support is the degree to which the proposition is supported by evidence. It is expressed in percent, and ranges from 0 to 100%. The negation of a proposition can also receive support, which is independent of the support for the proposition itself. If there is support for *both* a class and its negation, this indicates a conflict in the evidence, which must be resolved by the reasoner to maintain consistency.

Initially, every proposition has zero support. Each data expert, when responding to queries from the controller, attempts to determine the degree to which its data support a particular proposition. A response from a data expert of "AROMATIC 50" is interpreted to mean "based on my data, there is evidence, whose value is 50%, supporting the proposition that the unknown is aromatic." This evidence is then used by the reasoner to increase its belief in the aromaticity of the unknown. If the new evidence contradicts existing beliefs, the originating data expert will be requested to reconsider it, and may withdraw it or lower its value.

The probability of a class is simply the probability that that class occurs in the unknown compound. This probability ranges from 0 to 1, and the sum of the probabilities of a class and its negation always equals 1. Initially, each class is assigned a "prior probability" determined from the statistical occurrence of the compound in the library. These prior probabilities may be adjusted to conform to the occurrence probability expected by a particular user. They do not depend, however, on any evidence pertaining to the particular unknown compound being analyzed. When evidence is obtained, it is used to condition the prior probabilities using Bayesian rules of conditioning, to produce "posterior" probabilities that reflect the impact of the new evidence.

Plato makes explicit use of both probabilities and support, because neither alone is adequate to express the impact of the data. Probabilities alone fail, as pointed out by Shafer (among others), because they cannot distinguish between a belief based solely on prior probabilities (with no supporting evidence) and one based on strong evidence supporting an inherently unlikely proposition. For example, the AROMATIC class (prior = 0.4) with slight evidence against it may have a probability equal to that of SILOXANE (prior = 0.0003) with strong evidence in favor of it. Yet we would wish to propose the siloxane and not the aromatic class as a candidate for our unknown. On the other hand, the probabilities of the classes have a bearing on how we should interpret ambiguous evidence, and on which class to suggest when there is a choice. For example, a strong IR band at 3040 cm^{-1} may be due to a C—H stretch of an aryl or vinyl group, but it could also be due to a terminal epoxy group. Since the epoxy group is a priori unlikely, we don't want to give it support equal to the more likely possibilities, even if it could (if present) explain the peak equally well. The IR expert therefore uses the current class probabilities to determine likely assignments for IR peaks.

When both positive and negative evidence from diverse sources is collected, it is inevitable that conflicts will arise. One of the major advantages of an expert system over a more conventional algorithmic interpretation scheme is that the expert system, which maintains a record of the reasoning steps leading to its conclusions, is able to intelligently resolve contradictions. Contradictions are detected first by the reasoner, when it records support for a proposition whose negation is already supported.

If the total of support for a proposition and its negation exceeds 100%, this is taken as a sign that some of the evidence has been misinterpreted. When the reasoner detects this situation, it requests that the conflicting evidence be revaluated by the data expert that originally proposed it. Sometimes such evidence is in error because of missing or faulty rules applied by one of the experts, or because of unusual perturbations obtaining in the particular compound being analyzed. The reevaluation may result in a reduction of the value of the evidence, in its elimination and replacement by new evidence supporting alternative propositions, or by no change in the evidence. If the conflict cannot be resolved by this procedure, it is stored as evidence against the current hypothesis.

8.4.2. Belief Systems

8.4.2.1. Dempster–Shafer Theory

In our discussion of beliefs we have adopted the terminology and philosophical outlook of the Dempster–Shafer theory of evidence. In this theory, a proposition is believed to the degree to which it is supported by evidence. In the absence of evidence, every proposition has zero support. As evidence accumulates, it is combined with previous evidence to modify the total support for each proposition. Evidence is combined using the rules of deduction shown in Table 8.2. The Dempster–Shafer theory is attractive because of its consistent treatment of belief functions and conflicts, its probabilistic combining rule, and its explicit relationship to Bayesian probability theory.

The probabilistic Dempster–Shafer combining rule assumes implicitly that the evidence being combined is independent. That is, membership of the unknown compound in a class does not affect its probability of membership in other classes (except when such dependence is explicit in the subset relationships expressed in the hierarchy). This assumption is clearly not fulfilled in all cases; for example, the probability that the compound lacks saturated C—H groups is clearly larger if it is known to be aromatic. As a rule, however, this is the most reasonable assumption to make. See Gaines[30] for a lucid discussion of

TABLE 8.2. Deductions Made by the Reasoner

$e1 \equiv$ the value of evidence $e1(A)$ supporting proposition A	
If $A \subset B$ then:	
$e1(A) \Rightarrow B$,	Support value $e1$
$e2(\bar{B}) \Rightarrow \bar{A}$,	Support value $e2$
$e1(A)$ and $e2(A) \Rightarrow A$,	Support value $e1 + e2 - e1 * e2$
$e1(A \vee B)$ and $e2(\bar{B}) \Rightarrow A$,	Support value $e1 * e2$
$e1(A)$ and $e2(\bar{A}) \Rightarrow$ FALSE,	Support value $e1 * e2$

the assumptions about the world involved in the choice of different combining rules.

Computational intractabilities in the original Shafer formulation were partially addressed by adopting the approximations proposed by Gordon and Shortliffe.[31] Further approximations were required for Plato because the chemical classes are not mutually exclusive. It can be shown that the rules of deduction and evidence combination shown in Table 8.2 completely describe the most restrictive system, consistent with the full Dempster–Shafer formulation, which avoids attributing belief value to undefined sets of comounds (i.e., to intersections of defined chemical classes). These rules guarantee that the final support given to any proposition is independent of the order in which the evidence is processed, and that no proportion receives more support than it would receive under the full Dempster–Shafer theory (although it might receive less).

8.4.2.2. Bayesian Probabilities

Besides building up a Dempster–Shafer support function, the reasoner maintains a probability for each class believed. These probabilities are to be interpreted as the probability of membership of the unknown in each class. They comprise the chemical context used by the data experts when interpreting their data. The probability function used by Plato bears the same relationship to a true Bayesian belief function as our support function does to a true Dempster–Shafer support function; namely, it is incomplete in that all primitive propositions are not defined, so that the probabilities of defined propositions do not sum to 1 as they would for a canonical probability function. Given the independence assumption, however, a true probability can be computed for each intersection of classes if required.

In contrast to the support function, $p(x) + p(\bar{x}) = 1$, and the probabilities are additive. Also, a probability can be computed for each subset of compounds, whether or not any evidence bearing on that subset has been collected. Each class starts with a prior probability, equal to the fraction of compounds in a large test library that are members of the class. This probability is updated as evidence is accumulated, using equations derived from Shafer's equation (9.4) relating support functions to Bayesian probabilities[29]:

$$p(c \mid e1) = p(c)/\{1 - e1 * [1 - p(c)]\} \tag{8.4}$$

$$p(c \mid e2) = p(c) * (1 - e2)/[1 - e2 * p(c)] \tag{8.5}$$

where $p(c \mid e)$ is the posterior probability of class c given evidence e, $p(c)$ is the prior probability, $e1$ is the value of evidence supporting c, and $e2$ is the value of evidence supporting the negation of c.

8.4.2.3. What is Evidence?

The value of a piece of evidence in the Dempster–Shafer theory is supposed to represent the degree of support the observed data lend to a proposition. Once evidence has been produced by a data expert, its use by the reasoner to modify beliefs and probabilities is straightforward. The more interesting question is: how are the support values to be generated in the first place? There is no "right" answer to this question, which leads to philosophical inquiries into the meaning of inductive reasoning. There are, however, reasonable empirical assumptions which can be shown to be self-consistent and to exhibit qualitatively correct behavior.

The assumptions used by the IR expert to determine evidence values are based on the discussion in Chapters 9–11 of Shafer.[29] The values of negative evidence are designed to be consistent with equation (8.5). When the IR expert fails to find a band predicted by rule r of class c, it produces evidence supporting the proposition that the unknown does not belong to class c. Because of the nature of infrared spectra, the value of this evidence is assumed to be independent of the presence or absence of other classes. The value given to the evidence against c is just the nprob of rule $r(c)$ from equation (8.1). This nprob is the support value which, when used in equation (8.5) to condition the Bayesian prior probability of class c, produces the actual conditional probability observed in the library. To show this, solve equation (8.5) for $e2$, the value of the negative evidence upon the observation of $\tilde{r}$ (the failure of rule r):

$$e = [p(c) - p(c \mid \tilde{r})]/p(c) * [1 - p(c \mid \tilde{r})] \tag{8.6}$$

Equation (8.6) can be rearranged using Bayes' rule

$$p(c \mid \tilde{r}) = p(\tilde{r} \mid c) * p(c)/p(\tilde{r}) \tag{8.7}$$

and the identity

$$p(\tilde{r}) = p(c) * p(\tilde{r} \mid c) + p(\tilde{c}) * p(\tilde{r} \mid \tilde{c}) \tag{8.8}$$

to produce equation (8.1), showing that the nprob directly expresses the evidential value of rule failure.

It is a bit more complicated to determine the support that should be given to the presence of a class when a band is observed that could be assigned to a vibrational band of the class. This is because what is observed is merely the presence of a band, and not its correct assignment. So, the effect of competition among the possible alternative assignments must be taken into account. We first invoke the closed-world assumption; that is, each significant IR band must be assignable to one of our known rules. The possible vibrational assignments are then ranked according to their probability of producing the band (in the universe of compounds defined by the current context). The assignment probabilities are used to generate a set of consonant evidence, as discussed in Chapters 10 and 11 of Shafer. For example, assume that an IR band has the three possible assignments: $r1$ ($p = 0.6$), $r2$ ($p = 0.35$), and $r3$ ($p = 0.05$). Under the closed-world assumption, these probabilities must sum to 1. Then, following Shafer [equation (11.3)], we generate three pieces of evidence:

$e1(r1 \vee r2 \vee r3) = 100\%$ (there is 100% support for one of the choices)

$e2(r1 \vee r2) = 1 - 0.05/0.6 = 92\%$

$e3(r1) = 1 - 0.35/0.6 = 42\%.$

Shafer, in his discussion, gives a compelling rationalization for this method of distributing support among competing explanations for an observation. An additional advantage of the method is that it allows for straightforward adjustment of the levels of support when the context changes (and thus the assignment probabilities change).

8.4.3. Explanations

The reasoner also includes an explanation module. The explanation module backtraces the lines of reasoning responsible for each conclusion, to the level of individual pieces of evidence as supplied by the data experts. It then calls the appropriate data expert to explain the data-specific aspects of how that evidence is supported by the input data. This capability is used to explain to a user the reasons for the system's conclusions. It is also used when conflicts are detected to assign blame for the contradiction.

8.5. CONCLUSIONS

We have designed an architecture we believe can support a powerful and flexible structure elucidation and confirmation system. Our design decisions have been based on considerations of the problems frequently faced by teams of human spectroscopists, and on the techniques used to solve those problems. The distributed model works well with the limited number of data experts we have currently implemented. Work is in progress to complete a rule-based mass spectrum interpreter and integrate it with the existing modules. We are also investigating the adaptation of interpreters developed by others to function as data experts. The results of this research will provide a stringent test of the capability of the system to scale up smoothly to handle larger numbers of interacting modules.

Acknowledgments

I would like to acknowledge the contributions of Reed Letsinger to the controller and reasoner, Craig James to the molecular structure editor, and David Stranz and Jerry James to the IR data expert. All of these individuals, and others, contributed to the current system design. Cathy Cummings compiled and tested a large fraction of the chemical classes and IR rules knowledge bases. John Michnowicz has provided consistent support and direction for this research.

REFERENCES

1. Z. Hippe and R. Hippe, "Computer Retrieval of Spectral Data," *Appl. Spectrosc. Rev.* **16,** 135–186 (1980).
2. J. Zupan, ed., *Computer-Supported Spectroscopic Databases,* Ellis Horwood, West Sussex (1986).
3. D. H. Smith, N. A. B. Gray, J. G. Nourse, and C. W. Crandell, "The DENDRAL Project: Recent Advances in Computer-Assisted Structure Elucidation," *Anal. Chim. Acta* **133,** Computer Techniques and Optimization, 471–497 (1981).
4. H. B. Woodruff and Graham M. Smith, "Computer Program for the Analysis of Infrared Spectra," *Anal. Chem.* **52,** 2321–2327 (1980).
5. F. W. McLafferty, B. L. Atwater, K. S. Haraki, K. Hosakawa, I. K. Mun, and R. Venkataraghavan, *Adv. Mass Spectrom.* **8,** 1564 (1980).
6. N. A. B. Gray, "Applications of Artificial Intelligence for Organic Chemistry: Analysis of Spectra," *Artificial Intelligence* **22,** 1–21 (1984).
7. J.-E. Dubois, M. Carabedian, and I. Dagane, "Computer-Aided Elucidation of Structures by C-13 NMR," *Anal. Chim. Acta* **158,** 217–233 (1984).
8. K. P. Cross, P. T. Palmer, C. F. Beckner, A. B. Giordani, H. G. Gregg, P. A. Hoffman, and C. G. Enke," Automation of Structure Elucidation from MS/MS Data," in: *Artificial Intelligence Applications in Chemistry* (T. H. Pierce and B. A. Hohne, eds.), pp. 321–336, ACS Symposium Series 306, Washington, D.C. (1986).
9. W. Bremser and W. Fachinger, *Magn. Res. Chem.* **23,** 1056–1071 (1985).
10. B. R. Kowalski, P. C. Jurs, T. L. Isenhour, and C. N. Reilley," Computerized Learning Machines Applied to Chemical Problems," *Anal. Chem.* **41,** 1945–1949 (1969).
11. L. Domokos, I. Frank, G. Matolcsy, and G. Jalsovszky, "Pattern Recognition Applied to Vapor-Phase Infrared Spectra," *Anal. Chim. Acta* **154,** 181–189 (1983).

12. R. E. Carhart, D. H. Smith, N. A. B. Gray, J. G. Nourse, and C. Djerassi, "GENOA: A Computer Program for Structure Elucidation Utilizing Overlapping and Alternative Substructures," *J. Org. Chem.* **46,** 1708–1718 (1981).
13. H. Abe, I. Fujiwara, T. Nishimura, T. Okuyama, T. Kida, and S. Sasaki, "Recent Advances in the Structure Elucidation System, CHEMICS," *Computer-Enhanced Spectrosc.* **1,** 55–62 (1983).
14. M. E. Munk, M. Farkas, A. H. Lipkus, and B. D. Christie, "Computer-Assisted Chemical Structure Analysis," *Mikrochim. Acta* (*Wien*) 1986 II, 199–215 (1987).
15. B. Curry, "An Expert System for Organic Structure Determination," in: *Artificial Intelligence Applications in Chemistry* (T. H. Pierce and B. A. Hohne, eds.), pp. 350-364, ACS Symposium Series 306, Washington, D.C. (1986).
16. D. A. Laude, Jr., J. R. Cooper, and C. L. Wilkins, "Analysis of GC/MS Library Search Results with Edited and Quantitative C-13 NMR Spectra," *Anal. Chem.* **58,** 1213–1217 (1986).
17. L. J. Bellamy, *The Infared Spectra of Complex Molecules,* Chapman and Hall, London (1975).
18. R. A. Nyquist, *The Interpretation of Vapor-Phase Infrared Spectra,* Vol. 1, Sadtler Research Labs, Philadelphia (1984).
19. M. F. Delaney, P. C. Denzer, R. M. Barnes, and P. C. Uden, "Pattern Recognition Approach to Vapor-Phase Infrared Spectra Interpretation for Gas Chromatography," *Anal. Lett.* **12**(A9), 963–978 (1979).
20. Y. Ishida and S. Sasaki, "Application of Membership Function to Automated Structure Analysis of Infrared Spectra of Organic Compounds," *Computer-Enhanced Spectrosc.* **1,** 173–184 (1983).
21. M. O. Trulson and M. E. Munk, "Table-Driven Procedure for Infrared Spectrum Interpretation," *Anal. Chem.* **56,** 2137–2142 (1983).
22. S. A. Tomellini, J. M. Stevenson, and H. B. Woodruff, "Rules for Computerized Interpretation of Vapor-Phase Infrared Spectra," *Anal. Chem.* **56,** 67–70 (1984).
23. H. E. Dayringer and F. W. McLafferty, "Computer-Aided Interpretation of Mass Spectra," *Org. Mass Spectrom.* **11,** 543–551 (1976).
24. F. W. McLafferty, *Interpretation of Mass Spectra,* University Science Books, Mill Valley, CA (1980).
25. F. W. McLafferty and R. Venkataraghavan, *Mass Spectral Correlations,* 2nd ed, Advances in Chemistry Series 40, American Chemical Society, Washington D.C. (1982).
26. M. Passlack and W. Bremser, in *Computer Supported Spectroscopic Databases* (J. Zupan, ed.), Chichester, U.K., Ellis Horwood Ltd., (1986), pp. 92–116.
27. M. E. Munk, R. J. Lind, and M. E. Clay, "Computer-Mediated Reduction of Spectral Properties to Molecular Structures," *Anal. Chim. Acta* **184,** 1–19 (1986).
28. K. S. Haraki, "The Self-Training Interpretive and Retrieval System: Improved Substructure Perception," Ph.D. Thesis, Cornell University (1980).
29. G. Shafer, "A Mathematical Theory of Evidence," Princeton University Press, Princeton, NJ (1976).
30. B. R. Gaines, "Fuzzy and Probability Uncertainty Logics," *Information and Control* **38,** 154–169 (1978).
31. J. Gordon and E. H. Shortliffe, "A Method for Managing Evidential Reasoning in a Hierarchical Hypothesis Space," Report HPP 84-35, Heuristic Programming Project, Departments of Medicine and Computer Science, Stanford University, Stanford, CA 94305 (1984).

Chapter 9

Computer-Aided Solutions to ^{13}C NMR Spectral Interpretation Problems

Gary W. Small, Scott E. Carpenter, and Malcolm K. McIntyre

9.1. INTRODUCTION

The difficult step in the solution of many structure-elucidation problems is the determination of an exact structure once an overall structural backbone is known. Carbon-13 nuclear magnetic resonance spectroscopy (^{13}C NMR) is a powerful tool in the solution of such problems due to its sensitivity to subtle differences in carbon atom environments. The use of ^{13}C NMR spectra to discern an exact structure is a time-consuming task, however, because small differences in chemical shifts can be measured more easily than they can be assigned to specific structural units. Alternative instrumental techniques such as mass spectrometry may or may not be of help, depending on the type of structure involved. To simplify the interpretation of complex ^{13}C NMR spectra, chemists most often turn to specialized NMR experiments designed to provide more specific structural information. In this regard, a variety of decoupling experiments and two-dimensional NMR techniques have proven to be of great use.[1]

Gary W. Small, Scott E. Carpenter, and Malcolm K. McIntyre • Department of Chemistry, The University of Iowa, Iowa City, Iowa 52242. All correspondence should be addressed to Gary W. Small.

A practical limitation of many *experimental* solutions to complex spectral interpretation problems is the requirement for extensive data collection. In many academic and industrial NMR facilities, the available spectrometers are in constant demand. Users may have difficulty obtaining enough instrument time to perform desired experiments. It would be of great use in such cases if alternative spectral interpretation tools were available that did not require extensive data collection.

The motivation for one such alternative strategy is found in the answer to the following question: Why can't sufficient information be obtained from a standard broadband-decoupled ^{13}C NMR spectrum to allow exact structural assignments to be made in complex cases? The answer lies in the realization that even the most experienced spectroscopist cannot in general assign exact structural significance to specific chemical shifts. General structural conclusions can be drawn, but it is often very difficult to distinguish between possible candidate structures that are highly similar. A strength of computers, however, is their ability to store detailed information and to make detailed comparisons. Thus, it can be argued that computer-based spectral interpretation tools have the potential for allowing the spectroscopist to extract more structural information from the results of simple ^{13}C NMR experiments than can be done manually.

This chapter describes the development of two computer-based strategies for aiding the interpretation of ^{13}C NMR data. Both techniques are designed to aid the spectroscopist in extracting the maximum amount of information from broadband-decoupled spectra, thereby decreasing the need for additional experiments. The techniques are very different, yet complementary. Each will be described in turn, followed by a comparison of their strengths and weaknesses.

9.2. OVERVIEW OF COMPUTER-BASED SPECTRAL INTERPRETATION STRATEGIES

Computer-based methods designed to help interpret complex spectra can be divided into two general categories. Both categories of methods make use of a data base of known spectra. In effect, this data base serves the same purpose as the experience and knowledge of the trained spectroscopist. It represents a collection of known relationships between chemical structures and their corresponding spectral representations. Computer-based spectral interpretation tools differ in the manner in which they use spectral data bases.

As their name implies, direct data base methods make direct use of the spectral data base. Figure 9.1 is a flow chart that describes how these

FIGURE 9.1. Outline of direct data base methods for automating spectral interpretation.

methods work. Employing the spectrum of the compound whose structure is sought, the data base is interrogated to find similar spectral information. Discovered spectral similarities are then used to suggest candidate structures or structural fragments. This type of approach is typically termed a spectral library search.

Indirect data base methods make use of the data base in generating a spectral interpretaion algorithm. Once this algorithm is developed, however, the actual spectral data base is no longer used. While interpretation algorithms can have many forms, each mimics in some manner the thought processes of an expert spectroscopist. Effectively, the indirect methods use the data base to encode general spectra–structure relationships. Figure 9.2 is a flow chart describing these methods. Inputs to the interpretation algorithm are the spectrum of the unknown compound and a set of structural hypotheses. These hypotheses are typically a list of possible candidate structures or structural fragments. The interpretation algorithm makes a recommendation regarding the validity of these hypotheses. Most indirect data base methods can be placed in the following categories: artificial intelligence,[2] pattern recognition,[3] and spectrum simulation.[4]

Direct and indirect data base methods are complementary because the output of the direct methods (i.e., candidate structures) can serve as input to the indirect approaches. Furthermore, the availability of both types of strategies overcomes weaknesses in each. Direct methods are limited by the size and composition of the spectral data base, while

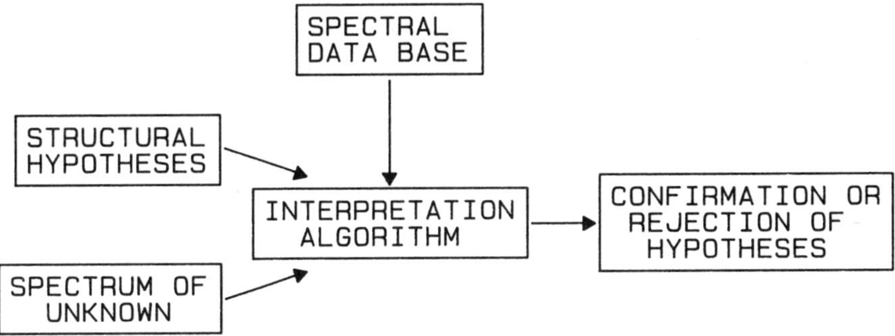

FIGURE 9.2. Outline of indirect data base methods for automating spectral interpretation.

indirect methods are limited by the requirement that candidate structures must be generated first. Thus, while the results of a library search may not specifically identify an unknown, structural features may be suggested that can lead to the generation of plausible candidate structures. The candidate structures can then be evaluated through the indirect data base approach. In the following sections, these concepts will be illustrated through a discussion of one direct data base method for ^{13}C NMR data and one complementary indirect method.

9.3. ^{13}C NMR LIBRARY SEARCHING

In a computer-based spectral library search, numerical procedures are used to compare the spectrum of an unknown to each member of a spectral data base. Those spectra judged most similar to that of the unknown are reported as candidate structures. The goals for a library search are straightforward. First, if the library contains the unknown, the corresponding spectrum should be retrieved as the closest match. Second, if the unknown is not contained in the library, the retrieved list of nearest matches should reflect, to the degree possible, the compounds in the library whose structures are most similar to the unknown.

9.3.1. Overview of Library Searching Methods

Library search methods have been used most often with infrared and mass spectral data. ^{13}C NMR searches have been limited by the lack of available spectral data bases and by inherent characteristics of ^{13}C NMR data that complicate library searching. Chemical shifts of identical carbon atoms can vary considerably between spectra due to differences in experimental conditions and solvent effects. Furthermore, the introduction or removal of structural symmetry can dramatically change the number of chemical shifts observed.

Despite these complications, several library searches have been described for ^{13}C NMR data. Binary (peak/no peak) searches have been used by Zupan et al.[(5)] and Uthman et al.[(6)] Clerc and coworkers have made use of a binary representation or "signature" in an attempt to represent underlying chemical structure in addition to shift information.[(7)] Mlynárik et al.[(8)] have reported a search that incorporates an adjustable shift tolerance value which improves shift matchings by accounting for movement of shifts due to structural, experimental, or solvent effects. Zippel et al.[(9)] have reported an algorithm that includes an adjustable shift tolerance, tolerance "windows" for the spectral matching procedure, and calculation of a matching factor that considers the relationship between the unknown spectrum and a library spectrum based on the number of matching signals.

While these previously reported procedures make some allowances for spectral variation, many do so only through manually adjustable tolerances. The motivation for our library search work is the realization that, as chemists, we automatically correct spectra for variations when making a visual spectral comparison. In effect, we give spectra the best opportunity to match by performing the best possible mapping of one spectrum onto another. The focus of our research, therefore, has been to develop an automated library-searching algorithm that encodes the same decision-making processes involved in visual ^{13}C NMR spectral matching.

9.3.2. Overall Search Design

A flowchart illustrating the overall design of our search algorithm is given in Fig. 9.3. A clustering algorithm is employed to separate the lines of an unknown spectrum into subsets or clusters based on the number of lines and their relative proximity to one another. The clustering results are used to divide the unknown spectrum into partitions. If a library spectrum is similar to the spectrum of the unknown, it should have the same number of clusters with similar or identical partition boundaries. Therefore, before each comparison, each library spectrum is divided into the same number of partitions with exactly the same partition boundaries as the unknown spectrum.

The next step of the search is to compare the lines in each partition of the unknown spectrum to the lines in the corresponding partition of each library spectrum. Within each partition, a match tolerance, or slide factor, is computed. The slide factor is the maximum distance that two lines can be separated in order to be still classified as matching.

The line mapping section of the search algorithm provides information about all possible line matches within the corresponding unknown and library partitions. Every line of one is compared to every line contained in the other and is stored in a matrix format representing all possible match combinations.

The last phase of the search algorithm involves scoring each of the selected best matches. Penalties are assigned to matching lines based on the closeness of the matches. Missing and extraneous lines also acquire a penalty. Each comparison between corresponding partitions is scored separately. The partition scores are summed, producing an overall score that describes the degree of similarity between the compared spectra. A score of zero indicates an exact spectral match. Increasing magnitude in the score value indicates increasing spectral dissimilarity.

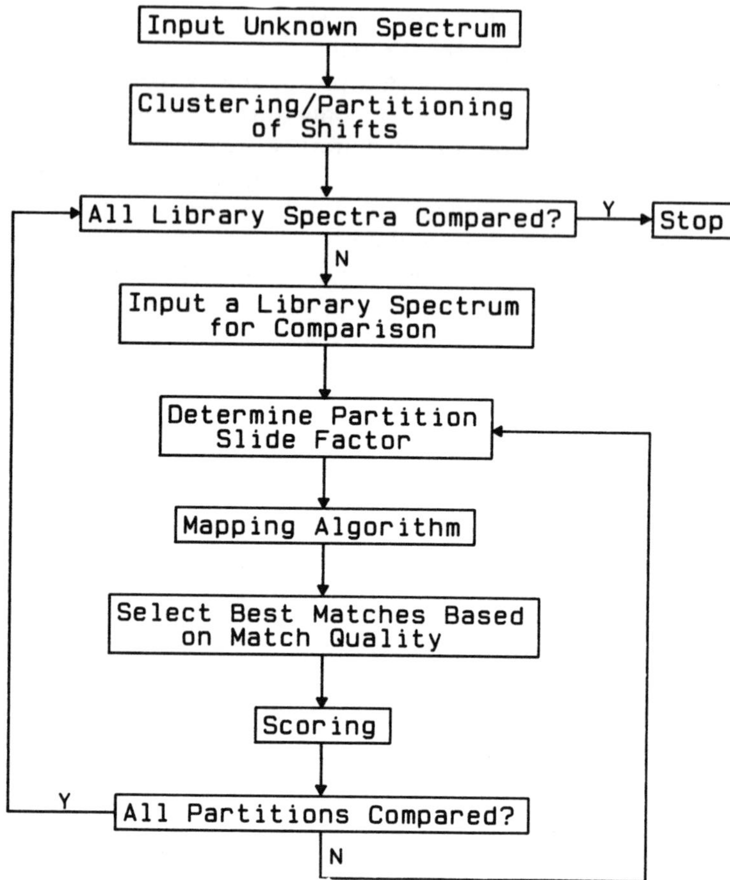

FIGURE 9.3. Flow chart illustrating search design.

9.3.2.1. Clustering and Partitioning

Clustering methods are numerical procedures for grouping similar objects. The algorithm developed by W. D. Fisher and described by Hartigan[10] was found to be particularly well-suited to the sorted one-dimensional data that comprises a ^{13}C NMR spectrum. The ^{13}C NMR spectrum of 1-hexyne contains six chemical shifts (84.5, 68.1, 30.7, 21.9, 18.1, 13.5 ppm), and serves as a useful example in explaining the principles of the Fisher algorithm. For this spectrum, the maximum number of possible clusters is equal to six, where each cluster contains one spectral line. However, this arrangement offers no information concerning the degree of similarity between neighboring lines. The lines may also be grouped into five clusters. There are five possible ways to group six objects into five clusters. Each possible five-cluster arrangement is evaluated by use of an error function. That arrangement producing a minimum value of the error function is taken as the best. The error function employed is the summation of the sums of squares calculated for each cluster. The sum of squares for a given cluster is simply the sum of the squares of the deviations of the lines within the cluster from the cluster mean. Analogously, all groupings of the six lines into four, three, and two clusters must be investigated. Upon completion, the Fisher algorithm provides a list of the best arrangements of the six spectral lines into one through six clusters.

To compare each clustering arrangement, a value representing the percent reduction of the total sum of squares can be computed using the standard percent-of-total calculation and the sum of squares of the one-cluster arrangement as the total sum of squares value. For many example spectra, it was experimentally found that the first encountered clustering arrangement representing a reduction in the sum of squares of approximately 85% or greater was the one that provided a result similar to that obtained by visual grouping of the spectral lines. Application of this procedure to the 1-hexyne chemical shifts reveals there are two clusters of spectral lines. The upfield cluster contains four lines corresponding to the four sp^3-hybrid ^{13}C atoms, and the downfield cluster contains two lines corresponding to the two sp-hybrid ^{13}C atoms. This represents the same clustering arrangement that would be naturally applied in a visual clustering of the chemical shifts.

Only the unknown spectrum undergoes cluster analysis. The resulting clusters are used to form partitions. A partition is a spectral region with assigned upper and lower boundaries which contains one cluster. Partitions are adjacent and consecutively cover the nominal ^{13}C NMR spectral range of -100 to 250 ppm. Setting the partition boundaries first involves calculating the mean and standard deviation of each cluster. A distance equal to one standard deviation is subtracted from the first shift of the cluster and added to the last shift of the cluster, forming an initial set of boundaries. Since it is desirable to have one boundary between two adjacent clusters, overlaps or spaces between the initial boundaries of the two are resolved, assigning the shared boundary to be the average of the initial boundary values. The lower boundary of the first partition is defaulted to be -100 ppm, and the upper boundary of the last partition is defaulted to 250 ppm.

9.3.2.2. Line Mapping

As mentioned previously, each spectral line in one partition is matched to every line in the corresponding partition. A record of each match and information about the match is stored as an element in a matrix containing information about all possible match combinations.

The quality of the match between two spectral lines is judged by comparing the distance between the lines to some value defining an acceptable match. This value is the partition slide factor and is determined based on two factors: (1) the number of spectral lines within each of the two corresponding partitions; and (2) the diameter of the cluster within each partition. Each partition has a separate slide factor which is continually adjusted before each library comparison. The slide factor is calculated using equation (9.1):

$$\text{Slide} = \frac{\text{largest cluster diameter (ppm)}}{\text{least number of lines}} \tag{9.1}$$

As all possible match combinations are recorded during the mapping procedure, a quality index is assigned to each based on the slide factor. This quality index is used later by the scoring section of the search. If the distance between two compared spectral lines is less than or equal to 0.10 ppm, the match is assigned a quality index of 1, indicating an exact match. The ± 0.10 ppm tolerance was incorporated to account for slight instrumental variations. If the distance between two compared spectral lines is less than or equal to the slide factor, the match is given a quality index of 2, indicating a "good" match. If the distance between the two spectral lines being examined is greater than the slide factor, the match has a quality index of 3, indicating a mismatch.

The format in which the mapping matrix is constructed depends on the number of spectral lines contained in each of the two partitions being compared. The maximum number of matches that can be made is given by the number of peaks in the partition containing the least number of peaks. Each peak in this partition will represent a row in the matrix. The partition with the least number of spectral lines is always mapped onto the partition with the greater number of lines. Each peak in the latter will represent a column in the match matrix.

The first upfield shift in the partition containing the least number of lines is consecutively matched, upfield to downfield, with each shift in the corresponding partition. Each match forms an element in row 1 of the matrix. Each element contains three numbers, termed row element indices. These are (1) match quality, represented by a 1, 2, or 3; (2) the distance (ppm units) separating the two lines; and (3) the column number identifying the match. After all possible match combinations have been examined and recorded, each row of the matrix is sorted from least to greatest distance, based on the second index of each row element.

A column pointer is associated with each row of the matrix, and the position of the pointer indicates the best match for a given row peak. After the matrix is sorted from least to greatest distance, the element in the first column of each row ideally represents the best match for a given row peak. Therefore, the column pointers are all initialized to be in column 1.

In most cases, the selection of best matches is much more complex. In many situations, more than one row peak will ideally match best with the same column peak, resulting in a situation termed a multiple match. When a multiple match occurs, the algorithm scans the multiple matching rows in search of some pattern based on the third index of each row element. The patterns sought are those that indicate a systematic shift of a group of lines. For example, if lines 2 and 3 both match best with line 5 in the corresponding partition, the second best matches for lines 2 and 3 are inspected. If the second best match for line 2 is line 4, a systematic shift is indicated. Thus, line 2 is matched with line 4, while line 3 is matched with line 5. Several other analogous patterns are also possible. If no pattern can be found, multiple matches are resolved by simply matching the line combination with the smallest chemical shift difference. The column pointers for the remaining participants in the multiple match are then incremented to allow matching with their next best matches. The pattern scanning procedure is continued until no multiple matches remain.

9.3.2.3. Scoring

The scoring procedure is divided into two sections: (1) the match penalty section, which assigns an appropriate penalty to "good" matches and mismatches; and (2) the structure penalty section, which assigns a penalty to extraneous peaks identified by the mapping algorithm.

As outlined, the position of the column pointer for each row indicates the best match for each row peak. Scores for each match are determined by examining the distance separating the two matched spectral lines. If this distance is less than or equal to 0.10 ppm, the match is considered to be exact, and no penalty points are assigned. If the match is a "good" match or a mismatch, the penalty is equal to the distance between the lines. The matches for each row peak are evaluated in this manner, resulting in an overall summation of match penalty points.

Extraneous peaks are assigned a penalty equal to the distance to the closest peak in the corresponding partition. The penalties for each extraneous peak are summed, resulting in an overall penalty score. This procedure minimizes the penalty in the case of an extraneous peak that results from differences in spectral resolution.

The match penalty score and the structure penalty score are then added to produce a score for the partition. Each partition in the unknown is evaluated using the same procedures as outlined above. The scores for each partition are summed to yield a total score for the overall spectral comparison, and these scores are sorted from smallest to largest in a continuously updated "hit list."

9.3.3. Evaluation of Search Results

The search algorithm described above was evaluated by use of a library of 7,197 spectra derived from the NIH–EPA database (Fein–Marquart Associates, Baltimore, MD). The results of two example searches are given below. The goal in evaluating search results is to determine the degree to which the structural integrity of the unknown is preserved and identified. In each of the results presented, a library spectrum was selected as the "unknown," thus permitting better evaluation of the resulting hit list of best matches.

The spectrum of 1-phenyl-1-propanone was selected as the target spectrum. The resulting hit list of the five best matches is presented in Table 9.1, and the corresponding spectra for the compounds are given in Fig. 9.4. Hits 1 and 2 are both spectra of 1-phenyl-1-propanone recorded under different experimental and solvent conditions. The first spectrum represents a neat sample, while the second spectrum was recorded in $CDCl_3$. Both spectra were collected on different instruments. Despite the

TABLE 9.1. Search Results

SEARCH 1

Target: 1-phenyl-1-propanone

Hit	Name	Score
1	1-phenyl-1-propanone	0.000
2	1-phenyl-1-propanone	9.630
3	1-(2,5-dimethylphenyl)-ethanone	19.700
4	2-methyl-1-phenyl-1-propanone	21.300
5	1-(2-methylphenyl)-ethanone	21.500

SEARCH 2

Target: 4-chloro-1-(phenylmethyl)-1H-pyrazole

Hit	Name	Score
1	4-chloro-1-(phenylmethyl)-1H-pyrazole	0.000
2	1-(phenylmethyl)-1H-pyrazole	11.000
3	3-chloro-1-(phenylmethyl)-1H-pyrazole	12.000
4	5-chloro-1-(phenylmethyl)-1H-pyrazole	14.200
5	(dimethoxymethyl)-benzene	18.000

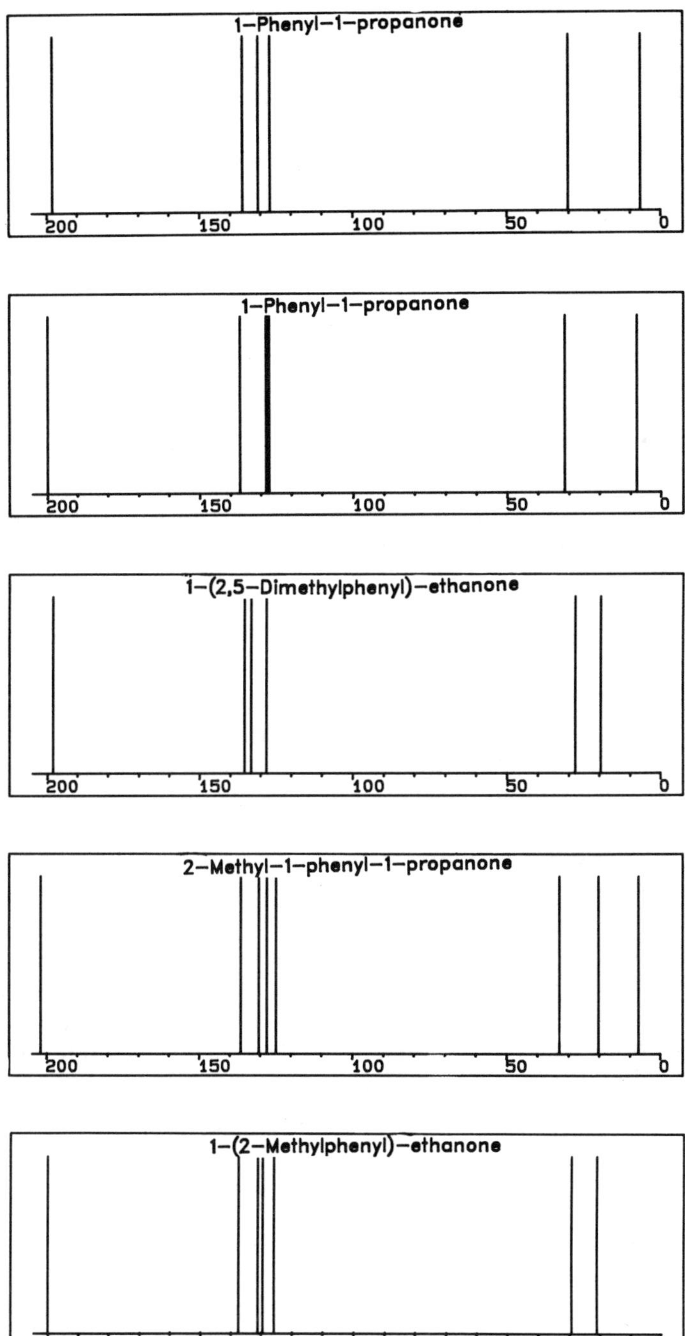

FIGURE 9.4. Broadband-decoupled ^{13}C NMR spectra of compounds on the hit list for Search 1.

differences in shift position and number of shifts, the search still accurately identifies the target compound. The remainder of compounds on the hit list are all 1-phenyl-1-ketones, correctly identifying the appropriate class of the target compound.

As a second test, the spectrum of 4-chloro-1-(phenylmethyl)-1H-pyrazole was selected as the target spectrum. The hit list of best matches is presented in Table 9.1, and the spectra of the compounds are illustrated in Fig. 9.5. The second best match for the spectrum of the target compound is the spectrum of 1-(phenylmethyl)-1H-pyrazole, the same compound minus chlorine substitution in the fourth position. The third and fourth best matches correctly identify compound class and also identify the presence of chlorine substitution. Hits 3 and 4 are the only other occurrences of chloro-substituted 1-(phenylmethyl)-1H-pyrazoles in the library. The fifth best match is useful in identifying the presence of a phenylmethyl structure in which the methyl ^{13}C atom is greatly deshielded.

9.3.4. Reverse Searching

Often an unknown spectrum is not contained in the library, and in some cases the library does not contain spectra of compounds similar to the unknown. Similarly, a spectrum that actually represents an isomeric mixture or which contains spectral components due to impurities will not likely match with any library spectrum, even if the appropriate compound class is well represented in the library.

In such cases, a reverse search strategy is useful. First developed for mass spectral library searches,[11] a reverse search looks for spectra that are subsets of the unknown spectrum. These subset spectra can then be used to identify specific substructural units of a complex molecule whose spectrum is not contained in the library, or the subset spectra may be used to identify mixture components and to analyze samples containing impurities. Laude and Wilkins have applied these concepts previously with good success to ^{13}C NMR library searching of mixture data.[12,13]

9.3.4.1. Reverse Search Design

In our approach, the procedural details of the reverse search differ only slightly from those of the normal forward search. The methodology used for slide factor calculation, line mapping, matrix construction, and selection of best matches remains exactly the same. The only two modifications are in the partitioning and scoring sections.

The Fisher clustering algorithm is performed on the unknown spectrum as before, but partition boundaries are assigned differently. A

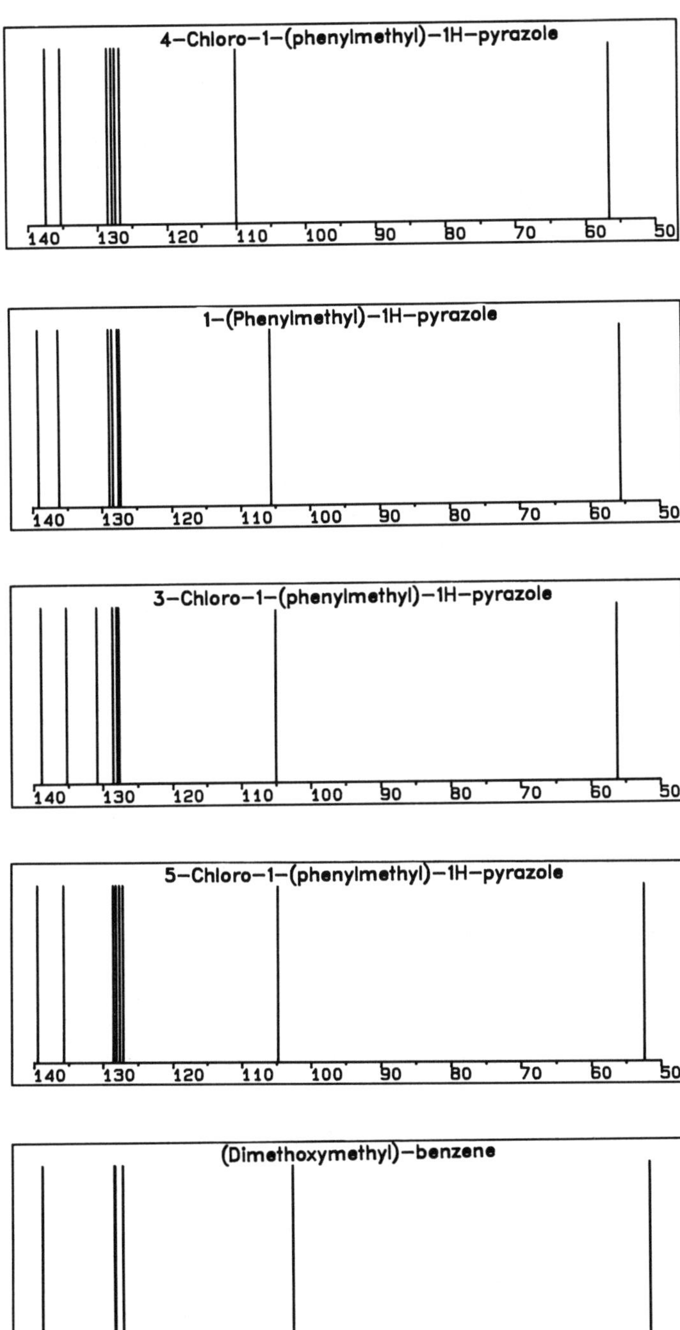

FIGURE 9.5. Broadband-decoupled ^{13}C NMR spectra of compounds on the hit list for Search 2.

distance of one standard deviation of the cluster is added to the first and last peak positions within the cluster to form initial partition boundaries. Overlapped boundaries are resolved by use of the mean of the two boundary values as the new shared boundary. Blank sections between partitions are not resolved, and the total spectral region examined is not defaulted to −100 to 250 ppm. This method of partitioning requires a library spectrum to have peaks within one or more of the distinct partitions to be classified as a valid subset.

The reverse search scoring procedure consists of three sections: (1) a match evaluation section for selecting the best matches; (2) a subset evaluation section which assigns a penalty to each extraneous peak in the library partition being compared to the corresponding partition of the unknown spectrum; and (3) a section for assigning a penalty to each extraneous peak of the library spectrum lying outside the specified partition boundaries. These three scoring features, working in combination, assure the best selection of library spectra that are true subsets of the unknown spectrum.

The scores from the match and subset evaluation sections of each partition are summed to produce a tentative final score. After all partitions in the unknown spectrum have been compared sequentially to the corresponding partitions within a given library spectrum, the last section of the scoring procedure checks for the presence of extraneous peaks lying external to the specified partition regions. These extraneous peaks acquire the most severe penalty because their presence clearly indicates that the given library spectrum is not a true subset of the target spectrum.

The best subset spectrum of a given target spectrum is itself. Therefore, if the unknown spectrum is contained in the library, that spectral comparison will have the lowest score, and the unknown will be positively identified. As before, a continuously updated hit list is kept during the search, and smaller scores are given to spectra that are better subsets of the unknown spectrum. It is important to note that the score is largely based on the number of matching lines contained in each subset spectrum. Thus, if the target spectrum contains ten shifts, a subset spectrum corresponding to seven matches (a 7-line subset) will result in a better score than a subset spectrum containing only four matches (a 4-line subset). However, both subset spectra may be equally important in the evaluation of the target spectrum. Therefore, given an unknown spectrum consisting of n shifts, $(n - 1)$ target lists of the best 2 through n-line subsets are constructed in addition to the hit list of best overall matches.

9.3.4.2. Evaluation of Reverse Search Results

A 1:1 mixture of 2-aminobenzoic acid and 4-aminobenzoic acid was prepared to test the ability of the reverse search to resolve the mixture spectrum into its component spectra. Upon examination of the reverse search hit lists, the spectrum of 2-aminobenzoic acid (7 lines) appeared as hit 1, and the spectrum of 4-aminobenzoic acid (5 lines) appeared as hit 4 on the list of best overall matches. Each appeared as hit 1 on their respective 7- and 5-line subset hit lists. After examining the results of the hit lists visually, it was evident that these two spectra were clearly the most accurate matches.

The reverse search can also provide information about substructures of complex molecules. As structural complexity of a compound increases, the likelihood that the library contains the spectrum of the compound decreases. However, the library may contain spectra of compounds that are substructural units of the more complex target compound. The spectrum of amygdalin, a substituted disaccharide ([(6-O-β-D-glucopyranosyl-β-D-glycopyranosyl)oxy]benzeneacetonitrile), was used to test the ability of the reverse search to identify substructures. An additional set of 46 spectra of various disaccharides was added to the library of 7,198 spectra to provide the reverse search with more information about the appropriate compound class. Since the spectrum of amygdalin (18 lines) was contained in the library, the target spectrum could be positively identified. The next two matches on the list of overall best matches were β-gentiobiose [6-O-β-D-glucopyranosyl-β-D-glucose (11 lines)] and α-gentiobiose [6-O-β-D-glucopyranosyl-α-D-glucose (11 lines)]. These two matches also appeared as the best matches on the list of best 11-line subset spectra. Upon inspection of the spectra and comparison of structures, it is found that the gentiobiose structure is the main structural backbone of amygdalin, and this is clearly indicated by the reverse search results.

9.3.5. Characterization of Library Searching Performance

The success of a search clearly depends on the size, quality, and diversity of the library used. For a given library, it is difficult to determine what the best hit list should be for a specific target compound. Based on visual results, therefore, as well as on knowledge about the various compound classes represented in the library used here, the search is judged to be highly successful in identifying the structural characteristics of a given target compound. The reverse search provides similar capabilities for the case of impure or mixture spectra. While we have no statistical basis for judging search performance, the search results presented here are typical of those obtained from a large number of test cases.

9.4. ^{13}C NMR SPECTRAL SIMULATION

No matter how large and varied spectral data bases become, they will never contain the spectrum of every unknown. Given an effective spectral comparison algorithm, however, it will usually be possible to use the search results in conjunction with other evidence to construct possible candidate structures for the unknown. The structure elucidation problem then focuses on ways to verify the validity of the candidate structures.

Spectrum simulation methods offer an attractive solution to this problem. If effective procedures were available for simulating the ^{13}C NMR spectra of the candidate structures, the simulated spectra could then be compared to the spectrum of the unknown. Judgements regarding the correctness of each of the candidate structures could be made through this procedure. The keys to the practicality of spectrum simulation, however, are both the ease with which the technique can be performed and the accuracy of the resulting simulated spectra.

9.4.1. ^{13}C NMR Spectra of Carbohydrates—A Simulation Example

The interpretation of ^{13}C NMR spectra of complex carbohydrates has always proven challenging because the subtle structural differences among the compounds produce spectra that are highly similar. These differences are largely stereochemical changes in the saccharide ring systems. If spectrum simulation techniques are to be successful in helping to solve structure elucidation problems involving carbohydrates, the simulation methodology must be sensitive to both the steric and inductive effects that determine carbohydrate chemical shifts.

A set of 40 disaccharides will be used to illustrate the two-step process that comprises our approach to spectrum simulation. The application of the simulation methodology to an example structure elucidation problem will be used to demonstrate the practicality of the approach.

9.4.2. Spectrum Simulation Based on Parametric Modeling

The spectrum simulation approach used here is based on the derivation of a set of linear models that relate numeric structural parameters to ^{13}C NMR chemical shifts. These models have the form

$$S_j = b_0 + b_1X_1 + b_2X_2 + \cdots + b_nX_n \tag{9.2}$$

where S_j is a predicted chemical shift for a given carbon atom j, the X_i are computed parameters that encode aspects of the chemical environment of

j that influence its chemical shift, and the b_i are coefficients determined through a multiple linear regression analysis in which the dependent variable is a set of known and correctly assigned chemical shifts. Thus, this approach to spectrum simulation is a two-step process. First, a spectral data base is used to assemble a set of chemical shifts that will form the basis for one or more simulation models. Once the appropriate models are derived, the second step is the use of the models to predict the spectra of candidate structures. The approach is an indirect data base method, as the actual application of the procedure does not involve direct use of the data base.

When structural geometries greatly influence chemical shifts, molecular mechanics techniques are useful in developing parameters that describe these geometrical effects. The efficient development of such parameters, however, requires computer-based structure handling, as well as a variety of other data processing utilities. Chemical structures used in this work were input in two-dimensional form into computer disk files via a graphical procedure developed by Brügger and Jurs.[(14)] No hydrogens were attached to the structures at this time. Other routine, but essential, tasks such as perception of structurally unique carbons, storage of computed models, and so on, were performed by use of a set of interactive software tools developed by Small and Jurs.[(15)]

9.4.3. Assembly of Data

In the development of chemical shift models for disaccharides, it was desirable to employ a spectral data base that was as large as possible. Data from five literature sources were used,[(16–20)] resulting in a set of 40 disaccharides and their corresponding ^{13}C NMR spectra. Table 9.2 lists each compound, along with an identifying number and the literature reference for the corresponding spectral data. Subsequent references to specific compounds will refer to the identifying numbers in the table.

A serious problem encountered when using data from multiple sources is the lack of a consistent reference for the chemical shifts. Carbohydrate spectra are particularly sensitive to changes in experimental conditions.

To overcome this problem, one data source was used as a base, and a calibration model was constructed to adjust each set of chemical shifts to that base. For the work reported here, the Usui et al. data[(16)] were chosen to define the base, because this collection contained the largest number of spectra. Each calibration model has the form

$$S_{\text{base}} = C_0 + C_1 S_{\text{expt}} \tag{9.3}$$

where C_0 and C_1 are parameters that convert an experimental chemical shift, S_{expt}, to a new shift, S_{base}, that represents the equivalent shift under the experimental conditions employed when the data chosen as the base were collected.

Simple linear regression was used to calculate C_0 and C_1 for each source of data, thereby allowing each set of chemical shifts to be adjusted to the Usui data. For the regression calculation to be performed on a given source of data, spectra had to be present that also occurred in the Usui collection. This allowed the dependent and independent variables in equation (9.3) to be constructed. For the Colson and King,[(17)] Pfeffer et al.,[(18)] Voelter et al.,[(19)] and Dorman and Roberts[(20)] data sources, 12, 36, 24, and 26 chemical shifts, respectively, were present that duplicated shifts in the Usui collection. The resulting calibrations were highly successful on statistical grounds, producing R^2 values of 99.99%, 99.99%, 99.80%, and 99.99%, respectively. The corresponding t values for the significance of C_1 were 423.00, 548.00, 100.00, and 455.00, respectively. As a final step, the addition of higher order terms to equation (9.3) was investigated. No such terms could be found that were statistically

TABLE 9.2. Compounds Used for Modeling and Testing

No.	Compound name	Ref.[a]
1	α-D-glucopyranosyl-β-D-glucose	P
2	α-D-glucopyranosyl-α-D-glucose	P
3	β-D-glucopyranosyl-β-D-glucose	U
4	2-O-β-D-glucopyranosyl-α-D-glucose	U
5	2-O-β-D-glycopyranosyl-β-D-glucose	U
6	2-O-β-D-glucopyranosyl-methyl-α-D-glucopyranoside	U
7	2-O-α-D-glucopyranosyl-α-D-glucose	U
8	2-O-α-D-glucopyranosyl-β-D-glucose	U
9	2-O-α-D-glucopyranosyl-methyl-β-D-glucopyranoside	U
10	3-O-α-D-glucopyranosyl-β-D-glucose	U
11	3-O-α-D-glucopyranosyl-α-D-glucose	U
12	3-O-β-D-glucopyranosyl-β-D-glucose	U
13	3-O-β-D-glucopyranosyl-α-D-glucose	U
14	4-O-α-D-glucopyranosyl-α-D-glucose	U
15	4-O-α-D-glucopyranosyl-β-D-glucose	U
16	4-O-α-D-glucopyranosyl-methyl-β-D-glucopyranoside	U
17	4-O-β-D-glucopyranosyl-α-D-glucose	P
18	4-O-β-D-glucopyranosyl-β-D-glucose	P
19	4-O-β-D-glucopyranosyl-methyl-β-D-glucopyranoside	U
20	6-O-β-D-glucopyranosyl-α-D-glucose	U
21	6-O-β-D-glucopyranosyl-β-D-glucose	U
22	6-O-β-D-glucopyranosyl-methyl-β-D-glucopyranoside	U
23	6-O-α-D-glucopyranosyl-α-D-glucose	U
24	6-O-α-D-glucopyranosyl-β-D-glucose	U

TABLE 9.2. *Cont.*

No.	Compound name	Ref.[a]
25	4-O-β-D-galactopyranosyl-α-D-glucose	P
26	4-O-β-D-galactopyranosyl-β-D-glucose	P
27	4-O-β-D-galactopyranosyl-methyl-β-D-glucopyranoside	D
28	4-O-α-D-glucopyranosyl-β-D-mannose	V
29	4-O-α-D-glucopyranosyl-α-D-mannose	V
30	3-O-β-D-glucopyranosyl-β-L-rhamnose	C
31	4-O-β-D-glucopyranosyl-β-L-rhamnose	C
32	2-O-β-D-glucopyranosyl-β-L-rhamnose	C
33	2-O-β-D-galactopyranosyl-β-L-rhamnose	C
34	3-O-β-D-galactopyranosyl-β-L-rhamnose	C
35	4-O-β-D-galactopyranosyl-β-L-rhamnose	C
36	3-O-β-D-glucopyranosyl-α-L-rhamnose	C
37	4-O-β-D-glucopyranosyl-α-L-rhamnose	C
38	2-O-β-D-glucopyranosyl-α-L-rhamnose	C
39	2-O-β-D-galactopyranosyl-α-L-rhamnose	C
40	3-O-β-D-galactopyranosyl-α-L-rhamnose	C
	Testing	
41	6-O-α-D-galactopyranosyl-α-D-glucose	L
42	6-O-α-D-galactopyranosyl-β-D-glucose	L
43	α-[(6-O-β-D-glucopyranosyl-β-D-glucopyranosyl)oxy] benzeneacetonitrile	L
44	α-[(6-O-β-D-galactopyranosyl-β-D-glucopyranosyl)oxy] benzeneacetonitrile	*
45	α-[(6-O-β-D-allopyranosyl-β-D-glucopyranosyl)oxy] benzeneacetonitrile	*
46	α-[(6-O-β-D-mannopyranosyl-β-D-glucopyranosyl)oxy] benzeneacetonitrile	*
47	α-[(6-O-β-D-altropyranosyl-β-D-glucopyranosyl)oxy] benzeneacetonitrile	*
48	α-[(6-O-β-D-glucopyranosyl-β-D-mannopyranosyl)oxy] benzeneacetonitrile	*
49	α-[(6-O-β-D-glucopyranosyl-β-D-allopyranosyl)oxy] benzeneacetonitrile	*
50	α-[(6-O-β-D-glucopyranosyl-β-D-galactopyranosyl)oxy] benzeneacetonitrile	*
51	α-[(6-O-α-D-glucopyranosyl-β-D-glucopyranosyl)oxy] benzeneacetonitrile	*
52	α-[(3-O-β-D-glucopyranosyl-β-D-glucopyranosyl)oxy] benzeneacetonitrile	*
53	α-[(4-O-β-D-glucopyranosyl-β-D-glucopyranosyl)oxy] benzeneacetonitrile	*
54	α-[(2-O-β-D-glucopyranosyl-β-D-glucopyranosyl)oxy] benzeneacetonitrile	*

[a] Key to source of spectral data: C, Colson; D, Dorman; L, Local; P, Pfeffer; U, Usui; V, Voelter; *, no spectrum obtained.

significant. Thus, it appears that a simple linear calibration is effective in adjusting multiple sources of data to a common reference.

Given the presence of duplicate spectra, a decision has to be made regarding which of the duplicate spectra to include in the modeling study. In this regard, the Pfeffer data were used where available, because these spectra were assigned by use of a deuterium isotope shift procedure. The Usui spectra were assigned the next highest priority, given the large size of the collection. The Colson and King, Dorman and Roberts, and Voelter et al. spectra were used only in cases in which they were the only source of data for a particular compound.

9.4.4. Molecular Mechanics Computations

The MM2 procedure of Allinger[(21)] was used to build three-dimensional models for the structures. To prepare the input structures for MM2, an initial force-field procedure was applied to convert the two-dimensional structures into rough three-dimensional form. These calculations were performed on the hydrogen-suppressed structures by use of the force field described by Stuper et al.[(22)] Hydrogens were then attached to the structures by direct calculation, employing standard bond lengths and bond angles. At the same time, two electron lone pairs were explicitly associated with each oxygen at a distance of 0.5 Å from the atom center, and at angles of 103.26° to carbon and 101.01° to hydrogen. These positions for the lone pairs are identical to those used by Allinger in MM2.

The computation of final models for the structures was undertaken with care regarding the conformational preferences of the saccharide ring systems. For each ring, a study of possible conformations was undertaken. Sites of clear conformational flexibility (e.g., the O-glycoside bond) were investigated carefully. In each case the conformer with the overall lowest strain energy was used to approximate the structure. It is clear that, in actuality, there is no unique conformation for each disaccharide. Investigation and use of the lowest energy conformation does seem to be a workable strategy in modelling ^{13}C NMR chemical shifts in these systems, however. As an example of the results of modeling, Fig. 9.6 depicts the modeled three-dimensional geometry of compound 4. The chemical shifts of four atoms are labeled to provide an example of the effects of geometrical similarities and differences.

9.4.5. Definition of Atom Groups

Based on previous experience, the successful modeling of complex structural effects requires models that are highly detailed. By necessity,

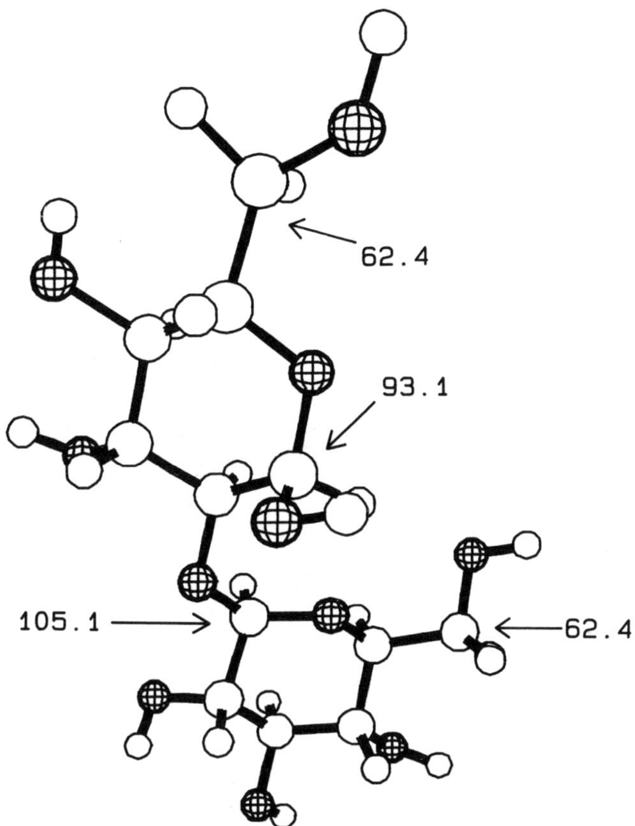

FIGURE 9.6. Three-dimensional plot of compound **4** based on the modeled atomic coordinates.

such models must be based on atoms in similar chemical environments. The standard procedure we have employed in previous studies has been to group atoms in similar environments together, and to compute separate models for each group. The task of how to group the atoms is somewhat based on trial and error. In all cases, models are sought that are as global as possible without sacrificing accuracy. The disaccharide ring atoms were found to require four groups: I, atoms 1 and 1′ (connected to the ring oxygen); II, atoms 2 and 2′ (two bonds from the ring oxygen); III, all other ring atoms; and IV, CH_2 carbons and CH carbons with no hydroxyl attached. For the 40 disaccharides, groups I–IV contained 78, 78, 222, and 78 carbons, respectively.

9.4.6. Design of Structural Parameters

An inspection of the modeled disaccharide structures revealed two key factors for the design of structural parameters for use in modeling chemical shifts. Structural effects on chemical shifts are clearly related to the distance from the carbon whose chemical shift is being simulated to the structural feature affecting the shift. Thus, parameters must be weighted by interatomic distance from the carbon of interest. Second, it seems reasonable to compute separate parameters that describe interactions of different types (carbon, lone pair, etc.). This conclusion results from observations that the only difference in some carbon environments is the presence of a particular species (e.g., oxygen atom) proximate to the carbon of interest. These guidelines result in parameters with great specificity. Such parameters have been found most useful in describing the highly detailed structural changes that induce chemical shift changes in the disaccharides.

In the investigation of the utility of specific structural parameters for the disaccharides, four additional points have been found significant. First, parameters based on distance alone are useful, but, by themselves, they are insufficient to model disaccharide chemical shifts accurately. These parameters are limited in that they consider atoms to be points in space, rather than spheres of charge density. Energy parameters based on van der Waals radii are of great use in overcoming this deficiency. Second, parameters based on interactions between hydrogens attached to the carbon of interest (the alpha hydrogen) and carbons, oxygens, hydrogens, or lone pairs in the molecule are of extreme value in constructing chemical shift models. Third, even crude attempts to encode the inductive effects of oxygens are beneficial in building accurate models. Last, in some instances, the most successful models must contain parameters derived simply from the topology of the molecule. To be useful for prediction, however, such parameters must be defined by continuous functions.

The observations noted above have led to the use of four families of structural parameters in modeling chemical shifts in the disaccharides. Each of these types of parameters is discussed below.

9.4.6.1. Distance Parameters

The distance parameters found useful have the form

$$D = \sum 1/d_{ij}^3 \tag{9.4}$$

where D is the computed parameter value, and the d_{ij} are computed distances between the target atom and other interacting atoms or lone pairs in the molecule. Distance computations employ the modeled atomic coordinates. The use of an inverse distance function allows the parameter value to approach zero at large distances. The choice of an inverse cubic form for the expression is based somewhat on empirical investigations, although it can be justified based on the inverse cubic distance relationship that determines the strength of the induced magnetic field in dipole interactions.[23] Individual parameters differ in three ways: (1) the type of the target atom, (2) the type of interaction forming the basis for the distance, and (3) the selection of the specific interactions to be entered into the summation.

The target atom is either the carbon whose chemical shift is being modeled or the alpha hydrogen, while the interaction type is either carbon, oxygen, hydrogen, lone pair, or carbon plus oxygen. The summation can be taken over all interacting species in the molecule or can be taken at a certain topological distance from the target atom. The latter procedure has the advantage of focusing the parameter at a specific bond distance from the target atom. The d_{ij} terms are still based on the modeled coordinates.

9.4.6.2. Energy Parameters

As noted previously, energy terms based on van der Waals radii have proven useful. Many potential functions can be used to define the interaction energy. We have chosen to implement the function used by Allinger in MM2, as our modeled atomic coordinates were clearly derived through a procedure in which this function was used prominently. This function has the form

$$E = \varepsilon(2.9 \times 10^6 e^{(-12.5/P)} - 2.25P^6) \qquad (9.5)$$

where

$$P = r^*/R \qquad (9.6)$$

In these equations, E is the computed interaction energy, ε is the depth of the potential well, r^* is the sum of the van der Waals radii of the two interacting atoms, and R is the computed interatomic distance based on the modeled coordinates. Allinger employs a composite well depth taken as the geometric mean of the well depths associated with the individual interacting atoms. As an example, Fig. 9.7 is a plot of the potential function generated for carbon–oxygen interactions. The minimum energy is found at 3.64 Å, the sum of the van der Waals radii of carbon and oxygen.

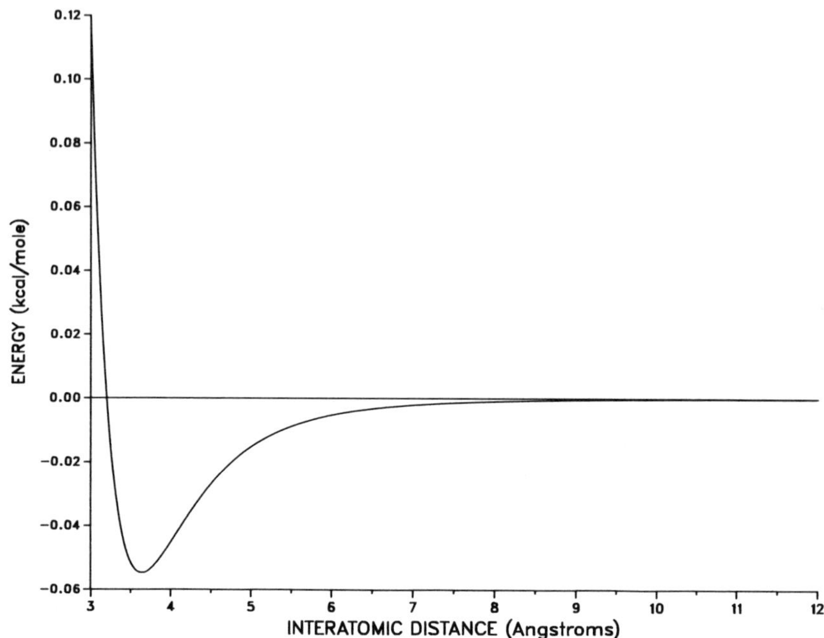

FIGURE 9.7. Potential function used in the generation of van der Waals energy parameters describing carbon–oxygen interactions.

Individual energy parameters differ by the choice of the target atom and the interaction type. Both carbon and alpha hydrogen-based parameters have proven useful. Parameters based on interactions with hydrogens, lone pairs, or carbons plus oxygens have found use. Typically, all interactions are used between three and seven bonds from the target atom. This designation treats lone pairs as pseudoatoms. Seven bonds represents a liberal estimate of the maximum distance at which an interaction can influence chemical shifts. Including interactions beyond seven bonds effectively adds "noise" to the parameter. In certain cases, a long-distance energy parameter has proven useful. Interactions between four and seven bonds from the target are used to compute this parameter.

9.4.6.3. Induction Parameters

To encode the fundamental inductive effects of oxygen atoms, the simple Linear Combination of Atomic Orbitals (LCAO) approach of DelRe[24] was used to compute partial sigma charges on the disaccharide atoms. At topological bond distances, specific parameters were computed as either the average charge per atom, the sum of charges, or the sum of absolute values of charges. While these parameters are admittedly crude, the best models for several of the atom groups require their presence.

9.4.6.4. Topological Complexity

The molecular connectivity index, developed by Randic[25] and modified by Kier and Hall,[26] was used to form the basis for a topological parameter encoding the complexity of the atomic environment at specific bond distances. As computed here, this parameter has the form

$$C = \sum 1/[(c_i c_j)^{1/2}] \tag{9.7}$$

where C is the computed index summed over a series of bonds. For a given bond between atoms i and j, the individual c terms specify the connectivity of atoms i and j (e.g., 1 for primary, 2 for secondary, etc.). The basic index was designed for carbon atoms only. Kier and Hall expanded the index to include specialized connectivity values for heteroatoms. The individual structural parameters found useful in the disaccharide work define the summation over all bonds at a particular topological distance from the target carbon. A related parameter also found useful is computed by dividing equation (9.7) by the number of bonds employed in the summation. Both types of parameters define a continuous variable that provides information about the structural complexity at a given topological distance from the target carbon.

9.4.7. Development of Models

Given the variety of structural parameters that can be computed, the model computation step involves the selection of which parameters best encode the chemical environments of the target group of atoms in terms of how the environments determine chemical shifts. For the disaccharide atom groups, no assumptions were made that a specific parameter must be used. It was considered preferable to investigate the utility of the computed parameters by employing automated selection procedures. A combination of stepwise regression and best subset regression[27] was used, producing a number of excellent models for each group of atoms. For each group, one model was selected for use in testing. The criteria used in this selection were the overall model statistics (standard error of estimate, F-value, correlation coefficient), the individual t-values for the significance of the regression coefficients, and the degree of collinearity existing among the parameters. Collinearity among the parameters was estimated by use of the method of Belsley et al.[28] Regression

TABLE 9.3. Summary of Model Statistics

Group	n^a	p^b	R^c	s^d	F^e
I	78	11	0.988	0.755	242
II	78	8	0.970	0.712	138
III	222	11	0.982	0.629	521
IV	78	9	0.997	0.301	1220

[a] Number of chemical shifts used to define model.
[b] Number of parameters in computed model.
[c] Correlation coefficient.
[d] Standard error of estimate in chemical shift units (ppm).
[e] F-value for significance of the model.

coefficients computed from a set of parameters exhibiting low collinearity are more precisely defined than those computed in the presence of high collinearity.

Table 9.3 provides a summary of the model statistics for each atom group. Included are the number of parameters in the model, the correlation coefficient, the standard error of estimate, and the F-value for the significance of the regression equation. Without exception, the models yield excellent statistics, with standard errors ranging from 0.30 to 0.76 ppm.

Within each compound, the individual predicted chemical shifts can be assembled to form complete simulated spectra. Since the individual chemical shifts were predicted by use of different models, perhaps the most valid statistic in assessing prediction accuracy is simply the average prediction error across the spectrum. For the 40 disaccharides, the mean of these average prediction errors was 0.45 ppm. The maximum and minimum of the average prediction errors were 0.64 ppm and 0.28 ppm, respectively. These results provide further indication of the excellent accuracy of the computed models.

9.4.8. Test of Predictive Ability

The derived chemical shift models were examined further by evaluating their ability to estimate chemical shifts in compounds not included in the model generation phase of the study. In a straightforward test, the models were used to predict the spectra of compounds **41** and **42**, commonly known as α- and β-melibiose. Table 9.4 is a comparison of our predicted and experimentally observed chemical shifts for the two compounds. To perform the comparison, a calibration model of the type described in equation (9.3) was derived to relate our local experimental conditions to those embodied in computed models. Local spectra were collected of compounds **14** and **15**, providing 24 chemical shifts for use in

TABLE 9.4. Comparisons of Experimental and Predicted Spectra of Compounds **41** and **42**

	Compound **41**			Compound **42**		
Atom	expt[a]	pred	error	expt	pred	error
1	100.73	101.16	−0.43	100.69	101.15	−0.46
2	71.04	72.99	−1.95	71.01	72.95	−1.94
3	71.76	73.19	−1.43	71.76	73.05	−1.29
4	72.03	69.82	2.21	72.03	69.88	2.15
5	73.48	73.07	0.41	73.48	73.05	0.43
6	63.67	63.45	0.22	63.67	63.37	0.30
1′	94.73	94.11	0.62	98.60	98.89	−0.29
2′	73.97	74.00	−0.03	76.60	75.67	0.93
3′	75.50	75.44	0.06	78.43	78.11	0.32
4′	72.13	72.08	0.05	71.98	72.72	−0.74
5′	72.63	73.59	−0.96	76.88	77.18	−0.30
6′	68.48	69.22	−0.74	63.38	69.25	−0.87
Average prediction error:			0.76			0.84

[a] All chemical shifts reported in ppm downfield from DSS or Me_4Si.

the simple regression analysis. As with the previous calibration examples, the regression yielded a calibration model with excellent statistics ($R^2 = 99.99\%$, $t = 482$).

The average prediction errors for the two compounds are significantly less than 1.0 ppm. An examination of the individual predicted chemical shifts does reveal a limitation of the derived models, however. The prediction errors for atoms 2, 3, and 4 in both compounds are significantly higher than the standard errors associated with the models used in predicting the shifts. This is due to the presence of the α-galactosyl moiety in compounds **41** and **42**. Eight compounds used in the model development work contained a β-galactosyl moiety, but none contained the corresponding α-substructure. This observation illustrates that a model cannot be used to extrapolate to the prediction of a significantly different chemical environment without some loss in prediction accuracy. In addition, this example illustrates the desirability of the use of a set of compounds in the model generation step that is as structurally diverse as possible.

In a second prediction test, a simulation was performed of the use of spectral simulation in structure elucidation studies. Compound **43** is a cyanogenic glycoside commonly known as amygdalin. Its experimental spectrum was collected under the same instrumental conditions used in the definition of the calibration model described above.

Compounds **44–54** represent structural modifications of compound **43**. Figure 9.8 depicts the structures of compounds **43–54**. The differences among the structures involve orientation changes among the hydroxyl groups, orientation changes at the ring connection point, or changes in the position of ring connection. For the prediction test, let us assume the structure of amygdalin is unknown and that compounds **43–54** represent

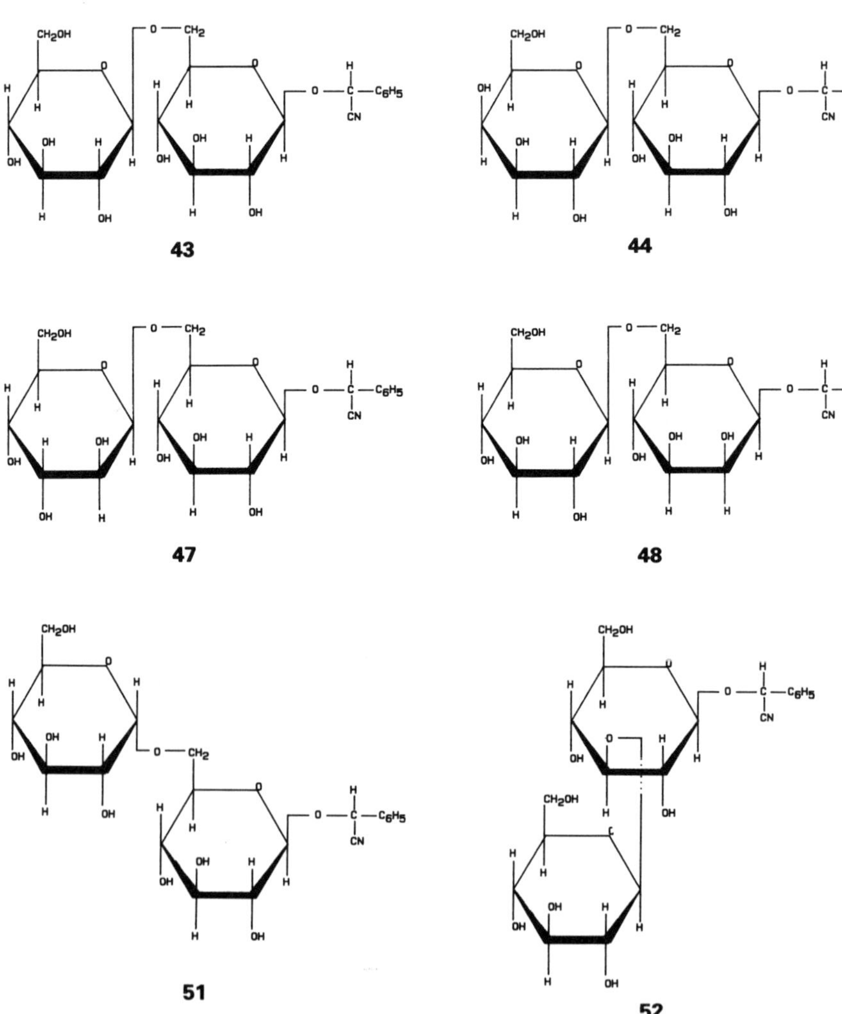

FIGURE 9.8. The twelve amygdalin candidate structures used in evaluating the prediction accuracy of the disaccharide models. Compound **43** is the correct structure.

possible candidate structures for the compound. The goal of the test was to determine if the derived chemical shift models could be used to confirm compound **43** as the correct structure, despite the presence of the benzeneacetonitrile group in all the compounds. Since none of the structures used in generating the models contained this side chain, it effectively served as the example of a potential interference to the performance of the models.

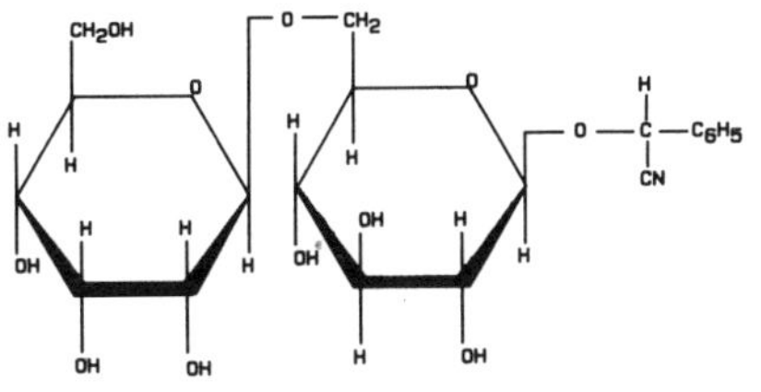

45

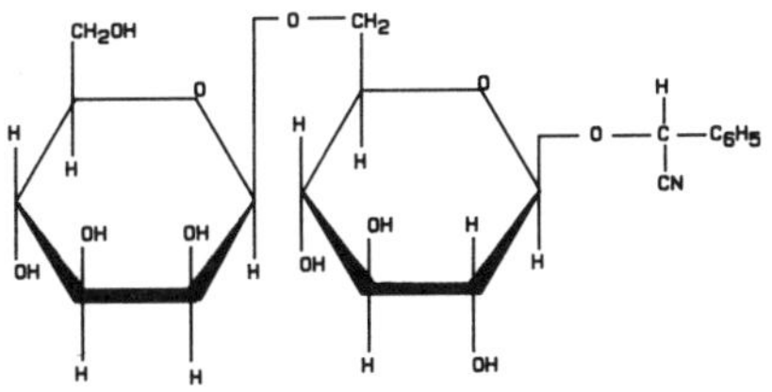

46

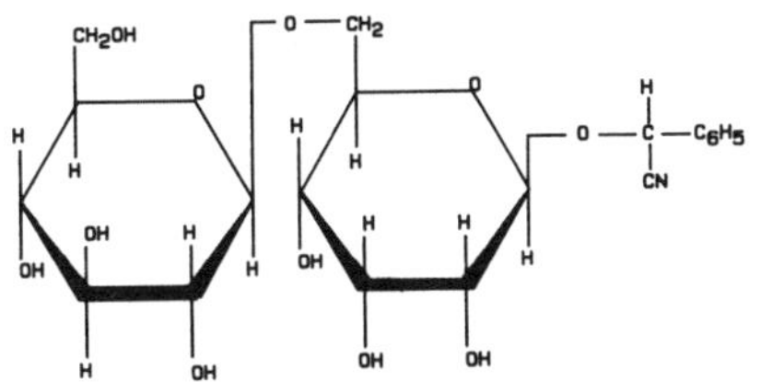

49

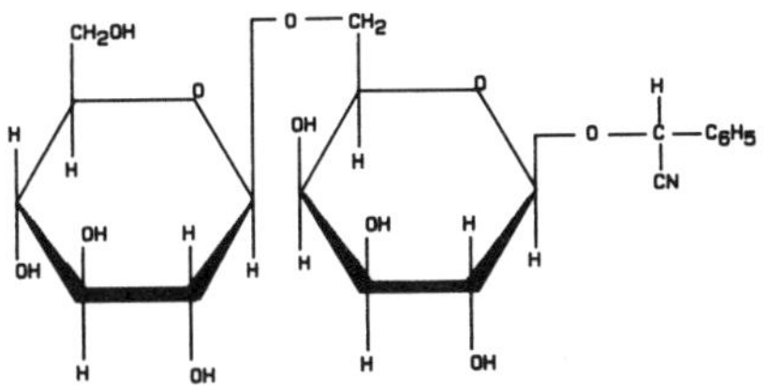

50

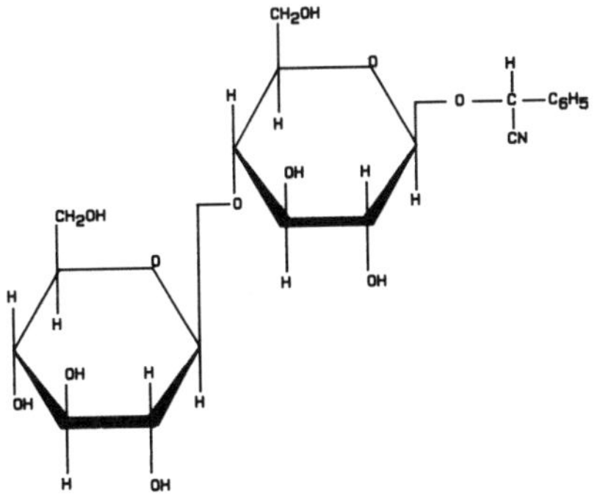

53

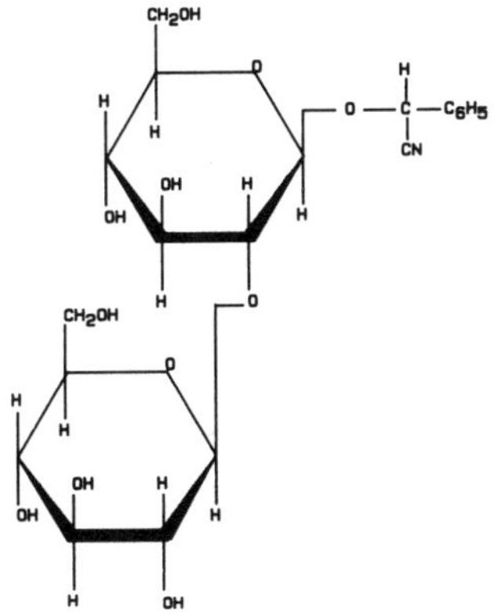

54

FIGURE 9.8. *Cont.*

The derived models were used to predict the chemical shifts of the twelve carbons associated with the saccharide ring systems in each of the twelve candidate structures. Lizotte and Poulton[29] have assigned the experimental spectrum of amygdalin previously. Employing these assignments, comparisons were made between the experimental chemical shifts of amygdalin and the predicted chemical shifts of the corresponding atoms in compounds **43–54**. For compounds **43–54**, the sum of squared differences between the experimental amygdalin chemical shifts and the candidate structure chemical shifts were 17.5, 41.5, 47.9, 53.7, 103, 55.4, 59.1, 53.6, 68.4, 145, 111, and 126, respectively. Despite the presence of the benzeneacetonitrile side chain, the derived models correctly predict compound **43** as the structure of amygdalin. Figure 9.9 is a comparison plot of the amygdalin experimental spectrum and the partial simulated spectrum of compound **43** (saccharide carbons only). The accuracy of the simulation is judged excellent, especially when the presence of the interfering side chain is considered.

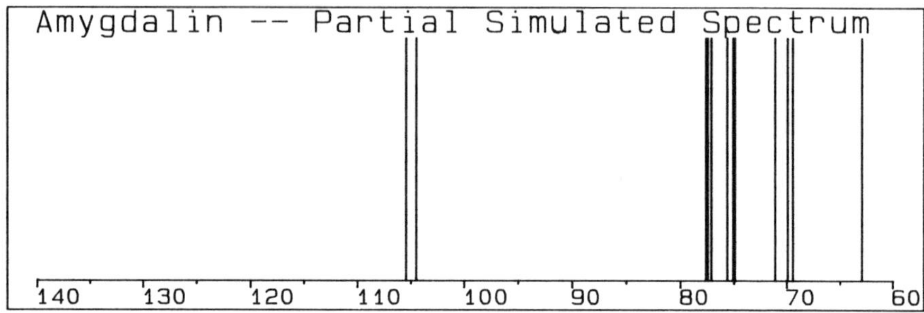

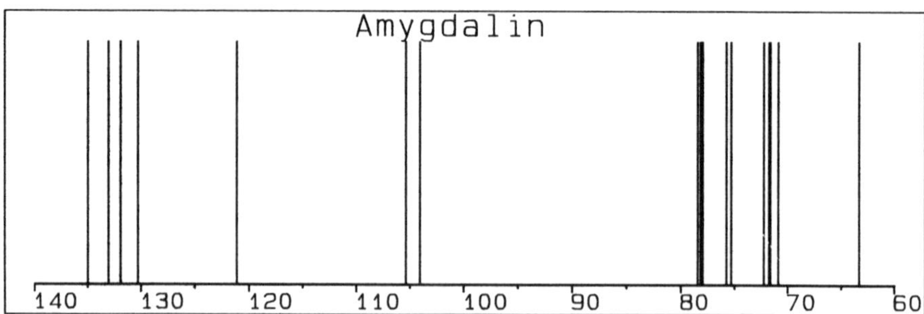

FIGURE 9.9. Comparison of the experimental spectrum of amygdalin with the partial simulated spectrum of compound **43**.

9.4.9. Characterization of Results

Based on the evaluation procedure outlined above, the chemical shift models of the disaccharides are judged highly accurate and useful for prediction. Potential application areas for the models lie in both structure elucidation studies and in the determination of chemical shift assignments.

The principal limitation of the models is the availability of only 40 disaccharide spectra for use in computing the models. As the prediction of the spectra of compounds **41** and **42** illustrates, the inclusion of more structural diversity in the model generation step would be expected to improve the versatility of the models for use in predictions.

9.5. CONCLUSIONS

Library searching and spectrum simulation are highly complementary methods for helping to automate the interpretation of ^{13}C NMR spectra. Each method can help to overcome limitations in the other. The availability of spectrum simulation eases the demand that the spectral library contain the spectrum of the unknown. A library search can be very useful, however, in helping the chemist to formulate candidate structures, an essential input to the spectrum simulation procedure.

The examples shown illustrate that no single computer-based technique can solve every spectral interpretation problem. A battery of methods must be available from which the spectroscopist can select the appropriate procedure for use with a given problem.

Both library searching and spectrum simulation are ultimately limited by the size and quality of spectral data bases. The major thrust of future work with these techniques will be the creation of larger and more consistent ^{13}C NMR data bases. Coupled with the increased performance characteristics of the typical laboratory computer, future prospects for the practical use of these procedures seem bright.

REFERENCES

1. A. E. Derome, *Modern NMR Techniques for Chemistry Research,* Pergamon, New York (1987).
2. D. H. Smith, N. A. B. Gray, J. G. Nourse, and C. W. Crandell, "Applications of Artificial Intelligence for Chemical Inference. Part XXXVIII. The DENDRAL Project: Recent Advances in Computer-Assisted Structure Elucidation," *Anal. Chim. Acta* **133,** 471–497 (1981).

3. K. Varmuza, "Pattern Recognition in Analytical Chemistry," *Anal. Chim. Acta* **122,** 227–240 (1980).
4. G. W. Small and P. C. Jurs, "Data Reduction in the Simulation of Carbon-13 Nuclear Magnetic Resonance Spectra of Steroids," *Anal. Chem.* **56,** 2307–2314 (1984).
5. J. Zupan, D. Hadzi, J. Marsel, and M. Penca, "Combined Retrieval System for Infrared, Mass, and Carbon-13 Nuclear Magnetic Resonance Spectra," *Anal. Chem.* **49,** 2141–2146 (1977).
6. A. P. Uthman, J. P. Koontz, J. Hinderliter-Smith, W. S. Woodward, and C. N. Reilley, "High-Throughput Microcomputer-Based Binary-Coded Search Systems for Infrared, Carbon-13 Nuclear Magnetic Resonance, and Mass Spectral Data," *Anal. Chem.* **54,** 1772–1777 (1982).
7. R. Schwarzenbach, J. Meili, H. Könitzer, and J. T. Clerc, "A Computer System for Structural Identification of Organic Compounds from ^{13}C NMR Data," *Org. Mag. Reson.* **8,** 11–16 (1976).
8. V. Mlynárik, M. Vida, and V. Kellö, Computer-Aided NMR Spectra Interpretation. Part 2. Minicomputer-Based ^{13}C/^{1}H-NMR File Search System," *Anal. Chim. Acta* **122,** 47–56 (1980).
9. M. Zippel, J. Mowitz, I. Köhler, and H. Opferkuch, "SPEKTREN—A Computer System for the Identification and Structure Elucidation of Organic Compounds," *Anal. Chim. Acta* **140,** 123–142 (1982).
10. J. A. Hartigan, *Clustering Algorithms,* Wiley, New York (1985).
11. F. W. Mclafferty and D. B. Stauffer, "Retrieval and Interpretative Computer Programs for Mass Spectrometry," *J. Chem. Inf. Comput. Sci.* **25,** 245–252 (1985).
12. D. A. Laude, Jr. and C. L. Wilkins, "Identification of Organic Mixture Components Without Separation: Quantitative and Edited Carbon-13 Nuclear Magnetic Resonance Spectrometry Data for the Analysis of Petroleum Distillates," *Anal. Chem.* **58,** 2820–2824 (1986).
13. D. A. Laude, Jr. and C. L. Wilkins, "Identification of Mixture Components in Organic Waste Materials by Carbon-13 Nuclear Magnetic Resonance," *Anal. Chem.* **59,** 576–581 (1987).
14. W. E. Brügger and P. C. Jurs, "Molecular Structure Input Program Using a Storage Cathode Ray Tube Terminal," *Anal. Chem.* **47,** 781–784 (1975).
15. G. W. Small and P. C. Jurs, "Interactive Computer System for the Simulation of Carbon-13 Nuclear Magnetic Resonance Spectra," *Anal. Chem.* **55,** 1121–1127 (1983).
16. T. Usui, N. Yamaoka, K. Matsuda, K. Tuzimura, H. Sugiyama, and S. Seto, "^{13}C Nuclear Magnetic Resonance Spectra of Glucobioses, Glucotrioses, and Glucans," *J. Chem. Soc. Perkin Trans.* **1,** 2425–2432 (1973).
17. P. Colson and R. R. King, "The ^{13}C–NMR Spectra of Disaccharides of D-Glucose, D-Galactose, and L-Rhamnose as Models for Immunological Polysaccharides," *Carbohyd. Res.* **47,** 1–13 (1976).
18. P. E. Pfeffer, K. M. Valentine, and F. W. Parrish, "Deuterium-Induced Differential Isotope Shift ^{13}C NMR. 1. Resonance Reassignments of Mono- and Dissacharides," *J. Am. Chem. Soc.* **101,** 1265–1274 (1979).
19. W. Voelter, V. Bilik, and E. Breitmaier, "Configurational and Mutarotational Effects on the ^{13}C-NMR Spectra of Some Disaccharides," *Collect. Czech. Chem. Commun.* **38,** 2054–2071 (1973).
20. D. E. Dorman and J. D. Roberts, "Nuclear Magnetic Resonance Spectroscopy. Carbon-13 Spectra of Some Common Oligosaccharides," *J. Am. Chem. Soc.* **93,** 4463–4472 (1971).
21. N. L. Allinger, "Conformational Analysis. 130. MM2. A Hydrocarbon Force Field Utilizing V_1 and V_2 Torsional Terms," *J. Am. Chem. Soc.* **99,** 8127–8134 (1977).

22. A. J. Stuper, W. E. Brügger, and P. C. Jurs, *Computer-Assisted Studies of Chemical Structure and Biological Function,* pp. 83–90, Wiley-Interscience, New York (1979).
23. D. E. Leyden and R. H. Cox, *Analytical Applications of NMR,* p. 28, Wiley-Interscience, New York (1977).
24. G. DelRe, "A Simple MO-LCAO Method for Calculating the Charge Distribution in Saturated Organic Molecules," *J. Chem. Soc.* **1958,** 4031–4040.
25. M. Randic, "On Characterization of Molecular Branching," *J. Am. Chem. Soc.* **97,** 6609–6615 (1975).
26. L. B. Kier and L. H. Hall, "Molecular Connectivity VII. Specific Treatment of Heteroatoms," *J. Pharm. Sci.* **65,** 1806–1809 (1976).
27. N. R. Draper and H. Smith, *Applied Regression Analysis,* Ch. 6, 2nd ed., Wiley-Interscience, New York (1981).
28. D. A. Belsley, E. Kuh, and R. E. Welsch, *Regression Diagnostics: Identifying Influential Data and Sources of Collinearity,* Ch. 3, Wiley-Interscience, New York (1980).
29. P. A. Lizotte and J. E. Poulton, "Identification of *R*-Vicianin in *Davallia trichomanoides* Blume," *Z. Naturforsch.* **41c,** 5–8 (1986).

Chapter 10

Expert System for Interpretation of the Infrared Spectra of Environmental Mixtures

Ying Li-shi, Steven P. Levine, and Sterling A. Tomellini

10.1. INTRODUCTION

To satisfy the requirements of hazardous waste analysis,[1–6] a program for automated waste mixture identification (PAWMI) through the interpretation of the infrared (IR) spectrum of the waste mixture was developed[7,8] and tested on hazardous waste drum samples.[9] Two limitations of PAWMI were that once a training set consisting of a reference library of spectra was defined, the rules for the inference engine (PAIRS)[10–16] had to be generated manually. The second limitation was that the PAWMI compound identification software uses only peak location information.

An approach to the automated generation of functional group interpretation rules for PAIRS was previously developed.[16] This system

Ying Li-shi and Steven P. Levine • School of Public Health, University of Michigan, Ann Arbor, Michigan 48109. Sterling A. Tomellini • Department of Chemistry, University of New Hampshire, Durham, New Hampshire 03824.

defined a value "occurrence," which was used to weight peak position information for the generation of expectation values for the presence of certain functional groups.

Efforts by other investigators have included the fuzzy data set,[17] as well as the hierarchic tree[18] and table-driven[19] programs developed by Munk and co-workers, and the pattern recognition approach of Frankel.[20] Most of these systems were primarily aimed at identifying functional groups in compounds, as was the original PAIRS program. A related work aimed primarily at identifying compounds in mixtures was that of Lowry and Huppler[21] which used a Boolean logic based search system. Many of these approaches owe their origins to earlier efforts that originated with Jurs, Isenhour, and co-workers.[22–24] Recent publications have included improvements in the PAIRS and hierarchic tree approaches, multispectroscopy expert systems, and various computer-aided spectral interpretation systems.[25–27]

This chapter describes a program for the identification of the principal components of mixtures based on computer-assisted interpretation of the mixture's infrared spectrum. This program (intIRpret) has five main subroutines: the interferogram processing and peak selection subroutine (PUSHSUB),[8] the automated knowledge acquisition subroutine (AUTOGEN),[16] the system optimization subroutine (STO), the interpretation subroutine (PAIRS),[7,10–16] and the final processing subroutine to subtract spectral similarity (PAIRSPLUS).[8]

Principal advantages of this system compared to the previously reported PAWMI system include speed (all spectral information is encoded automatically), flexibility (changes in the data base and in interpretation rules are readily accommodated), and accuracy (interpretation is based on peak position, frequency of occurrence, and peak size, each of which is weighted in an optimal fashion).

The method has been evaluated using the 62 most commonly identified organic compounds on hazardous waste sites.[3,9] IntIRpret was designed to be automatic, self-training, and self-optimizing.

IR spectra were generated using the following experimental procedures. All solvents were Aldrich Spectrophotometric Grade or equivalent. Mixtures were prepared on a weight basis. Film transmission spectra were acquired by placing a drop of sample between two KBr crystals. Spectra were acquired on a Nicolet 20-SX optical bench. Each spectrum was generated with a background and sample signal averaging of 128 scans. The number of data points collected was 16,384, resulting in a nominal spectral resolution of $2\,cm^{-1}$. All programming and spectral analysis, including rule writing, compiling and spectral interpretation was performed with a Nicolet 1280 computer.

10.2. COMPARISON OF intIRpret and PAWMI

IntIRpret has five main subroutines: the interferogram processing and peak selection subroutine (PUSHSUB),[8] the automated knowledge acquisition subroutine (AUTOGEN),[16] the system optimization subroutine (STO), the inference engine (PAIRS),[7,10–16] and the final processing subroutine which subtracts spectral similarity (PAIRSPLUS).[8] Figure 10.1 is a flow chart of the intIRpret process, where the logic of each of the five major subroutines is diagrammed. Emphasis is placed on describing STO, which is central to the operation of the self-training, self-optimizing mode of operation of intIRpret.

10.2.1. PUSHSUB

To automate PAWMI, a peak selection subroutine PUSHSUB, was developed that does not require the operator to set a peak selection threshold, and successfully follows nonlinear baselines.[8] PUSHSUB selects peaks by transforming the first 256 data points right of the centerburst from the original 16,384 data point sample interferogram into a threshold curve. PUSHSUB automatically calculates the threshold value from this file. This is described in detail in Puskar et al.[8] PUSHSUB stores the peak file in a format that can be used by AUTOGEN and STO.

10.2.2. STO

This subroutine accesses the peak tables generated by PUSHSUB. The peaks in a spectrum chosen for the purpose of decision making are called rule peaks. Not all spectral peaks are rule peaks. Each rule peak is assigned a "goodness value" that indicates the probable presence or absence of each compound in the training set. The question of "goodness" is discussed in detail elsewhere.[7,8] It is the purpose of the STO program to utilize the maximal amount of spectral information in an effort to enhance the predictive power of the goodness value.

Three factors are used to weight the goodness values assigned to each rule peak listed by AUTOGEN: $k1$ (frequency of occurrence), $k2$ (intensity), and $k3$ (frequency of occurrence × intensity) (Fig. 10.2). These three factors are designed to follow the logic used by an expert during the interpretation of the infrared spectra of mixtures. In this respect, the underlying intellectual framework is similar to that described in the work of McLafferty and co-workers in which match factors were automatically calculated for the interpretation of mass spectra.[28,29]

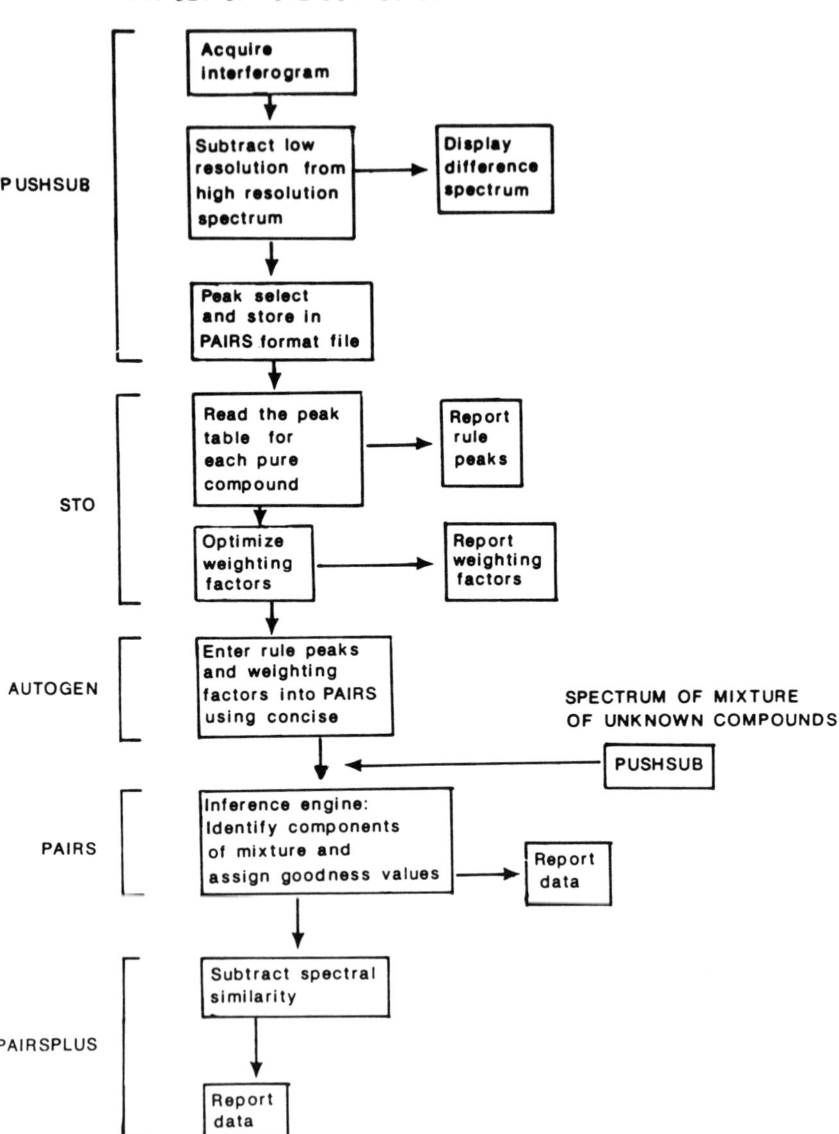

FIGURE 10.1. Flow chart of the intIRpret process, showing the logic of each of the five major subroutines.

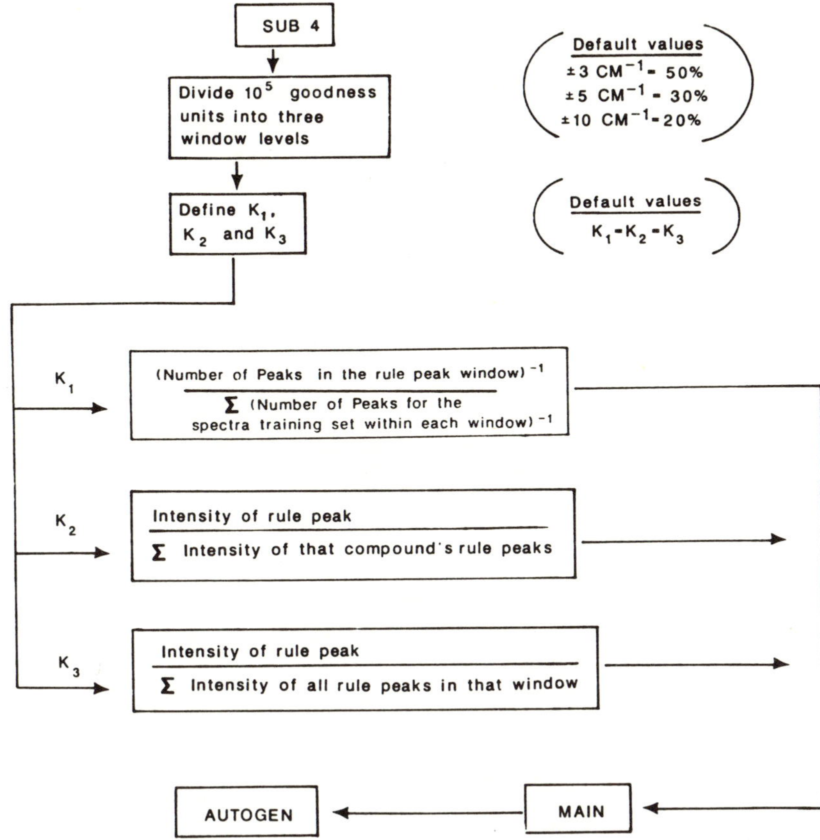

FIGURE 10.2. Flow chart of STO SUB 5, showing the relationship between *k*1, *k*2, and *k*3.

STO is structured around five subroutines, plus a "main," or driver, program. SUB 1 reads the peak table for each compound generated by PUSHSUB. The peak table is compared to the operator-defined window widths. If there is more than one peak in any given window, only the largest is retained. This results in the loss of potentially useful information, but it greatly simplifies later steps in the program with no apparent degradation of results.

SUB 2 reads the peak tables of all spectra in the training set and creates an array consisting of peak position and intensity information. This is used for calculations performed in SUB 4 and SUB 5.

SUB 3 decides which peaks in the peak table of each compound will be used for rule peaks for the PAIRS inference engine. This subroutine is designed to pick the largest peaks in the spectrum, up to a maximum of

20 peaks. If there are less than 20 peaks present using the highest threshold value, the threshold is automatically lowered incrementally until 20 peaks are chosen. In some cases, 20 peaks will not be present even at a low threshold, so the number of peaks necessary to satisfy this step of the program is lowered along with the threshold. If a minimum of three peaks are not present at a threshold of 3% or greater of the largest peak in the spectrum, then an error message is printed, and the spectrum of that compound in the training set must be reexamined by the operator. If the criteria for numbers of rule peaks and threshold are satisfied, the rule peak array is created from the information in SUB 2.

SUB 4 performs the frequency of occurrence and intensity analyses of data in the spectral array created by SUB 2 and SUB 3. The frequency of occurrence analysis counts the number of peaks within the window width surrounding each rule peak. The default value of the window widths was set at ±3, 5, and 10 cm^{-1}, which compensates for peak shifts expected in condensed phase mixtures.[7,8,10–17] For example, for a peak at 1,036 cm^{-1} in the spectrum of benzene, there are 11 other peaks for spectra in the training set within the tightest window of ±3 cm^{-1}, 19 other peaks present within the ±5 cm^{-1} window, and 26 other peaks within the ±10 cm^{-1} window. This information is utilized to assign weighted goodness values in SUB 5.

A similar calculation is performed for the peak intensity parameter. All peaks within the preset windows are not only counted, but their intensities are summed. This information is also used in SUB 5.

SUB 5 divides the total goodness between peaks and peak windows (Fig. 10.2). The first division of goodness is between windows, with the default value set at 50% for the tightest window, and 30% and 20% for the remaining two increasingly wide windows. These default values can be changed by the operator, if so desired.

Secondly, the factors $k1$, $k2$, and $k3$ are defined by the program. The goodness available to each peak window is divided between $k1$, $k2$ and $k3$, with the default value for the constants set equal. These default values can be changed by the operator.

Data generated by SUB 5 is accessed by the MAIN or driver program, which both calls SUB 1 to SUB 5 in sequence, and then creates a file for storing the goodness value for each window, peak, and compound in the training set. This data is stored in a form that is useable by AUTOGEN.

An example of the generation of the optimized goodness value for the rule peaks of benzene (Table 10.1). The values of $k1$, $k2$, and $k3$ are given for each of the peaks. For each compound, a total of 100,000 goodness units are allocated by STO. This is a change from the original PAIRS program in which 100 goodness units were allocated to the

TABLE 10.1. Comparison of $k1$, $k2$, $k3$ and Goodness Values for Six Peaks in the Spectrum of Benzene[a]

Peak position (cm^{-1})	674	1,036	1,479	3,036	3,071	3,091
Relative intensity (0–99)	99	9	28	18	5	9
Number of peaks in all spectra in						
±3 cm^{-1}	4	11	6	5	6	3
±5 cm^{-1}	8	19	10	8	10	8
±10 cm^{-1}	13	26	14	14	16	10
$k1$ (total for all 3 windows)[b]	6,659	2,701	5,024	5,882	4,883	8,175
$k2$ (total for all 3 windows)[c]	19,641	1,784	5,554	3,570	991	1,784
Total intensity of peaks in all spectra in						
±3 cm^{-1}	177	203	228	40	24	5
±5 cm^{-1}	335	254	368	80	39	391
±10 cm^{-1}	502	515	446	170	110	103
$k3$ (total for all 3 windows)[d]	10,682	1,009	2,726	7,763	3,828	7,317
$k1 + k2 + k3$ for						
±3 cm^{-1}	17,968	2,519	6,109	8,324	4,545	10,537
±5 cm^{-1}	11,324	1,786	4,145	5,681	3,383	3,678
±10 cm^{-1}	7,694	1,192	3,053	3,213	1,775	3,070
Total[e]	36,986	5,497	13,307	17,218	9,703	17,279

[a] Values associated with the peak at 674 cm^{-1} are discussed in the text.
[b] Apportioned based on 1/number of peaks in window in all spectra × 0.5, 0.3, and 0.2 for the three window widths.
[c] Apportioned based on 0.5, 0.3, and 0.2 for the three window widths.
[d] Apportioned based on 1/total intensity of peaks in window in all spectra × 0.5, 0.3, and 0.2 for the three window widths.
[e] Divide by 1,000 for percent contribution.

spectrum of each pure compound. For benzene, the allocation is made by apportioning the goodness between six rule peaks. The peak at 674 cm^{-1} is illustrative of the manner in which the system works. The peak at 674 cm^{-1} is the largest peak in the spectrum of benzene; therefore, the $k2$ value is the highest, with values of 9,821, 5,892, and 3,928, totalling 19,641 goodness units. The value 19,641 can be found in Table 10.1. These three values are for the ±3, 5, and 10 cm^{-1} windows, respectively, and represent an allocation of 50, 30, and 20% of the total $k2$ goodness, respectively.

The peak at 674 cm^{-1} has 4, 8, and 13 peaks in all of the other spectra of the compounds in the training set within the ±3, 5, and 10 cm^{-1} windows. Thus, the peak is in a window in which the frequency of occurrence of potentially interfering peaks is low, and the $k1$ values are correspondingly high. These are set at 3,450, 1,991, and 1,218, respectively, totalling 6,659, which is the value found in Table 10.1.

The total intensity, on a scale where the largest peak in a spectrum has an intensity of 99, of all other peaks in the spectra of the compounds in the training set, is 177, 335, and 502 for the three windows surrounding the 674 cm^{-1} peak. Thus, not only does this peak occur at a location where there are few other peaks in the spectra of other compounds in the training set, but those other peaks are relatively small. Therefore, the $k3$ values for this peak are set at the relatively high values of 4,696, 3,439, and 2,547 for the three windows, totalling 10,682, which is the value found in Table 10.1.

The total goodness assigned to the peak at 674 cm^{-1} is 36,986, or 37% of the goodness for all of the peaks in the entire spectrum of six rule peaks. Goodness is divided into 18% for $k1$, 53% for $k2$, and 29% for $k3$.

As stated earlier, the default values chosen for this study were window widths of ±3, 5, and 10 cm^{-1}; goodness values divided between these windows of 50%, 30%, and 20%, respectively; and $k1 = k2 = k3$. It is not known if these are the optimal values for this training set, for all possible mixtures that can be prepared for compounds in this training set, or for other training sets.

Using the STO portion of intIRpret allows the optimization of goodness values for each rule peak in each training set.

10.2.3. AUTOGEN

The automated generation of rules for a defined training set is essential to the success of this approach. Without AUTOGEN, PAIRS and PAWMI are hampered by the potential for errors that always occurs when data is manually encoded, and by the constraints imposed by the length of time it takes to enter data for new or modified training sets. Because of these problems, such a system is inherently inflexible. AUTOGEN solves these problems.

This subroutine has been modified from the program first reported.[16] The present version generates a three-level filter algorithm. The intensity algorithm has also been modified to generate information based on a scale of 0–99, rather than the previously utilized 0–9 scale.

At the completion of the running of AUTOGEN for a given training set, a complete set of three-level "if–then" rules has been generated for the PAIRS inference engine. If STO had not been used, goodness values, which are a measure of the probability of the presence of an unknown compound in a mixture, would be assigned on an equal basis to each peak in each spectrum of the training set. The use of STO allows the optimized goodness values to be entered in the rules by AUTOGEN for use by PAIRS.

10.2.4. PAIRS

As previously reported, PAIRS[10–16] was modified in the PAWMI program.[7,8] The goodness scale ranges from 0.001 for a complete mismatch to 0.999 for a complete match.

In the intIRpret program, peaks in the library spectra are picked by PUSHSUB, the goodness values are weighted by STO, and the three-level rules are written by AUTOGEN. A peak table is then created for the unknown mixture by PUSHSUB. PAIRS accesses that table and generate goodness values that indicate the probable presence of compounds in the mixture of unknowns.

10.2.5. PAIRSPLUS

As previously reported,[8] PAIRSPLUS was developed to limit the effect of spectral similarity. A detailed description of PAIRSPLUS can be found in Puskar et al.[8] PAIRSPLUS accesses both the complete array of known spectra and the PAIRS interpretation results, and subtracts the percentage of spectral similarly corresponding to the compound with the largest goodness value from all the remaining compounds' goodness values. Note that this is a subtraction of goodness values, not of actual spectra.

10.3. APPLICATION TO AIR MONITORING

A central focus of the work in our laboratory involves the interpretation of spectra of gases and vapors at concentrations of part-per-million to part-per-billion in ambient air. It was an objective of this research to apply intIRpret to these spectra.

The initial investigation in our group into the use of Fourier transform infrared (FTIR) for air monitoring was reported in 1986.[30] This was followed by application of the FTIR air monitoring technique to semiconductor device manufacturing process emissions[31–33] and to the general case of gases and vapors in air.[34–36]

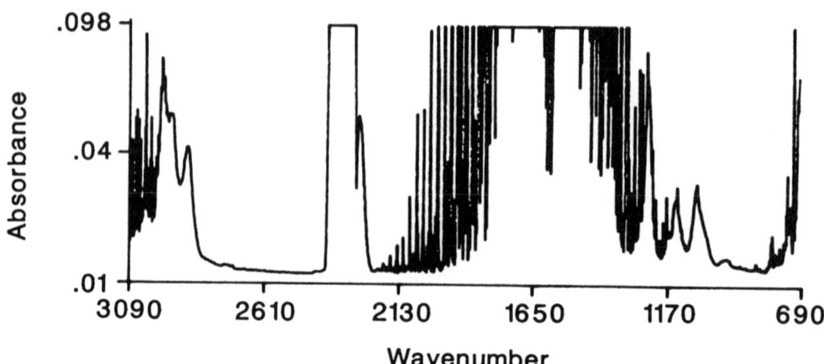

FIGURE 10.3. The spectrum of a 12-component solvent mixture at a concentration of 2 ppm each in ambient air.

During these investigations, it became clear that the use of intIRpret, and programs based on similar strategies, was not appropriate. Instead, optimal performance from the standpoint of a low incidence of false positive or false negative results at low concentrations could be obtained through the use of a least squares fit (LSF) program[33] written by Haaland et al.[37–39]

While the use of LSF under manual control yielded excellent results in terms of accuracy and limit of detection (LOD), it is unclear as to the best strategies for use of LSF under expert system control. An example of the difficulty is shown in Fig. 10.3.

This shows the spectrum of twelve common aliphatic and aromatic solvents from a typical paint spray formulation used in the automotive industry. The mixture is in ambient air at a concentration of 2 ppm for each component. At a concentration approaching the LOD for each solvent, the signal-to-noise ratio is seriously degraded from the spectrum shown here. Optimal peaks used in LSF quantitation include *o*-xylene at 720–760 cm^{-1}, 2-butanone at either 1140–1190 or 1350–1390 cm^{-1}, and ethylbenzene, and *n*-butanol at 2800–3200 cm^{-1}. The potential for interferences is very high, and the methods used by the human expert are, as yet, not well formulated. Writing expert system software for this application will be a challenge.

10.4. CONCLUSIONS

Results obtained through the use of PAWMI and intIRpret are shown in Table 10.2 for the training set of the spectra of 62 compounds frequently found at hazardous waste sites[3] and 67 four-component

TABLE 10.2. Results Obtained Using the PAIRS and PAIRSPLUS Subroutines of PAWMI and intIRpret[a]

PAIRSPLUS results obtained using	PAWMI[b]	intIRpret
Positives		
True	200	216
False	77	40
Improvement		40%
Negatives		
True	3,809	3,840
False	68	52
Improvement		24%
Total decisions	4,154	4,154

[a] The training set consisted of 62 compounds frequently found at hazardous waste sites.(3) The test mixtures consisted of 67 four-component mixtures of chlorobenzene, 1,1,1-trichloroethane (TCE), toluene, and benzene.
[b] These data do not match those previously reported(8) because the data set has been altered.

mixtures of those compounds. As stated previously, the difference between PAWMI and intIRpret are the subroutines STO and AUTOGEN, and a minor improvement in PAIRSPLUS. Thus, in PAWMI rule peaks are operator chosen and entered by hand into PAIRS using the subroutine CONCISE. All peaks are weighted equally, and a three-level logic structure is used to compensate for shifts of peak positions from the spectrum of the pure compound to the spectrum of the mixture.

In intIRpret, rule peaks are chosen by STO, and weighted for frequency of occurrence ($k1$), intensity ($k2$), and for the crossterm ($k3$). Rules are entered automatically by AUTOGEN and compiled into PAIRS. The software system is several orders of magnitude faster than when peaks were entered manually, immune from mistakes made when complex data is entered manually, and based on results that are consistently applied regardless of the operator or data set.

These data show a 40% decrease in false positive results and a 24% decrease in false negative results when intIRpret is compared to PAWMI. Some additional improvements in results can be expected after completion of a study of the optimal values of window widths, window weighting factors, and the relative weights of $k1$, $k2$, and $k3$. A certain degree of uncertainty, however, will remain in the direct interpretation of the infrared spectra of mixtures due to peak shifts in solution, the similarity of the spectra of structurally similar compounds, and the inability of the peak-picking routines to recognize the presence of peaks that appear as unresolved shoulders or in poorly resolved envelopes.

Factors that prevent the application of intIRpret for use in the interpretation of the spectra of mixtures of traces of gases and vapors in ambient air include potential interferences, high noise levels, and the poorly defined interpretive processes used by human experts for solving these problems.

ACKNOWLEDGMENTS

The authors would like to thank Greg Kinnes for his help in preparing the mixtures and acquiring the IR spectra, and to Mary Weed for preparation of manuscript figures. This work was supported by grants 1-R0-0H02066-0- and OH02404-01 from the National Institute for Occupational Safety and Health of Centers for Disease Control.

REFERENCES

1. M. A. Puskar, S. P. Levine, and R. Turpin, "Compatibility Testing and Materials Handling" in: *Protecting Personnel at Hazardous Waste Sites* (S. P. Levine and W. F. Martin, eds.), Ch. 6, Butterworths, Woburn, MA (1985).
2. D. F. Gurka, "Project Summary: Interlaboratory Comparison Study: Methods for Volatile and Semivolatile Compounds," Environmental Monitoring Systems Laboratory, Las Vegas, NV, EPA-600/S4-84-027 (June 1984).
3. P. A. Hallstedt, M. A. Puskar, and S. P. Levine, *J. Haz. Waste Haz. Mat.* **3**(2), 221–232 (1986).
4. W. P. Eckel, D. P. Trees, and S. P. Kovell, "Distribution and Concentration of Chemicals and Toxic Materials Found at Hazardous Waste Dump Sites," Proceedings of the National Conference on Hazardous Waste and Environmental Emergencies, Washington, D.C., (May 1985).
5. J. D. Mayhew, G. M. Sodaro, and D. W. Carroll, "A Hazardous Waste Site Management Plan," Chemical Manufacturers Association, Washington, D.C. (1982).
6. "The Hazardous and Solid Waste Amendments of 1984," *Congr. Rec.* H11103 Oct. 3, (1984).
7. M. A. Puskar, S. P. Levine, and S. R. Lowry, *Anal. Chem.* **58,** 1156–1162 (1986).
8. M. A. Puskar, S. P. Levine, and S. R. Lowry, *Anal. Chem.* **58,** 1981–1989 (1986).
9. M. A. Puskar, S. P. Levine, and S. R. Lowry, *Environ. Sci. Technol.* **21,** 90–96 (1987).
10. H. B. Woodruff and M. E. Munk, *J. Org. Chem.* **42**(10), 1761 (1977).
11. H. B. Woodruff and M. E. Munk, *Anal. Chim. Acta* **95,** 13–23 (1977).
12. H. B. Woodruff and G. M. Smith, *Anal. Chem.* **52,** 2321–2327 (1980).
13. H. B. Woodruff and G. M. Smith, *Anal. Chim. Acta* **133,** 545–553 (1981).
14. S. A. Tomellini, D. D. Saperstein, J. M. Stevenson, G. M. Smith, and H. B. Woodruff, *Anal. Chem.* **53,** 2367–2369 (1981).
15. S. A. Tomellini, J. M. Stevenson, and H. B. Woodruff, *Anal. Chem.* **56,** 67–70 (1984).
16. S. A. Tomellini, R. A. Hartwick, J. M. Stevenson, and H. B. Woodruff, *Anal. Chim. Acta* **162,** 227–240 (1984).

17. T. Blaffert, *Anal. Chim. Acta.* **161,** 135–148 (1984).
18. J. Zupan and M. E. Munk, *Anal. Chem.* **57,** 1609–1616 (1985).
19. M. O. Trulson and M. E. Munk, *Anal. Chem.* **55,** 2137–2142 (1983).
20. D. S. Frankel, *Anal. Chem.* **56,** 1011–1014 (1984).
21. S. R. Lowry and D. A. Huppler, **55,** 1288–1291 (1983).
22. P. C. Jurs and T. L. Isenhour, *Applications of Pattern Recognition,* Wiley, New York (1975).
23. G. T. Rasmussen, T. L. Isenhour, S. R. Lowry, and G. L. Ritter, *Anal. Chim. Acta.* **103,** 213–221 (1978).
24. J. A. de Haseth, H. B. Woodruff, S. R. Lowry, and T. L. Isenhour, *Anal. Chim. Acta* **103,** 109–120 (1978).
25. D. D. Saperstein, *Appl. Spectrosc.* **40**(3), 344–348 (1986).
26. *Computer Supported Data Bases,* J. Zupin, ed., Wiley, New York (1986).
27. P. C. Jurs, "Spectral Library Searching and Structure Elucidation," Ch. 16 in: *Computer Software Applications in Chemistry* (P. C. Jurs, ed.), Wiley, New York (1986).
28. K.-S. Kwok, R. Venkataragahaven, and F. W. McLafferty, *J. Amer. Chem. Soc.* **95,** 4185–4194 (1983).
29. B. L. Atwater, D. B. Stauffer, F. W. McLafferty, and D. W. Peterson, *Anal. Chem.* **57,** 899–903 (1985).
30. W. F. Herget and S. P. Levine, *Appl. Indus. Hyg.* **1,** 110 (1986).
31. C. R. Strang, S. P. Levine, and W. F. Herget, *Amer. Indus. Hyg. Assoc. J.* **50,** 70–77 (1989).
32. C. R. Strang and S. P. Levine, *Amer. Indus. Hyg. Assoc. J.* **50,** 78–83 (1989).
33. L. S. Ying and S. P. Levine, *Anal. Chem.* **61,** 677–683 (1989).
34. L.-S. Ying, S. P. Levine, C. R. Strang, and W. F. Herget, *Amer. Indus. Hyg. Assoc. J.* **50,** 354–359 (1989).
35. S. P. Levine, L.-S. Ying, and C. R. Strang, *Appl. Indus. Hyg.* **4,** 180–187 (1989).
36. L.-S. Ying and S. P. Levine, *Amer. Indus. Hyg. Assoc. J.* **50,** 360–365 (1989).
37. D. M. Haaland and R. G. Easterling, *Appl. Spectrosc.* **34,** 59 (1980).
38. D. M. Haaland, R. G. Easterling, and D. A. Vopika, *Appl. Spectrosc.* **39,** 73 (1985).
39. D. M. Haaland and R. G. Easterling, *Appl. Spectrosc.* **36,** 6 (1982).

Chapter 11

Approaches to Pyrolysis/Mass Spectrometry Data Analysis of Biological Materials

Kent J. Voorhees, Peter B. Harrington, Thomas E. Street, Stephen Hoffman, Steven L. Durfee, Joseph E. Bonelli, and Cynthia S. Firnhaber

11.1. INTRODUCTION

The earliest method of data analysis for pyrolysis/mass spectrometry (Py/MS) was the direct visual comparison between two or three spectra. This was the approach used by Zemany[1] whose major task was simply to decide if the mass spectra of pepsin and albumin were different. This method was successful as long as the data set was small and the

Kent J. Voorhees, Peter B. Harrington, and Thomas E. Street • Department of Chemistry and Geochemistry, Colorado School of Mines, Golden, Colorado 80401. Stephen Hoffman, Steven L. Durfee, and Cynthia S. Firnhaber • Somatogen, Inc., Broomfield, Colorado 80020. Joseph E. Bonelli • Health Science Center, University of Colorado, Denver, Colorado 80262.

differences between spectra were large. Unfortunately, there is only a small number of problems that fit into this category. In most examples, Py/MS experiments produce large amounts of data with the differences between samples being as subtle as minor changes for two or three of the mass spectral peaks. Both of these data characteristics demand a statistical comparison from several replicates for differentiation.

Pattern recognition techniques were first applied to data from pyrolysis/gas chromatography (Py/GC). Sekhon and Carmichael[2] used a taxonometric map in their work on fungi; Vincent and Kulik[3] employed a similarity index for *Penicillium* classification; Merritt and Robertson[4] described a set theory method for classifying various fingerprints and, finally, Reiner and co-workers[5] developed a computer-based library search program to aid in the identification of bacteria.

11.2. MULTIVARIATE Py/MS DATA ANALYSIS

Meuzelaar and his group at the F.O.M. Institute in Amsterdam were the first to apply pattern recognition techniques to Py/MS data.[6,7] The F.O.M. group showed that multivariate statistics tremendously expanded the usefulness of Py/MS. Their early procedures autoscaled the normalized data (each peak represented as a percentage of the total ion intensity) according to

$$x_i' = (x_i - \bar{x}_i)/s_i \tag{11.1}$$

where x_i' is the autoscaled intensity at mass i, x_i is the intensity at mass i, $\bar{x}_i$ is the mean intensity of mass i, and s_i is the standard deviation of the ith mass. If the information concerning the group membership was understood, Fisher weighting[8] was used

$$(WF)_{j,m,n} = \frac{(\bar{x}_{m,j} - \bar{x}_{n,j})^2}{\sum_{k=1}^{Nm} \frac{(x_{k,m,j} - \bar{x}_{m,j})^2}{Nm} + \sum_{k=1}^{Nn} \frac{(x_{k,n,j} - \bar{x}_{n,j})^2}{Nn}} \tag{11.2}$$

where $(WF)_{j,m,n}$ is the Fisher weight of the jth mass for categories m and n, $\bar{x}_{m,j} - \bar{x}_{n,j}$ is the difference between the means of the two categories m and n, and Nn and Nm are the number of spectra in categories n and m, respectively.

Following these calculations, a point in multidimensional space was calculated for each spectrum, and the distance between points measured.

TABLE 11.1. Distance Table for Py/MS Data from Air Particles

	Hanksville	SLC	Smelter	DWPP[a]	UWPP[a]
Hanksville	3.3	8.0	6.1	6.3	5.8
SLC		4.3	9.2	4.8	6.2
Smelter			3.1	7.6	6.6
DWPP				2.5	5.5
UWPP					3.8

[a] Represents downwind and upwind collection at same power plant.

The following formula was used for this step

$$d_{i,j}^2 = \sum_{k=1}^{N} (x_{i,k} - x_{j,k})^2 \tag{11.3}$$

where $d_{i,j}^2$ is the squared distance between mass spectra i and j with the number of mass spectral peaks equal to N. Table 11.1 represents a distance table.[(9)] The results of the distance calculation were displayed in matrix form, which are difficult to interpret and summarize. To overcome this problem, the Kruskal method of nonlinear mapping was used.[(10)] Nonlinear mapping is a display technique that allows the data in multidimensional space to be displayed graphically in a lower data space. The mathematical procedure involves an interative process that attempts to order the data points in a lower data space with the same relative order while maintaining the distance between points as nearly as possible. Figure 11.1 presents the nonlinear map corresponding to Table 11.1.

Although the nonlinear mapping technique allows for convenient differentiation between samples, it does not provide a means of determining the chemical difference responsible for the observed separation on the map. Extraction of the chemical information has been accomplished using factor analysis.[(11,12)] The factor–factor plots (K–L plots) produced from the analysis show the same relationship between samples as observed for the nonlinear map and, at the same time, processes the original data set to allow for extraction of the chemical information. Initially, the results from factor analysis are abstract in that it would be fortuitous if the data were separated into meaningful chemical categories. However, through a graphical rotation procedure,[(13,14)] the axes can be rotated to reveal the chemical information. The following equations illustrate the mathematical expressions

$$\begin{aligned} F_x' &= F_x \cos\alpha + F_y \sin\alpha \\ F_y' &= -F_y \sin\alpha + F_x \cos\alpha \end{aligned} \tag{11.4}$$

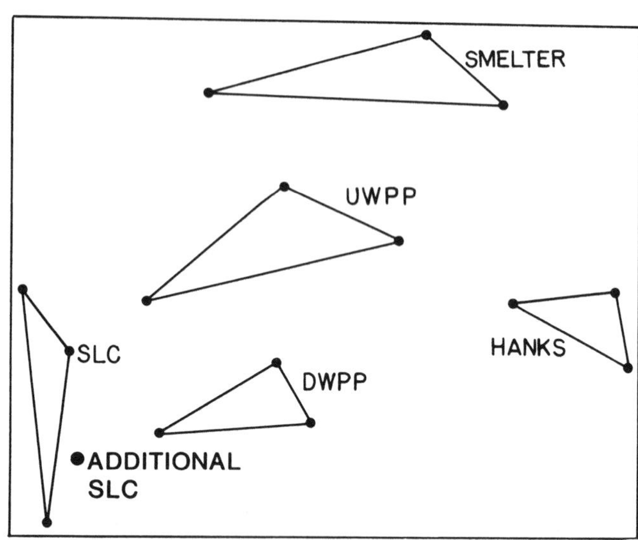

FIGURE 11.1. Nonlinear map of Py/MS air particulate data.

where F' is the original eigenvector (F) rotated to an angle α in the x,y-plane. A factor spectrum can be generated from the loadings calculated from this procedure. Factor spectra have been used successfully for identifying components in mixtures.[15–18]

11.3. APPLICATIONS OF MULTIVARIATE ANALYSIS

This general procedure is demonstrated by the following two examples. The first differentiates a series of Mycobacteria and the other distinguishes between two forms of *M. kansasii.* The classification of Mycobacteria is a clinical area where Py/MS could be routinely applied.[19] In the following example, seven Mycobacteria—*M. intracellulare, M. tuberculosis, M. kansasii, M. chelonae, M. fortuitum, M. smegmatis,* and *M. avium*—were investigated. In this particular study, we were interested in separating *M. tuberculosis* from the remainder of the Mycobacteria. Figure 11.2 is the histogram of the scores on a rotated factor in which separation of *M. tuberculosis* from the other bacteria was obtained. Another interesting aspect of this study is that *M. avium* and *M. intracellulare* were separated using the appropriate factors. Classical biochemical approaches do not adequately distinguish between these two organisms and usually report them as the A–I complex. Although identification of these organisms have not been as important in

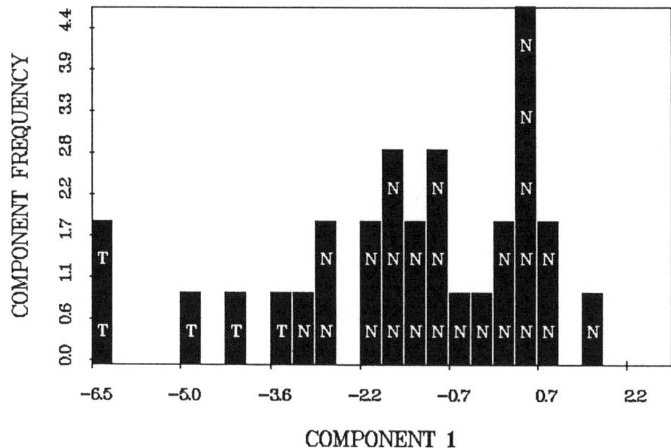

FIGURE 11.2. Histogram of rotated factor scores for Mycobacteria data.

the past as other Mycobacteria, the present-day problem of AIDS places higher demands on the identification procedures for all Mycobacteria.

The use of factor spectra to aid in the interpretation of the factor–factor plots is illustrated by the determination of the chemical differences between two strains of *M. kansasii* (Fig. 11.3). In this example, one of the strains possessed a rough colony morphology, while the other strain had a smooth colony morphology. A clear separation is observed for the organisms in factors 1 and 3. The factor spectrum shown in Fig. 11.4(a) was generated by rotating these factors for maximum

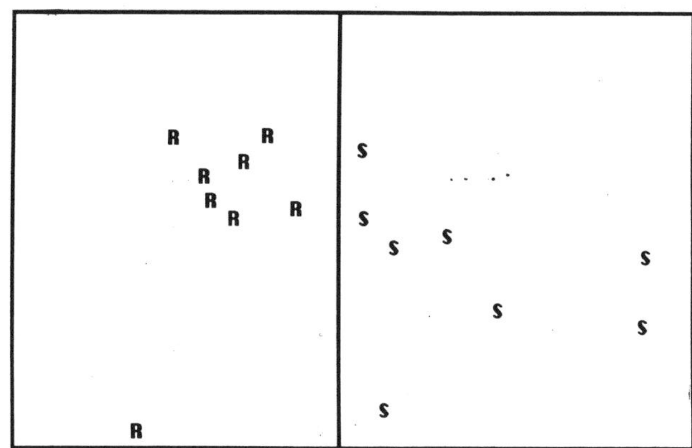

FIGURE 11.3. *K–L* plot for two strains of *M. kansasii.*

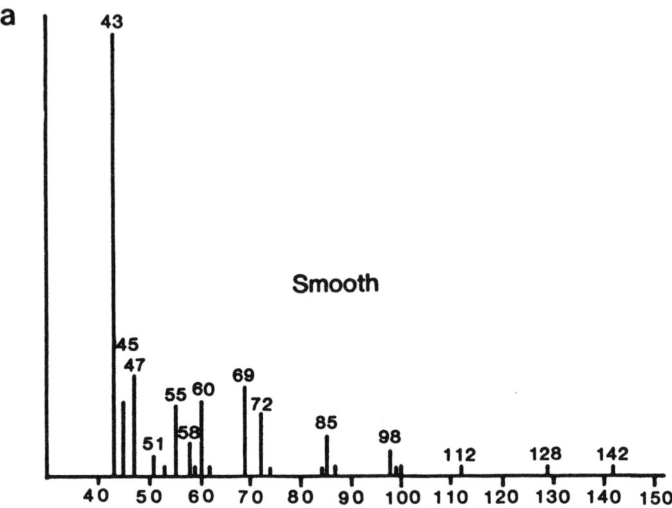

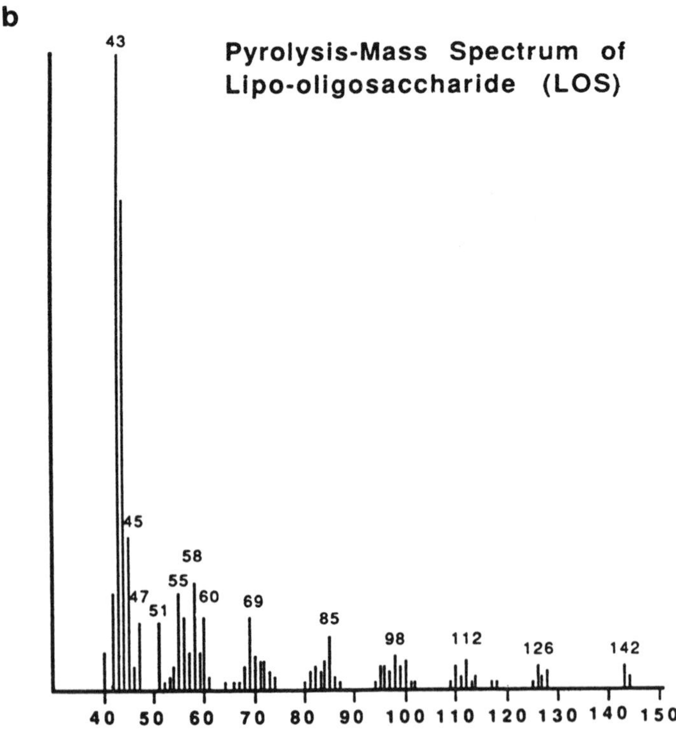

FIGURE 11.4. (a) Factor spectrum showing the chemical differences between smooth and rough colonies for *M. kansasii.* (b) Mass spectrum of lipooligosaccharide.

separation between the two strains of bacteria. Comparison of this spectrum to the mass spectrum [Fig. 11.4(b)] of a lipooligosaccharide isolated from the smooth colony material by extraction and liquid chromatography, indicates that the two spectra are of the same compound. Although biomarkers have been identified in Py/GC work,[20] this represents the first reported biomarker identified by Py/MS using factor analysis.

This same approach has been applied to several bioengineering problems. For example, the incorporation of plasmid is usually detected by the plasmid expression. Because of a change in overall structure after the introduction of the plasmid, it was speculated that Py/MS should be able to detect the introduction. The system used for this study was wild-type *Escherichia coli* with two plasmids known to introduce antibiotic resistance. Figure 11.5(a) shows the *K–L* plots of factor 1 and 2 for

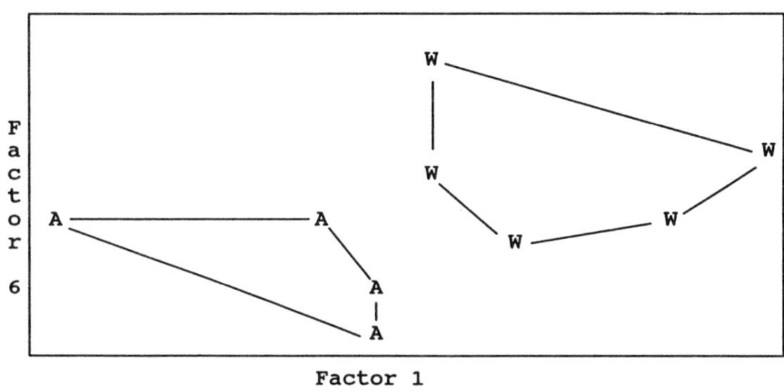

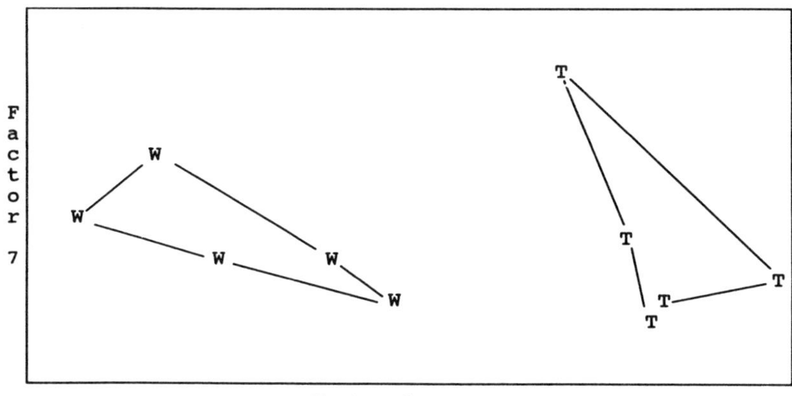

FIGURE 11.5. *K–L* plots for wild-type *E. coli* and plasmid reacted *E. coli* (*W* = wild type; *A* = ampicillin; and *T* = tetracycline).

both the wild-type *E. coli* and *E. coli* that had been transformed with the ampicillin resistant plasmid. Figure 11.5(b) shows a similar plot for wild-type and tetracycline-resistant *E. coli*. Both figures clearly show a difference between the wild type and the transformed organism. These differences have been studied for the two systems by other procedures. In the case of tetracycline, the expression is due to changes in the bacterial cell wall. For ampicillin the organism produces β-lactamase which modifies the structure of the antibiotic.

The study of the activity of collagenase produced from mammalian and bacterial cells offers additional insight into the potential of Py/MS for the analysis of bioengineering samples. Figure 11.6 shows the *K–L* plot of factors 4 and 5 for active and nonactive collagenases produced from both mammalian and bacterial cells. Examination of this plot suggests that activity, as measured by the degradation of collagen, is indicated on factor 5 and that factor 4 distinguishes the differences between mammalian and bacterial collagenases.

Two supervised procedures, discriminant analysis[21] and SIMCA,[22]

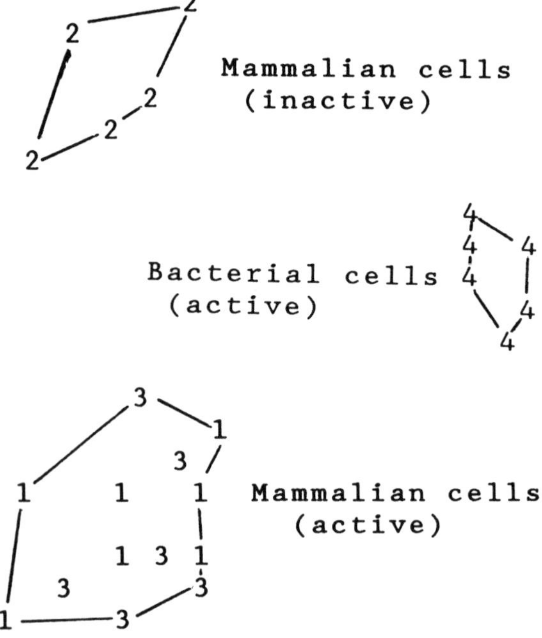

FIGURE 11.6. *K–L* plot showing activity for collagenase.

have also been applied to Py/MS data. Discriminant analysis assumes that a spectrum has been generated from materials associated with a number of known populations. By generating linear functions that can be used to separate known training sets, an unknown sample can be classified as belonging to one of the training sets.

The SIMCA method calculates a principal components model for each training set. An unknown is then related through a Euclidean distance to each principal components axis of the training sets. If the distance is within a specified standard deviation calculated for the training set, then the sample is assigned to that particular training set.

Classification procedures, when combined with unsupervised techniques, provide a powerful approach to data analysis. In a study designed to differentiate *Xanthomonas* bacteria from *Pseudomonas* bacteria, a combination of factor analysis and discriminant analysis was used for identification of each bacteria. The *Xanthomonas* bacteria are plant pathogens which can cause serious damage to a number of plants including citrus trees. It is therefore necessary that these two bacterial species be rapidly and reproducibly differentiated. Unfortunately, biological procedures are not very reliable and require up to four weeks to obtain results. Table 11.2 lists the bacteria used in the study.

Figure 11.7 shows spectra of *Pseudomonas* and *Xanthomonas* bacteria. The gross structure of the two spectra is very similar. Figure 11.8 shows the K–L plots for factors 5 and 8. This plot shows that a general separation trend exists for the two bacteria, and that a supervised technique, such as discriminant analysis, could be applied. By generating

TABLE 11.2. Summary of *Xanthomonas* and *Pseudomonas* Bacteria Used in Study

Pseudomonas
P. aeruginosa
P. fluorescens
P. alcaligenes
P. floridana
P. pseudoalcaligenes
Xanthomonas
X. campestris pv. citri
X. campestris pv. citri (Asia)
X. campestris pv. citri (Argentina)
X. campestris pv. citri (Brazil)
X. campestris pv. citri (Florida)
X. campestris pv. citri (unknown)
X. campestris pv. phaseoli

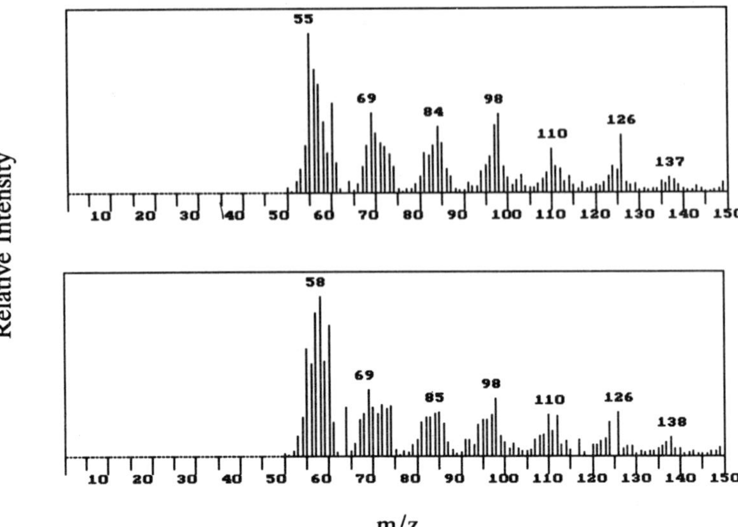

FIGURE 11.7. Py/MS spectra of *Pseudomonas* and *Xanthomonas*.

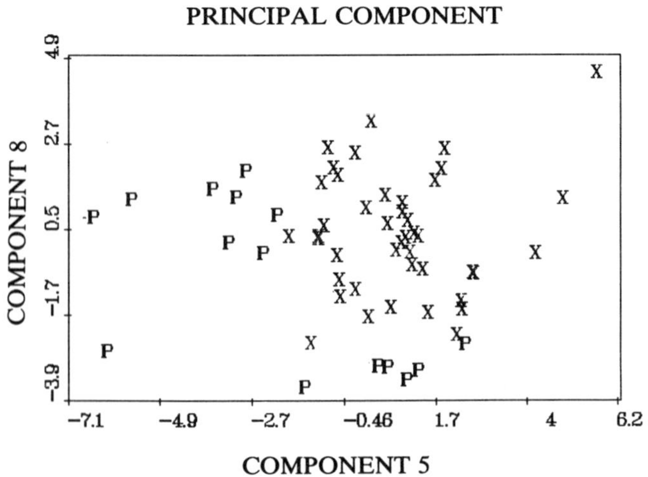

FIGURE 11.8. *K–L* plot showing separation of *Xanthomonas* and *Pseudomonas* bacteria (X = *Xanthomonas* and P = *Pseudomonas*).

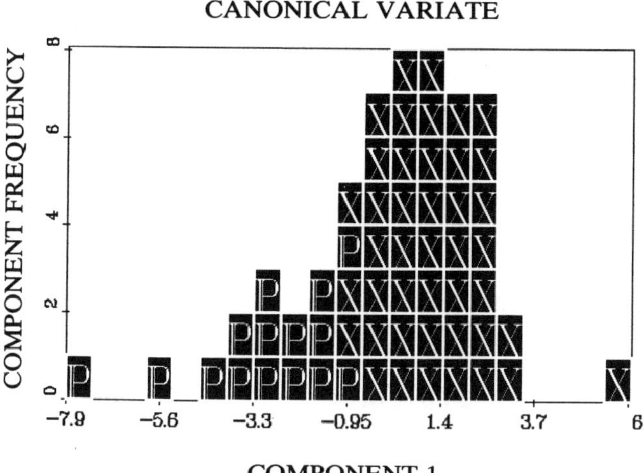

FIGURE 11.9. Histogram of scores for linear discriminant analysis for analysis *Xanthomonas* and *Pseudomonas*.

linear functions that can be used to separate known training sets, an unknown can be classified as belonging to one of the training sets.

The *Xanthomonas* and *Pseudomonas* data sets included 58 spectra and were treated as two known populations. Discriminant analysis was performed on this data and was evaluated using a cross validation procedure. Cross validation removes one sample at a time from the training set and treats it as an unknown. After cross validating all the samples in the data set, linear discriminant analysis correctly classified 77.6% of the samples. Figure 11.9 is a histogram of the scores on the linear discriminant which did not separate all the spectra. This level of performance is not acceptable for most applications and requires that improved methods be explored.

11.4. A RULE-BUILDING EXPERT SYSTEM

Expert systems are alternative methods that do not require linear separability or known probability distributions. These systems are usually composed of three parts: (1) the knowledge base, which is composed of facts and heuristics relating to the problem domain; (2) the inference engine, which decides on which facts and heuristics are to be used; and (3) the global data base, which does the accounting of the current state of the problem.

Extracting knowledge from a human expert in a consistent and efficient form is a difficult and tedious task. This problem is bad enough to be named the knowledge acquisition bottleneck and causes practical expert systems to take man-years to develop.

Commercial rule-building expert systems are available.[23] These systems use the ID3 algorithm to acquire a knowledge base of rules from symbolic or categorical data[24,25] and do not directly use numerical data. A disadvantage with commercial systems is that a source code is frequently unavailable and the users are uninformed regarding a program's algorithms and idiosyncracies.

11.5. Py/MS EXPERT SYSTEM

A rule-building expert system was devised for pyrolysis/mass spectra classification.[26] The rule-building expert system derives its classification rules from training sets of data. An expert system that builds its own rules eliminates the knowledge acquisition bottleneck. This expert system also devises its own certainty factors. The advantage of using the expert system over traditional classification methods such as linear discriminant analysis is that no assumptions requiring known probability distributions or linear separability are required.

The first step in any classification process is to convert the data into a suitable format for classification. Pyrolysis/mass spectra typically consist of intensity values obtained over a mass range of about 200 amu. The number of spectra in a data set may vary between ten spectra and several hundred spectra. For supervised methods of classification, the number of spectra should exceed three times the number of masses to obtain an accurate classification function.[27] An accurate classification function classifies by meaningful features in the spectra instead of by random features (noise). A classification function will converge to its true value as the number of samples is increased in the training set.

The pyrolysis/mass spectra are transformed into data that has three or more times the number of observations than variables by an eigenvector transformation, which maximizes the retained variance (information) while reducing the number of variables. A preselected number of eigenvectors (principal components) is calculated, and the data matrix is projected onto the basis set of eigenvectors to obtain transformed variables (scores). This procedure is also used in the traditional methods of Py/MS data analysis.

After compression, the data are used to build the classification rules and decision tree. A modified ID3 algorithm was devised for numerical data. This algorithm examines one variable at a time where each variable is a principal component score. The algorithm finds the best attribute to split the data based on the entropy of classification. Rules are constructed that determine whether a variable is greater or less than the attribute.

The data is sorted by each variable or score. The classes of the samples in the training set are known, and the algorithm selects attributes that divide the classes. An attribute is the median value between class changes of the sorted data. The attribute with the lowest entropy is selected for a given variable and is compared with the attributes of the other variables. The attribute with the overall lowest entropy is used in a rule that partitions the data.

The algorithm works by selecting attributes that decrease the entropy of the classification as it builds a tree of rules. The entropy of classification for class C and attribute A, $H(C \mid A)$, is a measure of uncertainty after a classification is made according to a decision criterion, and is given by the following two equations:

$$H(C \mid a_j) = -\sum_{i=1}^{N} p(c_i \mid a_j) \ln p(c_i \mid a_j) \tag{11.5}$$

$$H(C \mid A) = \sum_{j=1}^{M} p(a_j) H(C \mid a_j) \tag{11.6}$$

Equation (11.5) gives the entropy for an attribute. N is the number of different classes, and $p(c_i \mid a_j)$ is a probability obtained by counting the number of observations of class i and dividing that number by the total number of observations of the jth attribute. Usually entropies are calculated in units of bits using a base-2 logarithm, but the base of the logarithm only affects the units of the results. The same results are obtained if natural logarithms (ln) are used, except the units will be in nits instead of bits.

Equation (11.6) is the sum of the entropies for each attribute weighted by the prior probability that the attribute occurs. M is the number of attributes, and $p(a_j)$ is the number of observations with a given attribute divided by the total number of observations.

A zero entropy will be obtained when all the samples in the training set are classified correctly. The decision tree may be evaluated by examining the number of rules that are generated. Even random data will be 100% correctly classified by a decision tree; however, the tree structure will be deep. For a binary classification tree, the total number of decision rules required for a worst-case situation is simply the number of observations minus one. For a best-case situation, the number of rules will be the number of classes minus one. A simple measure of the efficiency of the decision tree is

$$\left(1.0 - \frac{\text{Number of rules} - \text{number of classes} + 1}{\text{Number of observations} - \text{number of classes}}\right) \times 100 = \text{eff\%} \tag{11.7}$$

Cross validation is another method for evaluating an expert system. It is useful because the goal of supervised classification techniques is to classify unknown samples based on a model developed from a training set of data.

Certainty factors are measures of the reliability of the results output from an expert system. Unfortunately, certainty factors are often arbitrary and are as difficult for a human expert to devise as the rules. It is important for expert systems to also generate their own certainty factors as well as rules. The certainty factors are generated for each attribute by

$$\mu_{a_j}(x) = e^{-|a_j - x|/\bar{a}} \tag{11.8}$$

where $\mu_{a_j}(x)$ is a certainty factor, a_j is the attribute for rule j, e is the exponential, x is the rule variable, and $\bar{a}$ is the average of all the variables in the training set possessing that attribute. The certainty factors vary between 0 and 1. If more than one rule is required for a classification, the minimum certainty factor is used.

11.6. CLASSIFICATION OF BACTERIA

The expert system was applied to a data set consisting of 58 pyrolysis/mass spectra obtained from *Pseudomonas* and *Xanthomonas* bacteria (see Table 11.2). These spectra contained a significant background from the growth media.

The rule-building expert system was compared to linear discriminant analysis for classifying the bacteria. Each method of analysis was evaluated by determining the percent of accurate predictions during cross validation.

Figure 11.10 is the decision tree obtained from the expert system. Two rules were created (rectangles). The first rule classifies all spectra with scores on the fifth eigenvector less than −2.0 (−2.0 is the attribute) as *Pseudomonas,* class P. The second rule classifies all scores on the eighth eigenvector less than −2.4 as *Pseudomonas*. Figure 11.8 is a *K–L*

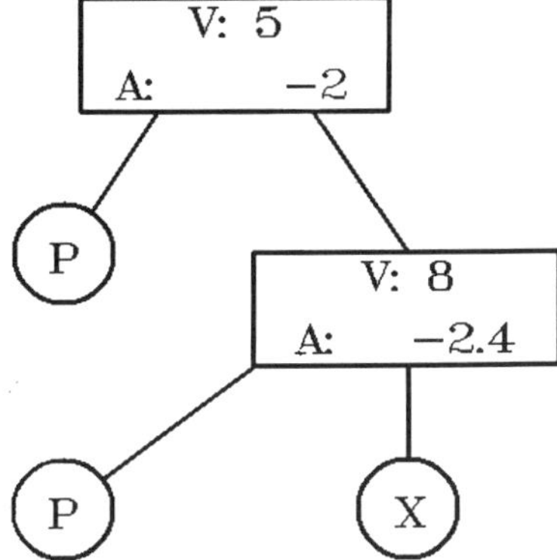

FIGURE 11.10. The expert system classification tree. The rectangles are rules, the circles are the classes (P = *Pseudomonas* and X = *Xanthomonas*), V are the variables, and A are the attributes.

plot of the scores on eigenvectors 5 and 8. With only two rules the entire data set is correctly classified, the entropy is zero, and the decision tree is 98.2% efficient. Using cross validation, the expert system correctly classified 98.3% of the spectra. The spectrum that was misclassified had a low certainty factor of 12.3%. The average certainty factor for the entire set of spectra was 47 ± 23%.

11.7. CONCLUSIONS

Pyrolysis/mass spectrometry in the 1950s and 1960s had to rely on visual inspection for interpretation. The 1990s offer a boon of computational power coupled to the mass spectrometer. Information is increasingly available to the spectroscopist in the form of mass storage and on-line data bases. The application of knowledge-intensive methods (artificial intelligence) will depend on the automated acquisition, storage and inference from Py/MS data. The future of Py/MS data analysis is bright and will combine knowledge intensive methods with traditional multivariate analysis.

ACKNOWLEDGMENTS

The authors wish to thank Dr. Stephen J. DeLuca, Mr. Steven M. Muskal, Mr. Douglas A. Wiesner, and Dr. Lewis Fink for their assistance and helpful comments. Support of this work by the Army Research Office under grant #DAAG29-85-K-0199 and Somatogen, Inc. is gratefully acknowledged.

REFERENCES

1. P. D. Zemany, "Identification of Complex Organics by MS Analysis of Their Pyrolysis Products," *Anal. Chem.* **24,** 1709–1713 (1952).
2. A. S. Sekhon and J. W. Carmichael, "Classification of Some Gymniasaceae by Py/GC Using Added Marker Compounds," *Sabouraudia,* **13,** 83–88 (1975).
3. P. G. Vincent and M. M. Kulik, "Differentiation of Species and Strains of Several Members of the *Aspergillus flavus* Group," *Appl. Microbiol.* **20,** 957–963 (1970).
4. C. Merritt, Jr. and D. H. Robertson, "Qualitative Analysis of GC Eluates by Means of Vapour Phase Pyrolysis. II. Classification by Set Theory," *Anal. Chem.* **44,** 60–63 (1972).
5. F. M. Menger, G. A. Epstein, D. A. Goldberg, and E. Reiner, "Computer Matching of Pyrolysis Chromatograms of Pathogenic Organisms," *Anal. Chem.* **44,** 423–24 (1972).
6. W. Eshuis, P. G. Kistemaker, and H. L. C. Meuzelaar, "Some Numerical Aspects of Reproducibility and Specificity," in: *Analytical Pyrolysis* (C. E. R. Jones and C. A. Cramers, eds.), pp. 151–166, Elsevier, Amsterdam (1977).
7. H. L. C. Meuzelaar, P. G. Kistemaker, E. C. Beuvery, P. M. Boonekamp, and R. H. Tiesjema, "Characterization of *Neisseria meningitides* Capsular Polysaccharides Containing Sialic Acid by Pyrolysis-Mass Spectrometry," *Anal. Biochem.* **104,** 407–418 (1980).
8. R. A. Fisher, "The Use of Multiple Measurements in Taxonomic Problems," *Annals of Eugenics* **7,** 179–88 (1936).
9. K. J. Voorhees and F. D. Hileman, "Pyrolysis/Mass Spectrometry Studies of Atmospheric Particles," *J. Anal. Appl. Pyrol.* **3,** 151–60 (1981).
10. J. B. Kruskal, "Nonmetric Multidimensional Scaling: A Numerical Method," *Psychometrika* **29,** 1–27 (1964).
11. D. Child, *The Essentials of Factor Analysis,* Academic, New York (1970).
12. E. R. Malinowski and D. G. Howery, *Factor Analysis in Chemistry,* Wiley, New York (1980).
13. R. J. Rummel, *Applied Factor Analysis,* Northwestern University Press, Evanston, IL (1970).
14. R. B. Catell, *Factor Analysis,* Harper & Row, New York (1976).
15. K. J. Voorhees and R. Tsao, "Smoke Aerosol Analysis by Pyrolysis/Mass Spectrometry/Pattern Recognition for Assessment of Fuels Involved in Flaming Combustion," *Anal. Chem.* **57,** 1630–36 (1985).
16. W. Windig and H. L. C. Meuzelaar, "Nonsupervised Numerical Component Extraction from Pyrolysis/Mass Spectra of Complex Mixtures," *Anal. Chem.* **56,** 2297–2303 (1984).
17. W. Windig, J. Haverkamp, and P. G. Kistemaker, "Interpretation of Sets of Pyrolysis/Mass Spectra by Discriminant Anaysis and Graphical Rotation," *Anal. Chem.* **55,** 81–89 (1983).
18. L. V. Vallis, H. J. MacFie, and C. S. Gutteridge, "Comparison of Canonical Variates Analysis with Target Rotation and Least Squares Regression as Applied to Pyrolysis/Mass Spectra of Simple Biochemical Mixtures," *Anal. Chem.* **57,** 704–709 (1985).
19. G. Wieten, J. Haverkamp, H. L. C. Meuzelaar, H. W. B. Engel, and L. G. Berwald. "Pyrolysis/Mass Spectrometry: A New Method to Differentiate Between the Mycobacteria of the Tuberculosis Complex and other Mycobacteria," *J. Gen. Microbiol.* **122,** 109–118 (1981).
20. L. W. Eudy, M. D. Walla, J. R. Hudson, S. L. Morgan, and A. Fox, "Gas

Chromatography/Mass Spectrometry Studies on the Occurrence of Acetamide, Propioamide, and Furfuryl Alcohol in Pyrolysates of Bacteria, Bacterial Fractions, and Model Compounds," *J. Anal. Appl. Pyrol.* **7,** 231–247 (1985).
21. G. L. Ritter and H. B. Woodruff, "Dimensionality and the Number of Features in Learning Machines," *Anal. Chem.* **49,** 2116–2118 (1977).
22. S. Wold, "Pattern Recognition by Means of Disjointed Principal Components Models," *Pattern Recogn.* **8,** 127–139 (1976).
23. M. P. Derde, L. Buydens, C. Guns, D. L. Massart, and P. K. Hopke, "Comparison of Rule-Building Expert Systems with Pattern Recognition for the Classification of Analytical Data," *Anal. Chem.* **59,** 1868–1871 (1987).
24. J. R. Quinlan, "Learning Efficient Classification Procedures and Their Application to Chess End Games," *Machine Learning: An Artificial Intelligence Approach* (R. S. Michalski, J. G. Carbonell, T. M. Mitchell, eds.), p. 463, Tioga Publishing Co., Palo Alto, CA (1983).
25. B. Thompson, and W. Thompson, "Finding Rules in Data," *Byte,* **11,** 149–158 (1986).
26. P. B. Harrington and K. J. Voorhees, "A Rule-Building Expert System for the Classification of Pyrolysis/Mass Spectra," presented at the 1988 Pittsburgh Conference, New Orleans, March 1988.
27. K. Varmuza, *Pattern Recognition in Chemistry,* Springer-Verlag, New York (1980).

Chapter 12

Theory and Application of the CIRCOM Software for Quantitative Spectroscopic Analysis

Gregory L. McClure and
Craig A. Lehmann

12.1. INTRODUCTION

12.1.1. Historical Aspects

There has been a growing trend in the last decade to use more calculation-intensive methods in quantitative infrared (IR) applications. This can be attributed in large part to the evolution of powerful, relatively inexpensive desktop computers with large memories. In addition, the improvements in IR instrument design features, especially microprocessor control and digital registration of spectra, have played a major role in enhancing the reliability and reproducibility of the acquired spectral data.

The impact of these developments has gradually been making itself felt in the world of practical analytical chemistry. In particular, the use of

---

Gregory L. McClure • Perkin-Elmer Corporation, Norwalk, Connecticut 06859-0227. Craig A. Lehmann • Department of Medical Technology, School of Allied Health Professions, State University of New York at Stony Brook, Health Sciences Center, Stony Brook, New York 11794.

multicomponent analyses that involve matrix calculations has increased substantially in the last half of this decade. Commercially available software packages have liberated the analyst from a large portion of the tedious aspects of the computations.

Prior to the proliferation of computerized IR instrumentation during the 1980s, the general practice of IR quantitative analysis consisted usually of relatively simple evaluations of spectral features on the basis of local absorbance maxima, possibly modified by subtraction of "baselines" estimated from absorbance minima in a neighbouring spectral region. This procedure worked satisfactorily if the number of variable components in the sample was not large, and if a characteristic band of the component of interest could be found that was not significantly overlapped by other compounds. Although matrix methods were well known in the 1970s as a means of handling multiple varying components, their use was less prevalent, partly because of limitations on the availability of commercially produced hardware and software to carry out the calculations.

12.1.2. Current Trends

When dedicated microcomputers (and in some cases minicomputers) became more widely available, multicomponent analysis using multiple linear regression (MLR) began to be considered for a wider range of applications. Moreover, there began to arise a trend toward using a greater portion of the mid-IR spectral region for the quantitative calculation, because matrix methods could easily cater to the larger data sets involved.[1–3]

More recently, the level of computational complexity in quantitative spectroscopy software has risen another level with the trend toward the incorporation of routines from the field of factor analysis. The potential of factor analysis in chemistry was first broadly demonstrated by Malinowski and Howery.[4]

The major strength of the factor analysis techniques rests in their ability to aid the analyst in isolating the desired relevant information that may be hidden in relatively large amounts of data. In recent years, analysts who deal with IR applications have become faced with instrumentation that can generate megabytes of data in minutes. The interest in deriving meaningful information from large data sets has become part of the basis for the recent reorientation toward, and activity in, factor analysis related quantitative techniques.

The course of evolution of factor analysis based quantitative software has diverged into two similar but unique programs which are known by the names principal components regression (PCR) and partial least squares (PLS). Each of these programs includes a step to carry out eigenanalysis on the dependent variable (such as spectral data) to evaluate the number of principal components in the data. In addition, both programs use some form of MLR to reach the final analytical result. The intervening steps and procedures, however, vary between the two.

The terms eigenanalysis, abstract factor analysis, and principal components analysis (PCA) are often used interchangeably to refer to the same operations. When multiple linear regression is combined with principal components analysis on the data of the dependent variable, the result is called principal components regression (PCR).[5,6] In the approach known as partial least squares (PLS), developed by Wold and coworkers, information contained in both the dependent and independent variables play a role in the development of the calibration model.[7–9]

Fredricks et al. described an enhancement to the simple PCR approach that refines the calibration model for each property by retaining only those factors judged to be statistically significant for the prediction of the respective property.[10] The statistical significance of each factor for each property may be decided on the basis of the relative error associated in the respective regression coefficient. This algorithm was developed under the name CIRCOM, which is an acronym for computerized infrared calculations on materials. It should be noted that CIRCOM and PLS, but not PCR, take into consideration both the absorbance and concentration information in the development of the final calibration model used to predict the properties of unknown samples.

The work presented in this chapter will include an elaboration on the originally published description of the theoretical basis of CIRCOM,[10] and will illustrate the use of CIRCOM with applications of varying levels of complexity.

12.2. GENERAL ISSUES OF QUANTITATIVE INFRARED ANALYSIS

12.2.1. Suitability of Infrared Data for Quantitation

Among the variety of physical properties that can be used as a basis for quantifying a particular component in a chemical system, the light absorption spectrum in the mid-IR region possesses a number of positive attributes. Most compounds have an IR spectrum that tends to be unique for that material. The absorption coefficients are large enough that measurable spectral responses can be obtained with small amounts of materials—easily microgram quantities, and, in some applications, nanograms quantities. It is also usually satisfactory to assume that the magnitudes of the absorbances in the spectral data vary as the sum of the absorbances contributed by each component, and that the absorbance contributed by each component at each spectral data element varies directly with the concentration of the component in the material being analyzed, or, more specifically, as the number of absorbing centers varies in the beam. This amounts to assuming that Beer's law is obeyed.[11]

12.2.2. The Effect of Spectral Bandwidths

Bandwidths of absorptions in the mid-IR (4000–400 cm^{-1}) are generally a small fraction of the spectral range, on the order of 6 to 10 cm^{-1} on average. Therefore, there is room in the spectral range for a large number of individual spectral features. Consequently, it is useful to sample and record many points in the spectral range in order to take full advantage of the information content of the spectral data. Moreover, it can be shown that there is advantage in including many points in the spectral range to contribute information to the multicomponent problem, especially when there is significant colinearity among some of the components. Spectra are said to be colinear when the differences between them are constant or constant plus a scaling factor times one of the spectra.

12.2.3. Calibration Requirements of Indirect Methods

The IR spectrum of most materials is most difficult to calculate accurately starting from first principles of light absorption, because the number of interactions makes for a problem of extreme complexity. Consequently, IR is not a primary method of analysis. It must use some external reference standard to establish the relationship between the observed absorbances and the concentration or physical property of interest. The practical consequence of this is that IR quantitative applications, like those of many other techniques, depend on a calibration step performed with known samples before a determination step can be performed on unknown samples of the material under investigation.

12.2.4. Limitations of Multiple Linear Regression

The spectral data obtained from computerized IR instruments is not a continuous report of the value of a physical property (e.g., percent transmittance) with respect to wavelength, wave number, time, etc. Rather, it is a systematic sampling of transmitted intensities at a series of small frequency ranges of light. The result is generally referred to as a digitized spectral data file. In the simplest terms, a digitized spectrum is just an ordered list of absorbance values.

The calibration data set will contain the digitized IR spectra of M prepared standard sample mixtures, and each spectrum will be represented by W spectral data points. Therefore, the calibration data set may be signified by the matrix D_{WM}, constituted of a block of absorbance values with W rows and M columns.

With IR spectra as a data source, there are usually more data points W per sample spectrum than there are calibration mixtures M, and therefore D is typically oversquare (more rows than columns) as drawn. The traditional way of relating the spectral data matrix D to the sample concentrations C involved the matrix representation of Beer's law, as in equation (12.1).

$$\mathsf{D}_{WM} = \mathsf{K}_{WN} * \mathsf{C}_{MN} \tag{12.1}$$

The use of MLR alone to solve the Beer's law equation has certain unfavorable limitations. Recently, these limitations of MLR have been reviewed by Haaland and Thomas.[12]

12.3. THE COMPUTATIONAL STEPS IN CIRCOM

12.3.1. General Approach

The CIRCOM software described in the following discussion was adapted to operate on the Perkin–Elmer model 7700 computer, as an extension of the features of the CDS-3 software for IR spectroscopy. Method development and unknown sample analysis were carried out from within the framework of the CDS-3 software as CIRCOM-specific enhancements of the regular set of CDS-3 commands. Calibration was carried out by running CIRCOM programs directly from the IDRIS operating system.

Many of the limitations on multivariate analysis that are related to pure MLR can be reduced or minimized by introducing some of the techniques of factor analysis into the calculations. With a factor analytical approach, it is possible to perform a substantial compression of the number of parameters needed to represent the data. Then, one can use an inverse least squares technique (also called the P-matrix approach) to find the empirical correlation between the concentrations or properties of interest and the intermediate variables calculated from the factor analysis.[13,14] The major problems associated with matrix inversion are minimized because of the way the factor analysis handles the data. Not only can the standards be tested as unknowns against the calibration, but a number of other useful statistical parameters can be derived from the calibration calculations. Moreover, the compression of the information in the data into a smaller number of calibration parameters can be much more extensive than is usually achieved with techniques that incorporate only MLR. The number of parameters needed to predict the properties of interest in unknown samples can be reduced to a minimum number, and that number can be estimated fairly well, in many instances, on the basis of one or more types of statistical tests.

12.3.2. Data Compression into a Correlation Matrix

If the data matrix D is multiplied by its transpose D^T, the resulting matrix Z is the product moment matrix of D, as shown in equation (12.2). The matrix Z will be symmetric about the diagonal and square, with M rows and columns, where M represents the number of standards in the calibration set:

$$Z_{MM} = D^T_{MW} * D_{WM} \tag{12.2}$$

12.3.3. Eigenanalysis and Derivation of Factor Loadings

According to the theory of eigenanalysis, it is possible to derive an orthonormal matrix L that diagonalizes the product moment matrix Z. This operation is indicated in equation (12.3), and it involves premultiplying Z by the transpose of L and postmultiplying by L. In literal terms the eigenanalysis equation states that there is a matrix L that diagonalizes Z, yielding the eigenvalue matrix E. Note that the column index of L is designated by H rather than M in equation (12.3). This notation is intended to signify that the number of significant types of variation in the data, H, may be less than or equal to the number of spectra, M, used to construct the product moment matrix Z. H is sometimes referred to as the rank (or an estimate of the rank) of the matrices D and Z.

$$L^T_{HM} * Z_{MM} * L_{MH} = E_{HH} \tag{12.3}$$

The matrix E contains the eigenvalues of the product moment matrix Z as the diagonal elements, which are arranged in decreasing order from upper left to lower right. The off-diagonal elements of E are all zero. The magnitude of each eigenvalue corresponds to a factor, that is, an independent type of variation found in the spectral data. It is found that, in many cases, the magnitudes of the eigenvalues decrease rapidly for the first several factors, and then tend to level off as one procedes further down the diagonal. When the eigenvalues tend to level off, this suggests that the types of variation in the data due to systematic effects (i.e., the absorbance signal generated by the components of the samples) have become roughly equal to the types of variation in the data due to random effects such as spectral noise. Consequently, the eigenanalysis process has effectively provided a means to estimate the number of factors related to independent components in the analytical system, and to separate them from the factors dominated by the random error sources or noise in the system.

It is worth noting a few of the relationships that arise from the special properties of the matrix L. Since L is constructed to be orthonormal, the relationship in equation (12.4) must hold by the definition of orthonormality.

$$\mathsf{L}^T_{HM} * \mathsf{L}_{MH} = \mathsf{L}_{MH} * \mathsf{L}^T_{HM} = \mathsf{I} \tag{12.4}$$

The columns of L are called eigenvectors, and the rows of L are referred to as the factor loadings.

12.3.4. Derivation of Eigenspectra

If the original data matrix D is multiplied by L, the product is a set of abstract factors or eigenspectra, F, of D, as is shown in equation (12.5):

$$\mathsf{F}_{WH} = \mathsf{D}_{WM} * \mathsf{L}_{MH} \tag{12.5}$$

Because of the orthonormality of L, equation (12.5) can be transformed into equation (12.6), which states that the original data can be reconstructed from a linear combination of the eigenspectra in F. Moreover, the reconstructed spectra may be expected to contain essentially all of the information of the original data, but with a lower noise level:

$$\mathsf{F}_{WH} * \mathsf{L}^T_{HM} = \mathsf{D}_{WM} \tag{12.6}$$

It can be shown by combination of the above equations that the dot product of any eigenspectrum (a column of the F matrix) with another eigenspectrum is zero, and the dot product of any eigenspectrum with itself is a scalar equal to the corresponding eigenvalue, as is written in equation 7.

$$\mathsf{F}^T_{HW} * \mathsf{F}_{WH} = \mathsf{L}^T_{HM} * \mathsf{Z}_{MM} * \mathsf{L}_{MH} = \mathsf{E}_{HH} \tag{12.7}$$

Note that all of the equations above apply to the full L and F matrices, in which all M columns are retained, and to the reduced matrices L1 and F1, in which only the H significant factors have been retained. In CIRCOM the value of H is known as the factor compression cut-off value, which it is convenient to refer to as the FCCV.

12.3.5. MLR with Factor Loadings

The second step in calibration by CIRCOM involves the MLR step that seeks to relate the information in the factor loadings matrix L with the matrix C.

Elements of the C matrix may contain chemical concentrations or a wide variety of physical and chemical properties. Examples of the diverse property types that may be used include the ash content of coal, and the BTU content of fuel oil. Even salinity of water may be predicted, although there are no directly identifiable bands of salt itself in the spectral data. The matrix expression that relates C and L is shown in equation (12.8):

$$\mathsf{C}_{MN} = \mathsf{L}_{MH} * \mathsf{B}_{HN} + \mathsf{ERROR}_{MN} \tag{12.8}$$

The object is to solve for B in equation (12.8) in a manner such that the ERROR matrix is a minimum. The form of equation (12.8) has a similarity to the P-matrix, or inverse Beer's law statement of the relationship between absorbance and concentration, in that the parameters of interest, C, are solved for directly as a function of some other measured parameter. Only, in this case, there is an important advantage because all of the significant information in the spectral data can be included in the calculation, in the form of the L matrix with M rows and H columns, without making the solution for B depend on a large cumbersome matrix inversion. When L and B are reduced from their maximum size of M columns to the first H columns, the respective reduced matrices may be symbolized as L1 and B1, respectively. The matrix solution for B1 is written in equation (12.9):

$$\mathsf{B1}_{HN} = (\mathsf{L1}^T * \mathsf{L1})^{-1}_{HH} * \mathsf{L1}^T_{HM} * \mathsf{C}_{MN} \tag{12.9}$$

The matrix inversion indicated in equation (12.9) need not be carried out explicitly on the original factor loadings matrix L1, because the orthonormal properties of L1 would lead to the collapse of the expression in parentheses in equation (12.9) to just the identity matrix. It is useful, however, to modify the L1 matrix of equation (12.9) to allow for a nonzero intercept in the relationship between the property of interest and the set of factor loadings derived from the eigenanalysis step. This can be done by adding a column of unity values to the L1 matrix. The modified matrix, represented by **L2**, can be used to solve for a better set of linear regression coefficients, **B2**, as shown in equation (12.10).

Note that H in equation (12.10) is now one column larger than the original FCCV. The regression coefficients derived in equation (12.10) belong to what is called the full regression model in CIRCOM terminology. Note that the prediction results obtained with a CIRCOM full regression model are essentially the same as would be obtained by the usual implementations of principal components regression (PCR), except that PCR does not necessarily include provision for a nonzero intercept in the model. For simplicity, however, it will be assumed that the CIRCOM full regression model and the PCR model are equivalent in the subsequent discussion:

$$\mathbf{B2}_{HN} = (\mathbf{L2}^T * \mathbf{L2})^{-1}_{HH} * \mathbf{L2}^T_{HM} * \mathbf{C}_{MN} \tag{12.10}$$

Some of the coefficients found in **B2** will be negligibly small or will have a comparatively large relative standard error. The conditions under which there is advantage in excluding certain factors from the prediction phase of the calculations has been described by Hocking.[15] Such tailoring of the regression coefficients of the **B** matrix can be carried out on a property-by-property basis rather than on all of the collected properties in the **B** matrix at once. Therefore, different factors may be retained for prediction of one property than are retained for the prediction of any other.

The danger in excluding too many factors is that bias may be introduced into the predictions. In CIRCOM, elimination of variables is carried out conservatively by a critical residual sum of squares test.[10] Generally, only those factors are dropped that have a very low correlation of the property with the factor loading and a relatively low error calculated on the critical residual sum of squares test. This amounts to an F-test on the combined significance of the values of **B2** to be eliminated for each property of interest. The matrices based on this third type of reduction may be distinguished by an added 3. Specifically, matrix **L3** can be used to derive regression coefficient matrix **B3**, which will contain the set calibration parameters that compose what is referred to as the reduced regression model in CIRCOM. The form of the solution is still the same as that indicated in equations (12.9) and (12.10). It is the reduced regression model that is used by CIRCOM to predict the properties of unknowns, and this model may be referred to as the reduced regression model or, more conveniently, the CIRCOM prediction model (CPM).

12.3.6. Determination of Factor Loadings for an Unknown

The first step in the determination of the property values of an unknown sample is the calculation of the factor loadings associated with the spectrum of the unknown. This can be carried out by a linear least squares estimate, using the reduced eigenspectra matrix **F1**, as is shown in equation (12.11):

$$\mathbf{L1u}_{H1} = (\mathbf{F1}^{t} * \mathbf{F1})_{HH}^{-1} * \mathbf{F1}_{HW}^{T} * \mathbf{Du}_{W1} \tag{12.11}$$

Equation (12.11) is really just a rearrangement of equations (12.5) and (12.6) above to solve for the factor loadings matrix, with the added change that M reduces to unity for the single unknown sample spectrum under evaluation. A simplification of equation (12.11) is possible because of the relationship in equation (12.7), which leads to equation (12.12):

$$\mathbf{L1u}_{H1} = \mathbf{E}_{HH}^{-1} * \mathbf{F1}_{HW}^{T} * \mathbf{Du}_{W1} \tag{12.12}$$

Since **E** is a diagonal matrix, the inverse of **E** may be obtained simply by substituting in the reciprocals of the values of the individual diagonal elements of **E**.

12.3.7. Calculation of Property Values in Unknown Samples

The factor loadings matrix calculated in equation (12.12) can be modified as before by adding a unity element to allow for the possibility of a nonzero intercept in the relationship between the property values and the factor loadings. Therefore, **L2u** may be constructed from **L1u**, and it will have the dimensions of $(H + 1)$ rows and a single column. The final calculation of the properties of an unknown sample can be carried out according to equation (12.13):

$$\mathbf{C}_{N1} = \mathbf{B3}_{NH}^{T} * \mathbf{L2u}_{H1} \tag{12.13}$$

12.4. CIRCOM CALIBRATION OF A SYNTHETIC DATA SET

The ideas of the preceding mathematical discussion can be demonstrated by application to a simple physical example. Assume that there are four components named peak 1, peak 2, peak 3, and peak 4, each of which have only one band in their individual IR spectra, which is centered at 1550, 1500, 1440, and 1450 cm^{-1}, respectively. Figure 12.1 shows the series of computer-generated spectra that were constructed to

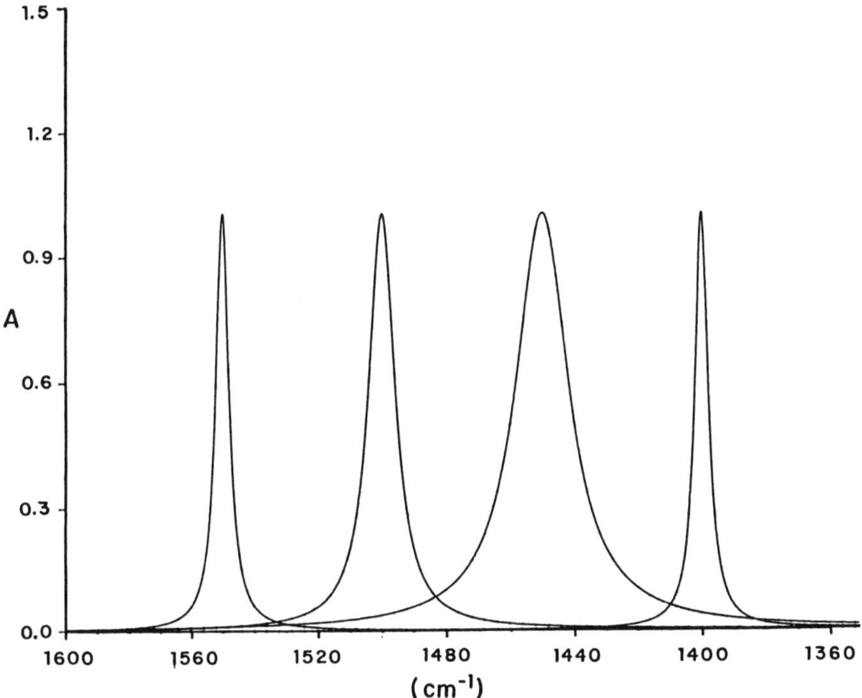

FIGURE 12.1. Overlay plot of synthetic spectral bands used to synthesize mixture spectra used in method pksc402.

represent a series of Lorentzian lineshapes of varying widths and frequency maxima. Figure 12.2 shows the group of spectra generated as linear combinations of the spectra in Fig. 12.1. Table 12.1 shows the nominal spectral parameters of the CIRCOM method along with the mixing coefficients of the original lineshape files used to generate the simulated mixture spectra. The mixing coefficients are equivalent to concentrations if one assumes, for simplicity, 1 unit of concentration corresponds to 1 unit of absorbance for each of the parent lineshapes. This data corresponds to the simple hypothetical example of a mixture of four materials, each of which has only one band in the spectral region considered.

Figure 12.3 shows a plot of the log of the reduced eigenvalue versus the factor number for the results of eigenanalysis of the data matrix formed by the twelve synthetic mixtures of the simulated spectra. The use of the reduced eigenvalue is preferable in this type of plot because it

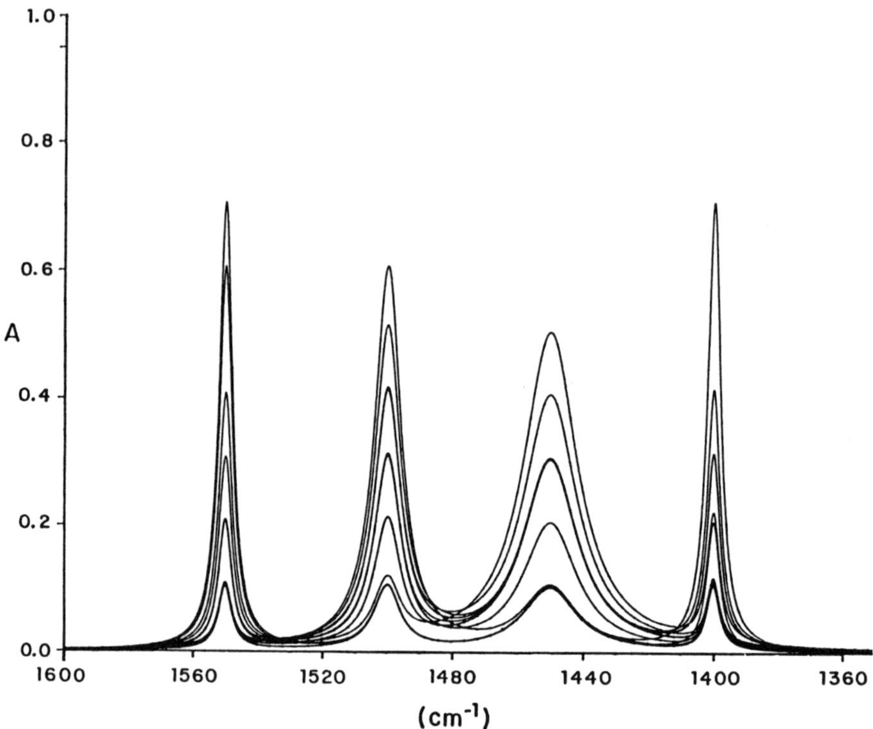

FIGURE 12.2. Overlay plot of synthetic spectral data used in method pksc402.

takes into account that there is one less degree of freedom as the eigenvalues for each successive factors are derived. The use of the reduced eigenvalue for this purpose has been described by Malinowski.[16] The eigenvalues of the first four factors are clearly much larger than the rest. Eigenvalues of factors 5–7 are smaller by about 8 orders of magnitude, and the remaining five eigenvalues are not significantly different from zero. The existence of four large eigenvalues is consistent with the fact that four major sources of spectral variance were used to construct the data set. The fact that there are three additional eigenvalues significantly different from zero in a computer-generated, supposedly noise-free, synthetic data set is fortuitous for illustration of the operation of the CIRCOM software. However, it is probably an inadvertent byproduct of the sequence of operations used to construct the twelve synthetic standards from the original four computer-generated spectral band contours (Lorentzian lineshapes).

Some idea of the information contained in the various factors can be

TABLE 12.1. CIRCOM Calibration Report

CIRCOM method name: pksc402

Method description: Synthetic Lorentzian bands
Peak 1 at 1,550
Peak 2 at 1,500
Peak 3 at 1,450
Peak 4 at 1,400

Starting wave number: 1,600.00
Finishing wave number: 1,350.00
Data interval: 0.50
Data points per spectrum: 501
Blanked region(s): 0
Calibration spectra: 12

	Expected property values			
Filenames	Peak 1	Peak 2	Peak 3	Peak 4
pks01	0.2000	0.2000	0.3000	0.3000
pks02	0.2000	0.1000	0.5000	0.2000
pks03	0.1000	0.5000	0.3000	0.1000
pks04	0.1000	0.4000	0.3000	0.3000
pks05	0.2000	0.6000	0.1000	0.1000
pks06	0.6000	0.1000	0.1000	0.2000
pks07	0.3000	0.3000	0.3000	0.1000
pks08	0.1000	0.2000	0.3000	0.4000
pks09	0.4000	0.3000	0.2000	0.1000
pks10	0.1000	0.4000	0.4000	0.1000
pks11	0.7000	0.1000	0.1000	0.1000
pks12	0.1000	0.1000	0.1000	0.7000

obtained by inspection of the corresponding eigenspectra, which are the columns of the **F** matrix. These are also sometimes referred to as abstract factors. Figure 12.4 shows the eigenspectrum corresponding to factor 1, and, as is usual, it resembles a scaled average of all of the spectra in the calibration set. Figures 12.5, 12.6, and 12.7 show the eigenspectra that correspond to the second, third, and fourth factors, respectively. Figures 12.8 and 12.9 show the eigenspectra for the fifth and sixth factors, respectively. In this simple example it is fairly obvious that the eigenspectra of the first four factors represent what appears to be linear combinations of the original pure component spectra. In contrast, the eigenspectra of the fifth and sixth factors look like noise spectra.

Table 12.2 shows the information used to determine the FCCV. Note especially the column labelled indicator function (IND). When the

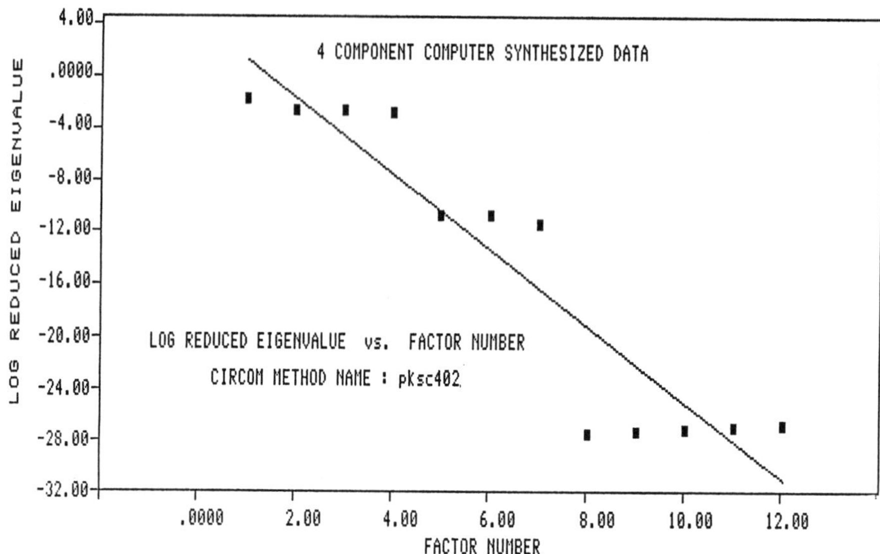

FIGURE 12.3. Log eigenvalue versus factor number for method pksc402.

error is normally distributed, it is generally expected that the indicator function will go through a minimum at the optimum number of factors for the construction of the full regression model. The properties of the indicator function and other criteria for determining the FCCV have been described by Malinowski.(17) In this synthetic data set the error is not due to absorbance noise, and indeed may not be normally distributed. Consequently, the FCCV is arbitrarily selected at 7, which is the point beyond which the real error can not be distinguished from zero.

Table 12.3 shows the results of using the first six factors as the full regression model (FRM) to predict the property called peak 3. The regression coefficients for peak 3, which are in fact the elements of the third row of **B2**, are large for the first four factors only. A similar trend is observed for the correlation coefficients of the property with the factor loadings. The ratio of the regression coefficients divided by the standard error in the regression coefficients is a t-value, which should ideally be as large as possible. The sign of the t-test can be ignored for present purposes. In Table 12.3, the t-values are large for the first four factors, but is very small for the fifth and sixth factors.

Table 12.4 shows the CPM obtained when the factors of the full regression model were evaluated systematically in a stepwise fashion for the property called peak 3. In the CPM, the F-test has increased, and the standard error of each of the regression coefficients is reduced over what was obtained in the full regression model. Additionally, the four

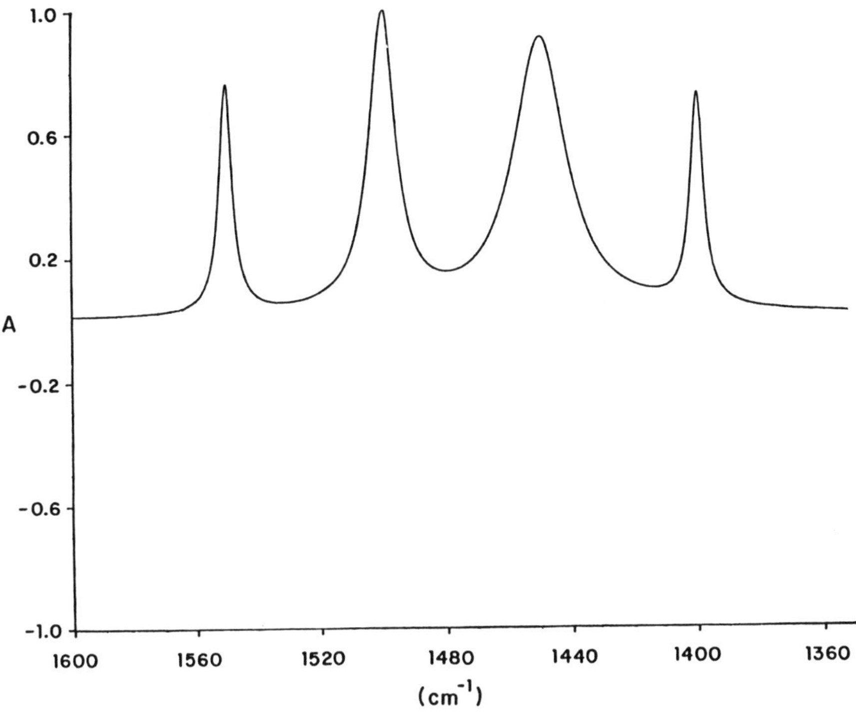

FIGURE 12.4. Eigenspectrum of factor 1 in method pksc402.

computed *t*-values are all relatively large, and the predicted values of peak 3 are correct to four places for all samples. The results are essentially equivalent for the other properties in the method.

12.5. CIRCOM CALIBRATION OF A SIMPLE CHEMICAL SYSTEM

Spectral data obtained from real samples are likely to contain significant levels of collinearity in the component spectra, some possible Beer's law nonlinearity, atmospheric absorption variations, other instrumental effects, and always some level of random noise. Such factors make the determination of property values more difficult. It is one of the advantages of factor analysis based techniques such as CIRCOM, however, that systematic effects unrelated to the property of interest can be largely isolated as separate factors. CIRCOM is then free to build a

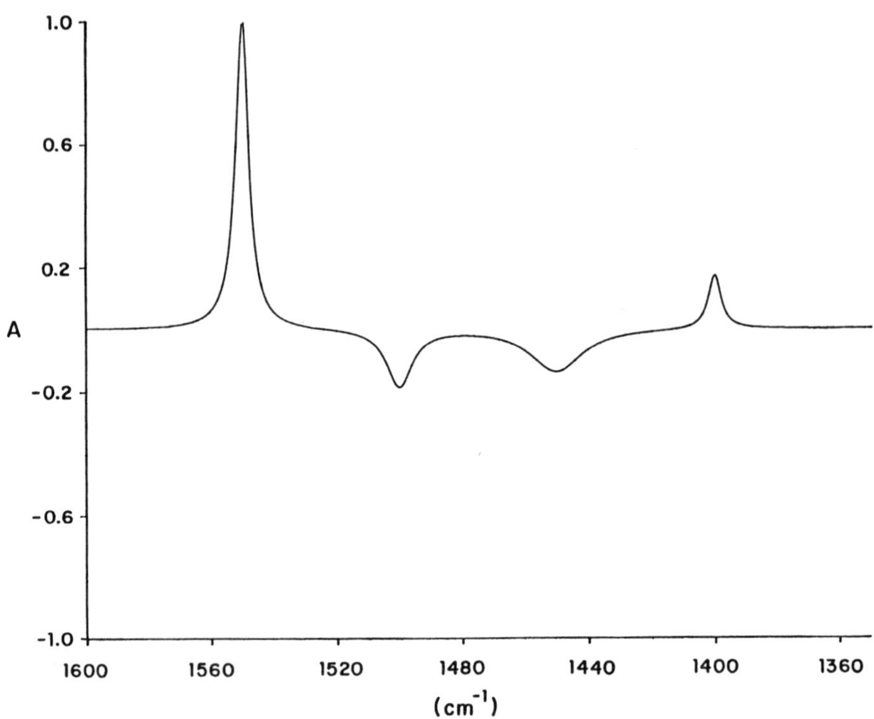

FIGURE 12.5. Eigenspectrum of factor 2 in method pksc402.

unique calibration model for each property which is based on the data features that are related to the respective property.

The three main types of lipids found in blood serum are triglycerides, phospholipids (particularly lecithins), and cholesteryl esters. Elevated levels of triglycerides are associated with a variety of diseases. High cholesterol levels are known to be associated with circulatory diseases, particularly atherosclerosis. While triglycerides and total serum cholesterol are determined on a routine basis, analysis of phospholipids is more difficult and costly. Consequently, phospholipids, particularly lecithins, are not screened as routinely as are triglycerides and cholesteryl esters.

The analysis of the three major types of lipids found in blood serum by IR spectrometry eluded clinical analysts until the age of microprocessor instrumentation and computerized data processing by matrix methods.[18,19] The use of MLR in the form of curve fitting has been shown to give good results on this analysis with laboratory samples.[20,21]

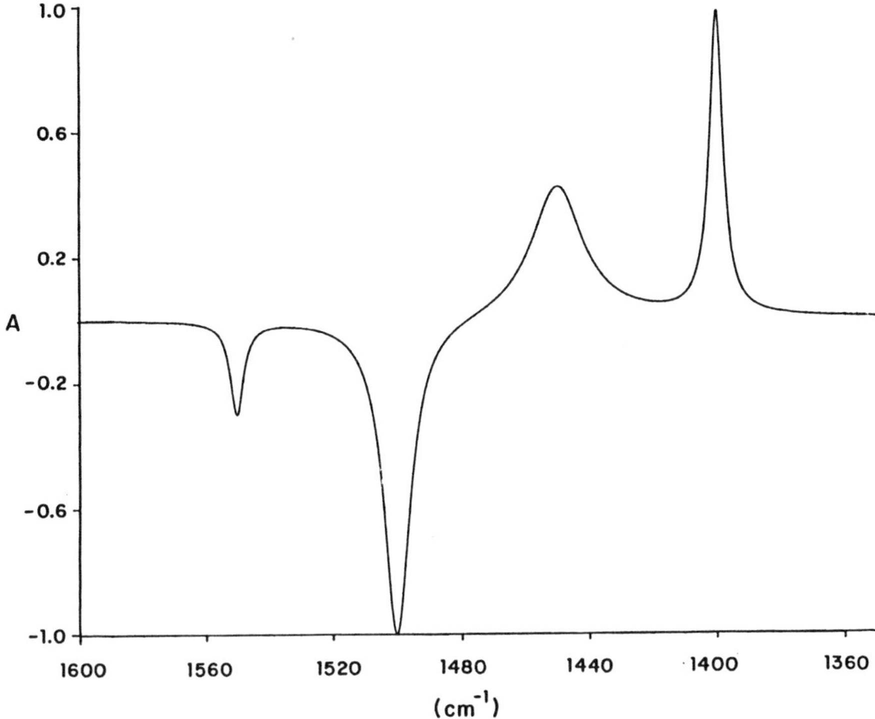

FIGURE 12.6. Eigenspectrum of factor 3 in method pksc402.

The generally accepted procedure for isolation of lipids involves extraction of the lipids with chloroform from serum mixed with aqueous methanol. Extracts are dried and redissolved in chloroform for subsequent analysis. Chloroform is an efficient solvent for these lipids and has large regions of relative transparency in the mid-IR.

The spectra of the three predominant lipids found in serum do not have unique bands that are not overlapped by bands of either of the other two materials. Rather, the spectral differences of each of the lipids amount to variations in relative intensity of absorbance across the spectral range. The spectral differences of the materials used as model compounds in this study—tripalmitin, dipalmitoyl lecithin, and cholesteryl palmitate—have been documented recently.[21] For this application, the region above 1900 cm^{-1} can be excluded from the calculations because of the high degree of colinearity of the spectra in this region. In the remainder of the mid-IR region, there are identifiable differences in

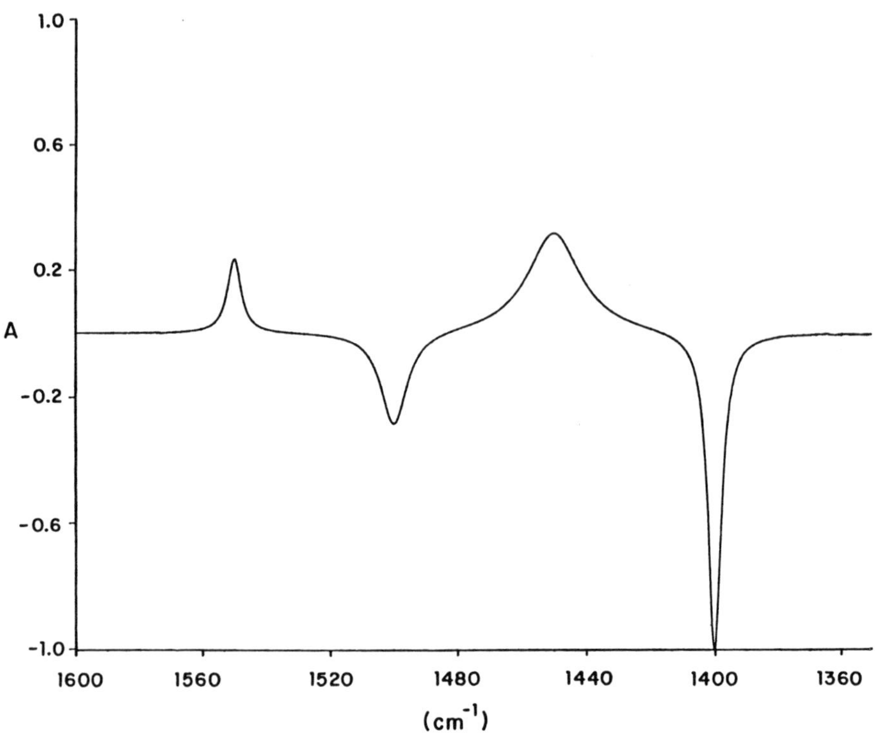

FIGURE 12.7. Eigenspectrum of factor 4 in method pksc402.

the pure component spectra, but the overlap of the bands of each component with those of the other two remains at a significant level. This situation is to be expected from the fact that each of these three materials has acyl groups of common fatty acids incorporated in their respective structures.

All samples of pure lipids were obtained from Sigma Chemical Company, St. Louis, MO. Spectra of weighted amounts of the lipids were dissolved in chloroform to make standard mixture samples. The spectra of all samples were recorded on the Perkin–Elmer model 1800 FT/IR spectrometer using a 0.2 sealed cell with KBr windows. The recorded spectra were stored on 5.25 in. floppy disks with a Perkin–Elmer model 7700 computer. The spectra are plotted in Fig. 12.10.

The experimental spectra are dominated by the features of the chloroform solvent. The selection of the size of the cell pathlength (0.2 mm) represented a trade-off between making the solute component

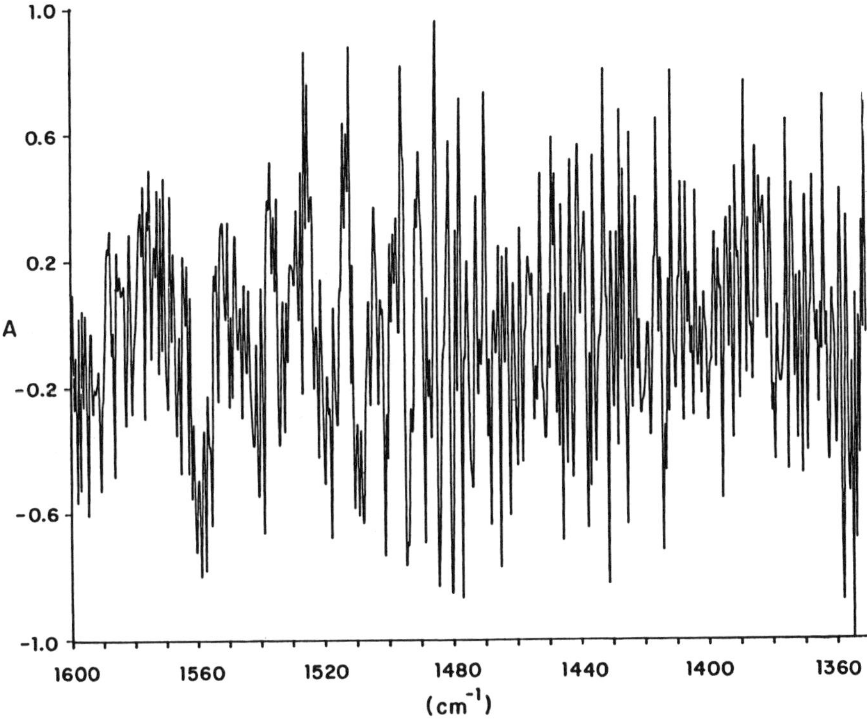

FIGURE 12.8. Eigenspectrum of factor 5 in method pksc402.

absorbances larger, while reducing the useable fraction of the spectral range because of the large solvent bands. Figure 12.11 shows the result of subtraction of the solvent spectrum from the solution spectra, and this plot shows essentially the signal on which the quantitation of the lipid components will depend. The noise in an absorbance measurement increases, however, as the absolute absorbance value increases. For mid-IR data, the relative noise in absorbance (i.e., the ratio of absorbance signal to absorbance noise) has a minimum in the range of 0.4–0.5 AU. Above the range of 1.0–1.5 AU, the magnitude of the absorbance noise tends to increase rapidly. Therefore, points in the spectrum were excluded from the calculation (as blanked regions) wherever the total absorbance tended to exceed an acceptable limit, for example, 1.0 AU. Table 12.5 shows the limits of the infrared data region used in the calculations, including the blanked regions.

To carry out the quantitative study of the three lipids, eighteen

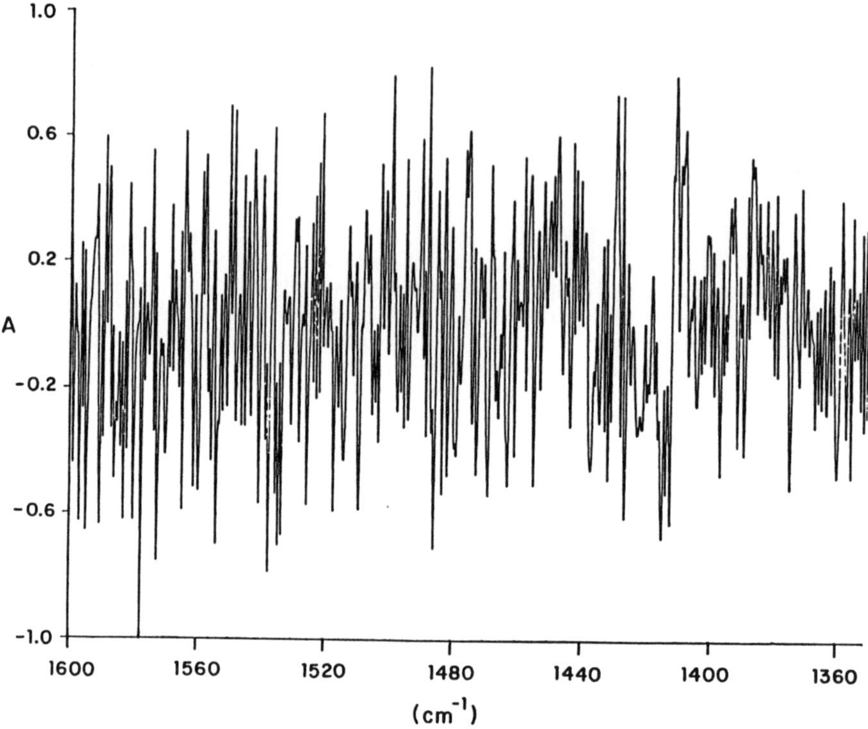

FIGURE 12.9. Eigenspectrum of factor 6 in method pksc402.

standards were prepared, all of which contained each of the three lipids at levels that could realistically be found in actual serum extracts (assuming a 1:1 volume/volume ratio between the original serum sample volume and the redissolved lipid extract). The concentration data on the eighteen mixture solutions are included in Table 12.5.

Table 12.6 shows the data from the eigenanalysis process on the set of standards. The full regression model has been formed on the basis of eight factors. A plot of log reduced eigenvalue versus factor number is shown in Fig. 12.12. The eigenvalue of factor 1 is significantly larger than the next three, which are in turn significantly larger than the remaining members of the set. The magnitude of the first eigenvalue may be understood in terms of the dominance of the average sample spectrum by the absorption bands of the chloroform solvent. The grouping of the next three eigenvalues at a lower magnitude is consistent with the idea that there are three distinguishable sources of variance in the data contributed

TABLE 12.2. CIRCOM Method Name: pksc402

Eigenanalysis based on covariance about the origin

Factor number	Eigenvalue	Reduced eigenvalue	Real error	Indicator function
1	0.5741×10^{2}	0.9549×10^{-3}	0.51501×10^{-1}	0.42563×10^{-3}
2	0.5971×10^{1}	0.1085×10^{-3}	0.41543×10^{-1}	0.41543×10^{-3}
3	0.5284×10^{1}	0.1058×10^{-3}	0.27307×10^{-1}	0.33712×10^{-3}
4	0.3362×10^{1}	0.7501×10^{-4}	0.42023×10^{-5}	0.65661×10^{-7}
5	0.3584×10^{-7}	0.9014×10^{-11}	0.31563×10^{-5}	0.64414×10^{-7}
6	0.2897×10^{-7}	0.8344×10^{-11}	0.14084×10^{-5}	0.39123×10^{-7}
7	0.5963×10^{-8}	0.2008×10^{-11}	0.00000	—
8	0.1000×10^{-23}	0.4049×10^{-27}	0.00000	—
9	0.1000×10^{-23}	0.5071×10^{-27}	0.00000	—
10	0.1000×10^{-23}	0.6775×10^{-27}	0.00000	—
11	0.1000×10^{-23}	0.1018×10^{-26}	0.00000	—
12	0.1000×10^{-23}	0.2041×10^{-26}	—	—

FCCV (factor compression cut-off value) selection:

Factor compression upper limit = 8
Indicator function minimum = *** (no minimum)
FCCV selected = 7

by the three serum lipids used to make up the standards. The issue is less clear as to how many of the remaining eigenvalues are significant for the prediction of the properties of interest, and how many represent mainly the effects of random error in the data. It is conceivable that other sources of variance in the spectral data could arise from atmospheric absorptions, instrument effects, and Beer's law nonlinearities. Each source of variation contributes another factor. At least the magnitude of the additional factors is relatively small, as evidenced by the fact that the magnitudes of the eigenvalues are substantially smaller after factor 4.

The calibration results for tripalmitin can be used to illustrate the differences in the full regression model and the CIRCOM prediction model (CPM). Table 12.7 shows the calibration statistics for tripalmitin when the complete set of the first eight factors are used. Table 12.8 shows the corresponding calibration statistics for tripalmitin in the CPM, in which only factors 2, 3, 4, and 5 have been retained. Table 12.8 also shows the results of using the CPM to predict concentrations of tripalmitin in set of calibration spectra. Note that the errors are relatively small for all standards. This suggests that the concentration errors in the standards are consistently small, and that the CPM may be capable of precise analysis. It is also worth comparing the increases in the value of the F-test and r-squared parameters between the full regression model

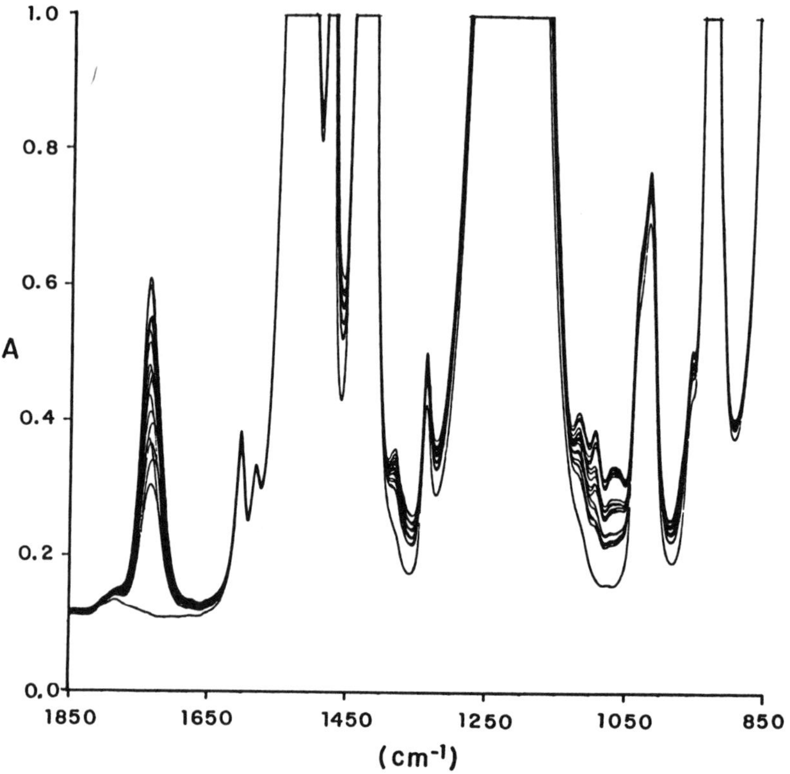

FIGURE 12.10. Overlay plot of spectral data used in method clip301.

and the CPM. These increases are consistent with the concept that reduction of the full regression model (analogous to the PCR model) (analogous to the CPM), which may lead to superior predictive capabilities. The results obtained for lecithin and cholesteryl palmitate were analogous.

12.6. CIRCOM CALIBRATION OF A DIFFICULT CHEMICAL SYSTEM

Since the calibration models for each property in a method are optimized independently in CIRCOM, it is possible to perform a rapid evaluation of which properties of interest may be predicted satisfactorily from a given calibration set, and to get an estimate of what kind of precision of prediction might be possible for each property. The following study illustrates a situation that was both successful and unsuccessful for

TABLE 12.3. Full Regression Model: Peak 3. Circom Method Name: pksc402

Regression results

Factor number	Correlation of property versus factor loading	Regression coefficient	Standard error of regression coefficient	Computed t-value
1	0.8709	0.9302	0.6648×10^{-1}	14.42
2	−0.6753	−0.1132	0.1342×10^{-1}	−8.439
3	0.3986	0.1829	0.1034×10^{-2}	176.8
4	0.6040	0.1683	0.7189×10^{-2}	23.41
5	-0.4414×10^{-1}	-0.6110×10^{-6}	0.1656×10^{-2}	-0.3689×10^{-3}
6	0.1527×10^{-1}	0.4351×10^{-6}	0.9148×10^{-3}	0.4765×10^{-3}
7	0.3230×10^{-2}	0.1622×10^{-6}	0.7675×10^{-3}	0.2113×10^{-3}

Intercept: 0.0000
Multiple correlation: 1.0000
r-squared: 1.0000
Standard error of estimate: 0.0008

Analysis of variance

Source of variation	Degrees of freedom	Sum of squares	Mean square	F-test
Regression	7	0.1900	0.2714×10^{-1}	0.4701×10^{5}
Deviation	4	0.2310×10^{-5}	0.5774×10^{-6}	
Total	11	0.1900		

deriving calibration models for the set of components entered into the method for evaluation.

Infrared spectra from 21 synthetic mixtures of 14 lipid materials found in serum at varying levels were used in the CIRCOM calibration. It is well known that the distribution of lipids in serum is more complicated than was suggested by the approximate three-component model used in the previously discussed data set. Consequently, the major objective of this experiment was to explore the potential limits of the analysis of chloroform solutions of lipid mixtures by IR spectroscopy. At the outset, it was not known how many of the serum components could be predicted from the data. This type of problem represents a typical example of the kind of problem often encountered in the process of developing quantitative methods in a new applications area.

In addition to lecithin, triglycerides, and cholesteryl esters, typical serum samples also contain varying amounts of unesterified cholesterol, free fatty acids, sphingomyelin, and a number of minor phospholipids. Table 12.9 shows the experimental concentrations of the lipids used in this study. Eigenanalysis by CIRCOM of the spectral data set led to the

TABLE 12.4. CIRCOM Prediction Model: Peak 3. Circom Method Name: pksc402

Regression results

Factor number	Correlation of property versus factor loading	Regression coefficient	Standard error of regression coefficient	Computed t-value
1	0.8709	0.9303	0.1028×10^{-1}	90.46
2	−0.6753	−0.1132	0.2154×10^{-2}	−52.57
3	0.3986	0.1829	0.2986×10^{-3}	612.5
4	0.6040	0.1683	0.1173×10^{-2}	143.5

Intercept: 0.0000
Multiple correlation: 1.0000
r-squared: 1.0000
Standard error of estimate: 0.0003

Analysis of variance

Source of variation	Degrees of freedom	Sum of squares	Mean square	F-test
Regression	4	0.1900	0.4750×10^{-1}	-0.6198×10^{6}
Deviation	7	-5364×10^{-6}	-0.7663×10^{-7}	
Total	11	0.1900		

Predicted property values: Peak 3

File name	Expected	Predicted	Percent residual	Residual
pks01	0.3000	0.3000	0.0	0.0000
pks02	0.5000	0.5000	0.0	0.0000
pks03	0.3000	0.3000	0.0	0.0000
pks04	0.3000	0.3000	0.0	0.0000
pks05	0.1000	0.1000	0.0	0.0000
pks06	0.1000	0.1000	0.0	0.0000
pks07	0.3000	0.3000	0.0	0.0000
pks08	0.3000	0.3000	0.0	0.0000
pks09	0.2000	0.2000	0.0	0.0000
pks10	0.4000	0.4000	0.0	0.0000
pks11	0.1000	0.1000	0.0	0.0000
pks12	0.1000	0.1000	0.0	0.0000

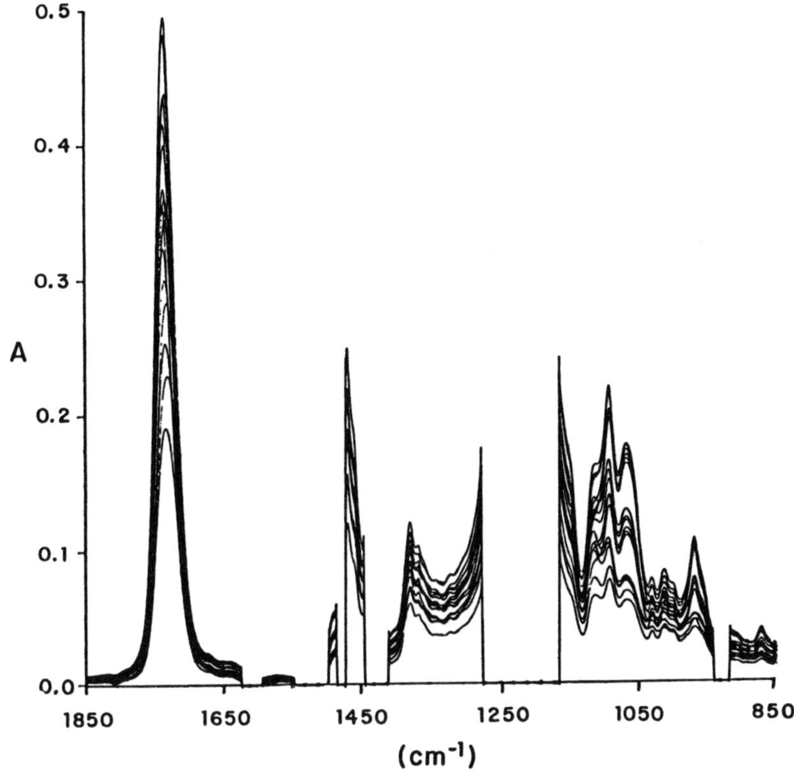

FIGURE 12.11. Overlay plot of solvent subtracted difference spectra used in method clip301.

estimation of the FCCV at ten factors for the full regression model, as shown by the data in Table 12.10. A plot of log reduced eigenvalue versus factor number is shown in Fig. 12.13, which shows no major breaks in the rather gradual decrease in the eigenvalues after the first factor.

With only ten factors considered significant in the spectra of mixtures composed of 14 known compounds, it is not unexpected that independent calibration models cannot be obtained for all components. Also, predictions based on the best models that can be derived will not be equally reliable for each component. Therefore, the problem is to evaluate which property prediction models will be likely to be useful and which will not. Table 12.11 shows the comparison of the *F*-test statistic for the full regression model and the CPM along with the number of factors used in the CPM. In some cases CIRCOM could not find a good model in the data, and reduced the number of factors down to one. In other cases

TABLE 12.5. CIRCOM Calibration Report

Method name: clip301

Method description: CIRCOM analysis of three-component lipid system:
Model 1800 data from DTGS detector;
chloroform solutions; transmission spectra;
0.2 mm barium fluoride cell;
data blanked in regions of
high solvent absorbance

Data region upper limit: 1,850.00
Data region lower limit: 860.00
Data interval: 1.00
Data points per spectrum: 991
Blanked region(s): 6

	Start	Finish
1	1,622.00	1,592.00
2	1,545.00	1,496.00
3	1,483.00	1,471.00
4	1,443.00	1,410.00
5	1,274.00	1,164.00
6	939.00	917.0

Calibration spectra: 18

Expected property values

File name	Lecithin	Tripalmitin	Cholesteryl palmitate
ltc201	59.8340	183.9700	64.6590
ltc202	57.0850	120.4000	70.8600
ltc203	58.6320	60.4610	128.6700
ltc204	56.7360	181.2400	133.7100
ltc205	45.0460	120.9500	208.3500
ltc206	59.1860	67.3560	195.8500
ltc207	119.3000	182.9400	70.7010
ltc208	115.5000	123.3000	68.2400
ltc209	119.2000	58.2750	131.8500
ltc210	120.0300	184.5000	126.6000
ltc211	115.7800	123.1400	188.8900
ltc212	119.0100	60.5460	196.9500
ltc213	179.9200	184.7100	68.1690
ltc214	181.5400	128.4900	70.1400
ltc215	191.9200	69.6170	126.8900
ltc216	177.9800	181.2200	129.2800
ltc217	183.5900	124.3900	204.1100
ltc218	177.1800	63.9110	209.8800

TABLE 12.6. CIRCOM Method Name: clip301

Eigenanalysis based on covariance about the origin				
Factor number	Eigenvalue	Reduced eigenvalue	Real error	Indicator function
1	0.2582×10^{4}	0.19677	0.17906×10^{-1}	0.61960×10^{-4}
2	0.3050×10^{1}	0.24645×10^{-3}	0.88975×10^{-2}	0.34756×10^{-4}
3	0.6648	0.57153×10^{-4}	0.48626×10^{-2}	0.21612×10^{-4}
4	0.2573	0.23627×10^{-4}	0.35311×10^{-2}	0.18016×10^{-5}
5	0.7326×10^{-3}	0.72177×10^{-7}	0.23870×10^{-3}	0.14124×10^{-5}
6	0.3149×10^{-3}	0.33457×10^{-7}	0.16039×10^{-3}	0.11138×10^{-5}
7	0.1277×10^{-3}	0.14719×10^{-7}	0.11016×10^{-3}	0.91040×10^{-6}
8	0.4771×10^{-4}	0.60073×10^{-8}	0.82488×10^{-4}	0.82488×10^{-6}
9	0.1631×10^{-4}	0.22621×10^{-8}	0.71231×10^{-4}	0.87940×10^{-6}
10	0.7657×10^{-5}	0.11816×10^{-8}	0.66295×10^{-4}	0.10359×10^{-5}
11	0.5948×10^{-5}	0.10340×10^{-8}	0.62108×10^{-4}	0.12675×10^{-5}
12	0.4798×10^{-5}	0.95463×10^{-9}	0.58339×10^{-4}	0.16205×10^{-5}
13	0.4143×10^{-5}	0.96304×10^{-9}	0.54291×10^{-4}	0.21716×10^{-5}
14	0.2934×10^{-5}	0.81955×10^{-9}	0.51752×10^{-4}	0.32345×10^{-5}
15	0.2854×10^{-5}	0.99790×10^{-9}	0.47603×10^{-4}	0.52893×10^{-5}
16	0.2173×10^{-5}	0.10144×10^{-10}	0.43688×10^{-4}	0.10922×10^{-4}
17	0.1773×10^{-5}	0.12433×10^{-10}	0.37221×10^{-4}	0.37221×10^{-4}
18	0.1010×10^{-5}	0.14185×10^{-10}	—	—

FCCV (factor compression cut-off value) selection

Factor compression upper limit = 12
Indicator function minimum = 8
FCCV selected = 8

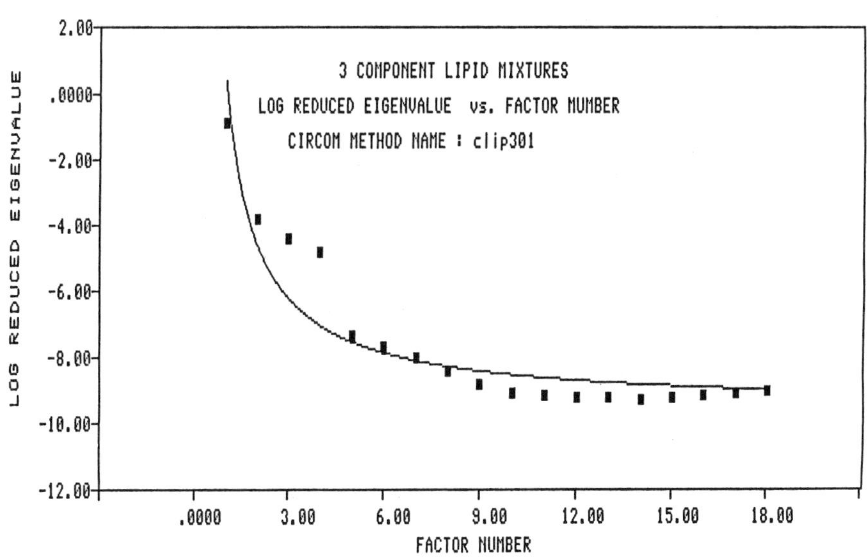

FIGURE 12.12. Log eigenvalue versus factor number for method clip301.

TABLE 12.7. Full Regression Model: TRIPALMITIN. CIRCOM Method Name: clip301

Regression results

Factor number	Correlation of property versus factor loading	Regression coefficient	Standard error of regression coefficient	Computed t-value
1	0.4387	−2,225.0	0.3535×10^{5}	-0.6294×10^{-1}
2	0.6085	194.1	1,080	0.1798
3	−0.7898	−149.6	227.4	−0.6577
4	-0.7759×10^{-1}	−24.61	129.9	−0.1895
5	0.3585×10^{-2}	1.565	13.56	0.1154
6	0.1793×10^{-2}	1.137	12.54	0.9070×10^{-1}
7	0.4577×10^{-3}	0.4633	6.170	0.7509×10^{-1}
8	-0.6893×10^{-3}	0.1751	5.573	0.3143×10^{-1}

Intercept: 648.6066
Multiple correlation: 0.9993
r-squared: 0.9986
Standard error of estimate: 2.6190

Analysis of variance

Source of variation	Degrees of freedom	Sum of squares	Mean square	F-test
Regression	8	0.4310×10^{5}	5,388	0.7854×10^{3}
Deviation	9	61.73	6.859	
Total	17	0.4316×10^{5}		

CIRCOM found the best model to include ten factors, and the CPM was equivalent to the full regression model. For a number of properties, satisfactory models were obtained with fewer than ten factors, and the F-test values were observed to increase on reduction of the number of factors used in the full regression model to form the CPM.

Table 12.12 shows the results of the F-test, coefficient of determination (r-squared), and standard error of calibration in both the full and reduced regression models for each of the properties of interest. Where there is a useful calibration model, the F-test is very large (>100), the r-squared is near 1.0, and the relative standard error is small. The relative standard error was found by dividing the standard error of estimate by the average value of the property in the set of calibration standards.

One can anticipate that triglycerides, lecithin, sphingomyelin, free cholesterol, and esterified cholesterol may yield satisfactory predictions over the concentration ranges included in the calibration. The results for

TABLE 12.8. CIRCOM Prediction Model: TRIPALMITIN. CIRCOM Method Name: clip301

Regression results

Factor number	Correlation of property versus factor loading	Regression coefficient	Standard error of regression coefficient	Computed t-value
2	0.6085	126.5	0.3286	384.8
3	−0.7898	−164.1	0.3284	−499.5
4	-0.7759×10^{-1}	−16.13	0.3284	−49.11
5	0.3585×10^{-2}	0.7458	0.3284	2.271

Intercept: 123.9762
Multiple correlation: 1.0000
r-squared: 1.0000
Standard error of estimate: 0.3284

Analysis of variance

Source of variation	Degrees of freedom	Sum of squares	Mean square	F-test
Regression	4	0.4316×10^{5}	0.1079×10^{5}	0.1000×10^{6}
Deviation	13	1.402	0.1079	
Total	17	0.4316×10^{5}		

Predicted property values: TRIPALMITIN

File name	Expected	Predicted	Percent residual	Residual
ltc201	183.9700	183.8895	0.0	0.0805
ltc202	120.4000	120.7202	−0.3	−0.3202
ltc203	60.4610	60.8913	−0.7	−0.4303
ltc204	181.2400	181.3125	0.0	−0.0725
ltc205	120.9500	120.8216	0.1	0.1284
ltc206	67.3560	66.9182	0.7	0.4378
ltc207	182.9400	183.1117	−0.1	−0.1718
ltc208	123.3000	122.9174	0.3	0.3826
ltc209	58.2750	58.1227	0.3	0.1523
ltc210	184.5000	184.3799	0.1	0.1201
ltc211	123.1400	123.0565	0.1	0.0835
ltc212	60.5460	60.7671	−0.4	−0.2211
ltc213	184.7100	184.5659	0.1	0.1441
ltc214	128.4900	128.0581	0.3	0.4319
ltc215	69.6170	69.8781	−0.4	−0.2611
ltc216	181.2200	181.7957	−0.3	−0.5758
ltc217	124.3900	124.2215	0.1	0.1685
ltc218	63.9110	63.9877	−0.1	−0.0767

TABLE 12.9. Serum Lipid Standards Concentration Data for 14 Component Serum Lipid Mixtures in Chloroform

Compound name	Concentration range (mg/dl)	Concentration average (mg/dl)
Cholesterol	40–160	90
Cholesteryl esters (cholesteryl palmitate)	75–300	193
Triglycerides (tripalmitin)	50–400	188
Lecithin (dipalmitoyl-)	25–200	110
Sphingomyelin	12–963	48
Palmitic acid	2–16	8
Oleic acid	3–24	12
Lysolecithin	4–32	16
Cardiolipin	0.5–4	2
Phosphatidyl-		
ethanolamine	3–24	13
inositol	3–16	8
glycerol	1–8	4
serine	0.5–4	2
Phosphatidic acid	0.5–4	2
Total		
Lipids	365–1064	695
Cholesterols	115–380	282
Lecithins	29–224	128
Fatty Acids	18–46	20

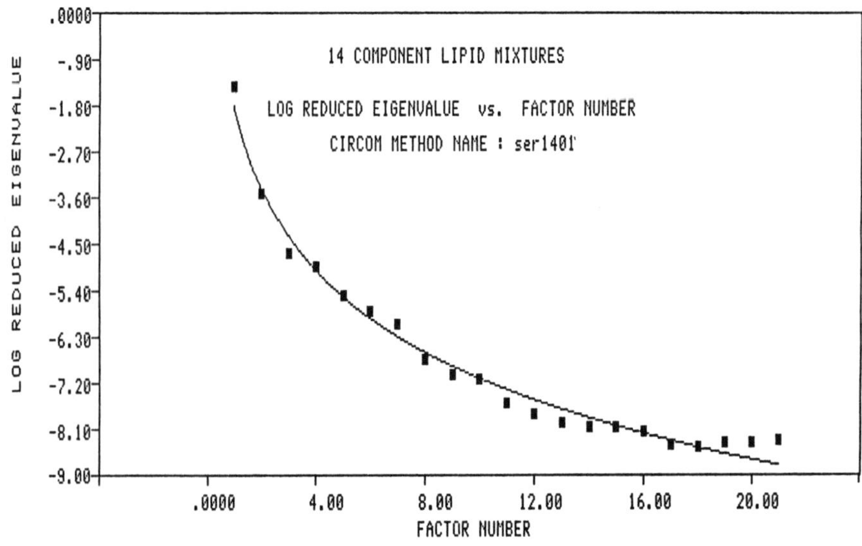

FIGURE 12.13. Log eigenvalue versus factor number for method ser1401.

TABLE 12.10. CIRCOM Method Name: ser1401

Eigenanalysis based on covariance about the origin

Factor number	Eigenvalue	Reduced eigenvalue	Real error	Indicator function
1	0.2140×10^{4}	0.56028×10^{-1}	0.22586×10^{-1}	0.56466×10^{-4}
2	0.1654×10^{2}	0.45640×10^{-3}	0.75409×10^{-2}	0.20889×10^{-4}
3	0.1109×10^{1}	0.32230×10^{-4}	0.51031×10^{-2}	0.15750×10^{-4}
4	0.5668	0.17397×10^{-4}	0.30302×10^{-2}	0.10485×10^{-4}
5	0.1538	0.50011×10^{-5}	0.21103×10^{-2}	0.82434×10^{-5}
6	0.7502×10^{-1}	0.25933×10^{-5}	0.14113×10^{-2}	0.62723×10^{-5}
7	0.3881×10^{-1}	0.14318×10^{-5}	0.77784×10^{-3}	0.39685×10^{-5}
8	0.7174×10^{-2}	0.28374×10^{-6}	0.58924×10^{-3}	0.34866×10^{-5}
9	0.3416×10^{-2}	0.14558×10^{-6}	0.46811×10^{-3}	0.32508×10^{-5}
10	0.2639×10^{-2}	0.12191×10^{-6}	0.32671×10^{-3}	0.27001×10^{-5}
11	0.7891×10^{-3}	0.39788×10^{-7}	0.27182×10^{-3}	0.27182×10^{-5}
12	0.4386×10^{-3}	0.24340×10^{-7}	0.23497×10^{-3}	0.29009×10^{-5}
13	0.2611×10^{-3}	0.16108×10^{-7}	0.21003×10^{-3}	0.32817×10^{-5}
14	0.1833×10^{-3}	0.12730×10^{-7}	0.18965×10^{-3}	0.38704×10^{-5}
15	0.1686×10^{-3}	0.13388×10^{-7}	0.16266×10^{-3}	0.45183×10^{-5}
16	0.1225×10^{-3}	0.11355×10^{-7}	0.13503×10^{-3}	0.54011×10^{-5}
17	0.5197×10^{-4}	0.57841×10^{-8}	0.12499×10^{-3}	0.78121×10^{-5}
18	0.4021×10^{-4}	0.55972×10^{-8}	0.11592×10^{-3}	0.12880×10^{-4}
19	0.3720×10^{-4}	0.69081×10^{-8}	0.99484×10^{-4}	0.24871×10^{-4}
20	0.2299×10^{-4}	0.64075×10^{-8}	0.84349×10^{-4}	0.84349×10^{-4}
21	0.1290×10^{-4}	0.71947×10^{-8}	—	—

FCCV (factor compression cut-off value) selection

Factor compression upper limit = 14
Indicator function minimum = 10
FCCV selected = 10

fatty acids are somewhat questionable, however, and predictions for the remaining lipids are not likely to produce satisfactory results with the present procedures used in sample preparation and spectral data collection.

The results summarized in Table 12.12 indicate the diversity of the quality of the calibration models for the different properties. The differences can be shown more dramatically in graphic form, such as in the comparison of plots of expected versus predicted concentrations of the properties in the calibration standards. Figures 12.14 and 12.15 show the plots of predicted versus expected concentrations for cholesterol and palmitic acid, respectively. It is easy to see that the cholesterol predictions are relatively good, and that the palmitic acid predictions are rather poor. It is useful in this case to take advantage of the fact that properties to be predicted by CIRCOM may be related to more than one chemical component. When palmitic acid (a saturated fatty acid) and

TABLE 12.11. Calibration Model Statistics Comparison of 14 Component Serum Lipid Mixtures in Chloroform

Compound name	Full regression model (PCR) (10 factors) F-test	CIRCOM prediction model (CPM) Number of factored retained	CIRCOM prediction model (CPM) F-test
Cholesterol			
Free	355	8	517
esterified (palmitate)	567	10	567
Triglycerides (tripalmitin)	550	7	853
Lecithin (dipalmitoyl-)	109	8	150
Sphingomyelin	141	5	266
Palmitic acid	2	2	7
Oleic acid	4	9	5
Lysolecithin	3	1	16
Cardiolipin	1	1	3
Phosphatidyl-			
ethanolamine	1	1	6
inositol	3	1	21
glycerol	1	1	2
serine	1	1	6
Phosphatidic acid	1	2	6
Total			
Lipids	374	7	1,808
Cholesterols	873	8	1,513
Phospholipids	11,620	10	11,620
Fatty acids	9	8	9

oleic acid (an unsaturated fatty acid) are combined and predicted as a single property (total fatty acids), the calibration statistics improve. The plot of predicted versus expected concentration for total fatty acids in Fig. 12.16 illustrates the improved precision of the calibration model for the combination property

Other combination properties can be formed in a similar fashion. Cardiolipin and lysolecithin have structures similar to that of lecithin. The combination of these three components can be predicted under the property name of total lecithins. Likewise, free and esterified cholesterol can be predicted under the property name of total cholesterol. The combination of all minor phospholipids can be predicted as trace phospholipids. Finally, the sum of all lipids can be combined under the heading of total lipids.

Table 12.12 includes a summary of the calibration statistics for the set of combined properties. The numbers suggest that the trace phos-

TABLE 12.12. CIRCOM Prediction Model Calibration Statistics for 14 Component Serum Lipid Mixtures in Chloroform

Compound name	Standard error of estimate	Relative standard error (%)	R-squared	CPM *F*-test
Cholesterol				
Free	3.1	3.4	0.9971	517
Esterified (palmitate)	5.0	2.6	0.9982	567
Triglycerides (tripalmitin)	6.8	3.6	0.9978	853
Lecithin (dipalmitoyl-)	8.6	7.9	0.9901	150
Sphingomyelin	3.8	7.8	0.9888	266
Palmitic acid	4.2	52.9	0.4277	7
Oleic acid	5.1	42.9	0.7707	5
Lysolecithin	8.4	52.7	0.4627	16
Cardiolipin	1.2	60.3	0.1417	3
Phosphatidyl-				
ethanolamine	7.3	56.0	0.2404	6
inositol	3.6	45.4	0.5279	21
glycerol	2.3	56.9	0.2290	6
serine	1.3	65.8	0.0741	2
Phosphatidic acid	1.1	53.1	0.5521	6
Total				
Lipids	7.6	1.1	0.9990	1,808
Cholesterols	3.6	1.3	0.9990	1,513
Phospholipids	1.2	0.9	0.9999	11,620
Fatty acids	4.7	23.7	0.8556	9

pholipids will not give satisfactory prediction results without some changes in sample preparation or data collection procedure. On the other hand, it is possible that the results for total fatty acids may be useful on a semiquantitative basis to detect the abnormally high levels of fatty acids found in some diabetic conditions. The calibration statistics suggest satisfactory predictions may be obtained for a number of other combination properties.

While it may not be possible to predict some of the properties included in this application on the basis of the current data set, the overall experiment can be judged productive and successful. It appears from this data that satisfactory prediction of up to six serum lipid components, as well as additional combination properties, may be obtained simultaneously, without any additional separation techniques other than those used traditionally to separate lipids from serum. This constitutes the first report of this capability known to these authors. In the further stages of this area of investigation, it remains to test the ability of the calibration model described to predict lipids isolated from

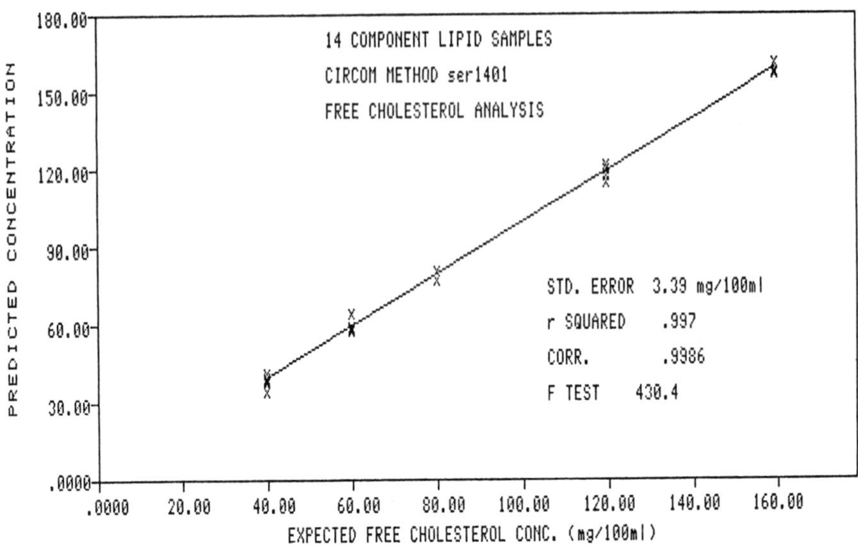

FIGURE 12.14. Predicted versus expected values of free cholesterol from method ser1401.

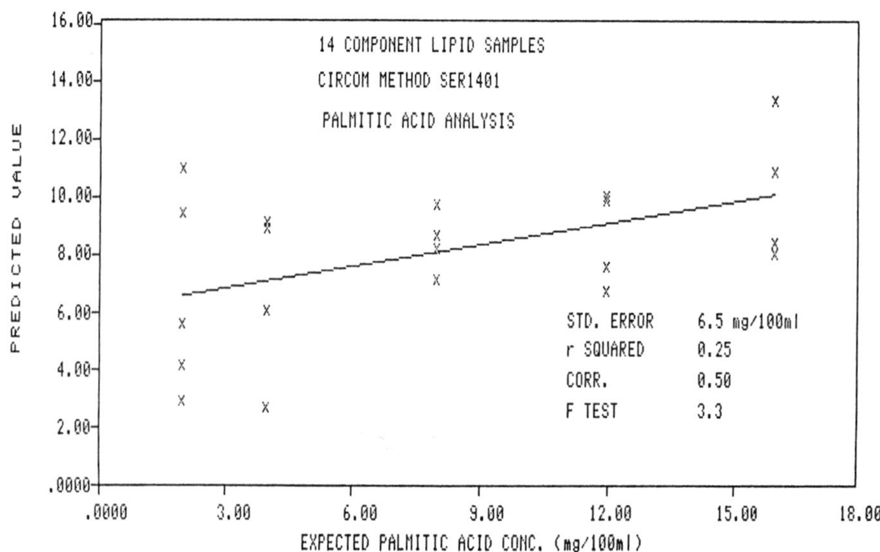

FIGURE 12.15. Predicted versus expected values of palmitic acid from method ser1401.

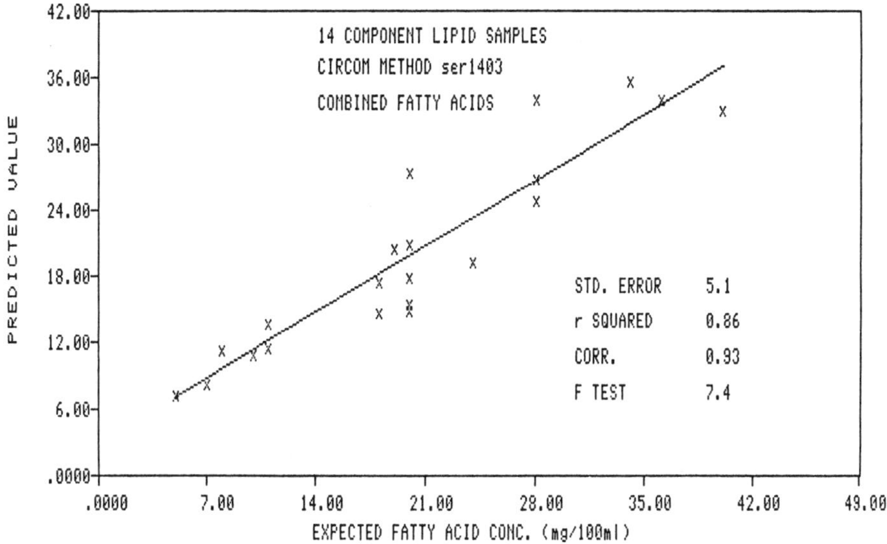

FIGURE 12.16. Predicted versus expected values of total free fatty acid from method Ser1403.

authentic serum samples, and, ultimately, to determine what adjustments must be made to adapt the multivariate spectroscopic method for general clinical use.

12.7. CONCLUSIONS

The mathematical processes involved in the CIRCOM software have been described. The operation of CIRCOM has been illustrated with both synthetic and authentic IR spectral data. The number of factors used in full regression models is similar to that which would be used in the technique of principal components regression (PCR). The CIRCOM prediction models have been shown to offer better calibration statistics for some properties in certain systems when the number of factors in the model is reduced below that which would be selected for CIRCOM full regression models. The set of calibration standards used to develop the prediction model should constitute a representative sampling of the materials, the properties of which are to be predicted in unknown samples. In such cases, the statistical parameters used to evaluate the calibration model may provide an estimate of the limits of the performance of the calibration model in the prediction of unknown materials. In some systems it may be found that some of the properties included in the method can be predicted well, while no satisfactory calibration model can be gleaned from the data for other properties. Included in this work is a study of the analysis of fourteen lipids found in blood serum. Good prediction models were obtained for cholesterol, cholesteryl esters, triglycerides, lecithins, and sphingomyelin. The prediction of fatty acids in the mixtures was less precise, but useable for assessment of high and low levels, which would allow differentiation of normal and potentially pathological conditions. In addition, combination properties may be predicted well, such as total cholesterol, total phospholipids, and total lipids. These results mark the beginning of a more extensive investigation of multiple lipid analysis on a variety of types of biological samples. The possibility of developing an IR-based lipid profile analysis is anticipated. The results described in this work constitute the first report known to these authors of the simultaneous independent determination of such a large number of lipid parameters from the measurement of each sample with only one instrumental method, and independent indirect aqueous methods.

REFERENCES

1. D. M. Haaland and R. G. Easterling, *Appl. Spectrosc.* **34,** 539–548 (1980).
2. D. M. Haaland and R. G. Easterling, *Appl. Spectrosc.* **36,** 665 (1982).
3. D. M. Haaland, R. G. Easterling, and D. A. Vopica, *Appl. Spectrosc.* **39,** 73 (1985).
4. E. R. Malinowski and D. G. Howery, *Factor Analysis in Chemistry,* Wiley, New York (1980).
5. I. T. Jolliffe, *Principal Components Analysis,* Springer-Verlag, New York (1986).
6. C. L. Lawson and R. J. Hanson, *Solving Least Squares Problems,* Prentice-Hall, Englewood Cliffs, NJ (1974).
7. W. Lindberg, J. A. Persson, and S. Wold, *Anal. Chem.* **55,** 643–648 (1983).
8. P. Geladi and B. R. Kowalski, *Anal. Chim. Acta* **185,** 1–17 (1986).
9. A. Lorber, L. E. Wangen, and B. R. Kowalski, *J. Chemometrics* **1,** 19–31 (1987).
10. P. M. Fredricks, J. B. Lee, P. R. Osborne, and D. A. J. Swinkels, *Appl. Spectrosc.* **39,** 311–320 (1985).
11. A. Beer, *Ann. Phys. Chem.* **86,** 78–88 (1852).
12. D. M. Haaland and E. V. Thomas, *Anal. Chem.* **60,** 1193–1202 (1988).
13. J. C. Sternberg, S. L. Stillo, and R. H. Schwendeman, *Anal. Chem.* **32,** 84–90 (1960).
14. C. W. Brown, P. F. Lynch, R. J. Obremski, and D. S. Lavery, *Anal. Chem.* **54,** 1472–1479 (1982).
15. R. R. Hocking, *Biometrics* **32,** 1 (1976).
16. E. R. Malinowski, *J. Chemometrics* **1,** 33 (1987).
17. E. R. Malinowski, *Anal. Chem.* **49,** 606–617 (1977).
18. H. J. Kisner, C. W. Brown, and G. J. Kavarnos, *Anal. Chem.* **54,** 1479–1485 (1982).
19. H. J. Kisner, C. W. Brown, and G. J. Kavarnos, *Anal. Chem.* **54,** 1703–1707 (1983).
20. L. L. Tyson, Y. C. Ling, and C. K. Mann, *Appl. Spectrosc.* **8,** 663–668 (1984).
21. G. L. McClure, P. B. Roush, J. F. Williams, and C. Lehmann, "Application of Computerized Infrared Spectroscopy to the Determination of the Principal Lipids Found in Blood Serum", in: *Computerized Quantitative Infrared Analysis* (G. L. McClure, ed.), ASTM Special Technical Publication 934, American Society of Testing Materials, Philadelphia, PA (1987).

Index